Megawatts and Megatons

Megawatts and Megatons

The Future of Nuclear Power and Nuclear Weapons

RICHARD L. GARWIN

— & —

GEORGES CHARPAK

This book belongs to
Alex Roth

THE UNIVERSITY OF CHICAGO PRESS

Published by arrangement with Alfred A. Knopf, a division of Random House, Inc.

The University of Chicago Press, Chicago, 60637
Copyright © 2001, 2002 by Richard L. Garwin and Georges Charpak
All rights reserved. First published in the United States
by Alfred A. Knopf, Inc. in 2001.
University of Chicago Press edition 2002
Printed in the United States of America
09 08 07 06 05 04 03 02   1 2 3 4 5
ISBN: 0-226-28427-1

CIP data have been requested from the Library of Congress.

This work is substantially based on the authors' French-language book, *Feux follets et champignons nucléaires,* with sketches by Sempé, originally published in France by Editions Odile Jacob, Paris. Copyright © 1997 by Editions Odile Jacob.

Grateful acknowledgment is made to Basic Books and I.B. Taurus & Co. Ltd. for permission to reprint excerpted text from *The Truth About Chernobyl* by Grigori Madvedev, English translation copyright © 1991 by Basic Books Inc. Reprinted by permission of Basic Books, a member of Perseus Books, L.L.C., and I.B. Taurus & Co. Ltd.

CONTENTS

ACKNOWLEDGMENTS

This book began as our volume in French, *Feux Follets et Champignons Nucléaires*, published by Editions Odile Jacob, Paris, in 1997, and subsequently translated into English by Robert and Ellen Chase. We were greatly aided by our physicist colleague Venance Journé in expanding, organizing, and presenting the material. Our longtime colleague Valentine Telegdi read the manuscript and provided invaluable information and advice, as did Patricia McFate. W. K. H. Panofsky was similarly helpful with two chapters. We acknowledge with gratitude the support of the Alfred P. Sloan Foundation in the research and writing of this book.

Richard L. Garwin
Georges Charpak

NOTE TO THE PAPERBACK EDITION

Richard L. Garwin

Megawatts and Megatons was printed in hardcover just before the September 11, 2001, terrorist attacks on the World Trade Center and the Pentagon. Nevertheless, we were well aware of the threat of terrorism when we wrote the book so that Chapter 12, "Current Nuclear Threats to Security," is largely devoted to a discussion of nuclear and biological megaterrorism. Improvised nuclear devices (nuclear explosives) are discussed on pages 349–50 and so-called dirty bombs (radiological dispersal devices) on pages 339–42. Anthrax and smallpox, also discussed in Chapter 12, have been much in the media since September 11 and the five deaths from anthrax-spiked letters during the fall of 2001.

We also wrote about threats to security in the March 2002 publication *Striking Terror: America's New War*, contributing "The Many Threats of Terror" and the epilogue. (This is available, along with other materials published by Richard Garwin, at *http://www.fas.org/rlg.*) Further, in a September 2002 article, we provide detailed calculations for casualties expected from a one-kiloton nuclear explosion in Manhattan: 30,000 killed by blast, 70,000 killed immediately by burns, and approximately 70,000 more killed by the initial radiation.

As for the megawatts — or nuclear power — side, on February 15, 2002, President Bush authorized the spent fuel repository at Yucca Mountain. Though the governor of Nevada protested, the House supported the President, and the approval expected from the Senate will move the controversy over the Yucca Mountain Repository to an almost certain argument before the Supreme Court.

A new type of problem was detected in one of the 103 nuclear reactors that produce 20% of U.S. electrical power. On March 6, 2002, severe corrosion in the form of a large conical pit penetrating six inches of steel was discovered in the reactor pressure head at the Davis-Besse plant near Toledo, Ohio. This was the unexpected effect of a modest leak of reactor coolant that was spiked with boric acid. Undetected, it had the potential of becoming a severe leak, although it probably would not have resulted in a significant dispersal of radioactive materials outside the reactor building. Industrywide inspections and corrective measures are under way.

On the megatons — or nuclear weapons — side, in December 2001 President Bush gave the requisite six-months notice that the United States would no longer be bound by the 1972 ABM Treaty. He did this without asking our Russian partners to accept interpretations of the treaty that would permit planned U.S. testing for a national missile defense and the deploying of a modest system for protection against threats from rogue nations. Since June 2002, the United States has not been limited by any of the testing or deployment constraints of the ABM Treaty.

Also in December 2001, Secretary of Defense Rumsfeld renamed the Ballistic Missile Defense Organization (BMDO) the Missile Defense Agency (MDA) and freed it from many of the normal requirements for development and deployment of Department of Defense systems. Operational tests and evaluations will no longer be required. If the need arises, the plan is to deploy a capabilities-based defense — that is, to dub test facilities "operational" and to use them to counter a nuclear-armed intercontinental ballistic missile (ICBM) threat from North Korea or other "rogue state." However, as we point out on pages 355–58 and has been stated in National Intelligence Estimates of recent years, a greater threat from these same states is short-range ballistic missiles or cruise missiles, armed with nuclear or biological payloads and launched from ships near U.S. shores.

On the side of rationality, on December 31, 2001, the BMDO/MDA issued a call for proposals for studies to identify systems which, by 2005, could demonstrate boost-phase intercept of ICBMs, using homing kill vehicles on the interceptors which will actually collide with and thus destroy an ICBM. This advance could lead to the deployment of a boost-phase intercept system by 2010.

Also in December 2001, the Rumsfeld Pentagon completed its Nuclear Posture Review. This introduced a new term, "operationally deployed strategic nuclear warheads," so that a planned limit results in more weapons deployed than were the goal of the Clinton administration. Furthermore, in addition to the 1700–2200 newly designated weapons, the Moscow Treaty, signed May 24, 2002, by Presidents Bush and Putin, allows augmenting the deployed force with 6000 or more similar but undeployed weapons. Such a loose agreement is quite contrary to our proposals on pages 373–74 and will only ensure that Russia maintains 10,000 or more strategic nuclear weapons, constituting a grave threat to U.S. security. Furthermore, the treaty limit of 1700–2200 operationally deployed strategic warheads applies only in 2012, at which time the treaty expires.

We trust that this volume provides the basis for understanding future events — even unforeseen future events — in the fields of nuclear power, nuclear weapons and arms control, and terrorism. Events and the actions of the Bush administration have focused public attention on the topics featured in this book. Input from the broader public, resulting in further analysis within government, is urgently needed to move toward a safe, secure, and prosperous society.

AN OPTIONAL REVIEW

OF UNITS AND DIMENSIONS

UNITS

As the reader will discover, this is neither a physics nor an engineering text. But, like any serious book on nuclear energy, it contains both physics and engineering. When we do physics or engineering we inevitably come up against the question of how big or small, how heavy or light, how fast or slow things are. We see here the need for units.

Units are standards. They are quite arbitrary, but are chosen in such a way as to maximize their convenience for whatever purpose is at hand, and so that they can be communicated readily to a wide range of users.

When the colonists came to the United States from England they brought with them the set of units that—at least in its essentials—we still use today. They included the yard, the foot, and the inch, along with less often used items such as the pole, the rod, and the perch. They also included the "wine gallon," which was a volume of 231 cubic inches. This is now our gallon. "A pint's a pound the world around" helps American children to recall that a pint of water weighs about a pound, but American cooks and beer drinkers are more or less shocked to find that in England the pint is based on the imperial gallon—277.42 cubic inches (20% larger than ours)—so that a British pint is not 16 but 19.2 U.S. fluid ounces (and 20 imperial fluid ounces!). And similarly for a "gallon" of gasoline (which is still called "petrol" in Britain).

But what is, say, a yard? Here the arbitrariness of these units becomes apparent. To illustrate, what is known as the British imperial yard was defined by an act of Parliament in the nineteenth century to be the distance between two gold studs placed on a bronze bar that had been cast by the firm of Troughton & Simms. Since bronze, like any metal, expands and contracts with the temperature, it was specified that this bar had to be kept at a temperature of sixty-two degrees Fahrenheit. Likewise, the so-called imperial pound was defined to be the weight of a particular platinum cylinder. It is clear, then, that we are not dealing here with laws of cosmic universality.

In the meantime, a parallel unit enterprise had been unfolding on the European continent. The French, who have always regarded themselves as more logical than anyone else, introduced what is known as the metric system. This system is logical in the sense that if, for example, you know what a meter is and you have been told that the prefix "kilo-" means a thousand, you will have no problem knowing what a kilometer is. Likewise, if you are told that the prefix "milli-" stands for a thousandth, you will surely know what a millimeter is. In contrast, unless someone hands you a guidebook you will not know that a yard is thirty-six inches — three feet — to say nothing of not knowing that a rod is sixteen and a half feet.

For this reason all scientific and engineering work is conducted in the metric system. Occasionally we torture freshman physics students by making them convert metric to British units, but that is only to see the expressions on their faces. Since we cannot see the expressions on the faces of the readers of this book, we will use metric units. But we will tell you how you can convert them into the units that you use around the house. Another advantage of using metric units is that you will be able, if so inclined, to read the reference material cited in the footnotes. These authors also use metric units. And with metric units it will be a lot easier to find your way in new fields of science, technology, and even commerce.

The British have moved to a decimal and metric system, although they still drive on the left side of the road. On the other hand, the metric system has been legal and standard in the United States since the Metric Act of 1866, but the replacement of the "English units" has been stalled by the U.S. Congress.

The metric system is constructed from three basic units of mass, length, and time — respectively the gram, the centimeter, and the second. This last, at least, is common to both the metric system and the common units of daily life. As you may imagine, the history of how the other two units came to be defined is a long one. Originally the meter — which is one hundred centimeters — was defined as precisely one part in ten million of the (polar) quarter-circumference of the earth — that is, the distance from the equator to one of the poles — and that is still a handy way to recall the size of the earth. Because you don't measure the earth's circumference every day, the meter then became the length of a standard bar made of platinum and iridium kept in France at zero degrees centigrade — thirty-two degrees Fahrenheit. But late in the nineteenth century it was related to a certain number of wavelengths of light. In 1927 this was fixed at 1,553,164.13 wavelengths of the red light emitted by cadmium — a standard that can readily be created in any laboratory. You should recall that a meter is a little longer than a yard—1.09 yards, roughly. On the other hand, one inch is 2.54 centimeters (now defined as precisely 2.54000000 centimeters). And now you know that the wavelength of cadmium's red line is about 1/1.55 = 0.7 micrometer.

When it comes to weights, the gram is a bit small for purposes of trade, so that what is usually defined is the kilogram—a thousand grams. This too is defined arbitrarily as the weight of a standard squat cylinder of platinum and iridium. Originally it was the weight of water in a cubic volume measuring a tenth of a meter on each of its sides—a volume that is known as a liter. But the weight of water is, if you want to measure it accurately, a pretty slippery quantity, so the basic unit of weight became a metal cylinder. Again, for our purposes, you should remember that a kilogram is about 2.2 pounds. If you have ever weighed yourself on a scale in Europe, you have no doubt experienced the shock of finding that you weigh something like half what you did before you got on the plane—only to realize that the scale is in kilograms.

When, as in many cases, the units are very small or very big compared to the measured quantity, the prefixes for units are commonly used:

1000	10^3	thousand	kilo	k
1,000,000	10^6	million	mega	M
1,000,000,000	10^9	billion	giga	G
1,000 billion	10^{12}	trillion	tera	T
1,000 trillion	10^{15}	quadrillion	peta	P
1,000 quadrillion	10^{18}	quintillion	exa	E

and for decimal fractions of a unit:

0.1	10^{-1}	tenth	deci	d
0.01	10^{-2}	hundredth	centi	c
0.001	10^{-3}	thousandth	milli	m
0.000 001	10^{-6}	millionth	micro	μ
0.001 millionth	10^{-9}	billionth	nano	n
0.001 billionth	10^{-12}	trillionth	pico	p

Using the symbol "≡" to mean "is defined as,"

$$1 \text{ cm} \equiv 0.01 \text{ m}; 1 \text{ bbl} \equiv 42 \text{ gal.}$$

$$1 \text{ metric ton} \equiv 1000 \text{ kilograms (kg)} \equiv 10^6 \text{ g} \equiv 1 \text{ megagram.}$$

The metric-ton unit is abbreviated as *t*.

Since 1 kg = 2.204 pounds, 1 t = 2204 pounds. In Britain and some other countries, the metric ton is known as a "tonne."

We use degrees centigrade in this book—specifically, the Celsius scale ("C" stands for Celsius, the scale having been proposed by the Swedish astronomer Anders Celsius in 1742), in which the melting point of ice is 0°C and the boiling point of water is 100°C at sea-level atmospheric pressure. Since the Fahrenheit scale (°F) has the ice point at 32°F and the boiling point at 212°F, it is easy to verify the simple formula that gives the Fahrenheit temperature T_f in terms of the corresponding Celsius temperature T_c:

$$T_f - 32 = ((212-32)/100) \times (T_c), \text{ or } T_f = 32 + (9/5) \times T_c$$

Some T_c and corresponding T_f used in the sections on reactor technology are:

T_c	−40	−18	0	15	37	100	300	500	°C
T_f	−40	−0.4	32	59	98.6	212	572	932	°F

We define specific units as they are introduced throughout the book; whether we use as a unit of energy the British Thermal Unit (BTU), the megajoule, the erg, the quad, the barrel-of-oil equivalent, the electron volt (eV), the calorie, or the Calorie is a matter of convenience. These particular quantities all refer to energy, and they all have the same "dimensions."

DIMENSIONS

Distance has the dimensions of length, whether you choose to measure length in feet, or meters/kilometers, or light-years—the distance light travels in a year. Other quantities obviously do not have the dimensions of length, for instance the day; the "length of the day" cannot be measured in feet or even in km—for this purpose we have other units such as seconds, days themselves, years, or centuries. Dimensions do not proliferate endlessly; as was already mentioned, we can measure almost all physical magnitudes with three such dimensions—length $[L]$, mass $[M]$, and time $[T]$.

Other physical quantities then have dimensions that are unique composites of these three, and you can work them out yourself:

Volume: $[L]^3$; area: $[L]^2$:

Note that it is all right to "cancel the units" to show that, for example,

1 hr = 60 min × 60 sec/min = 3600 seconds.

Many of the quantities that are used in physics, chemistry, engineering, etc. involve combinations of units. Their "dimensions" refer to these combined units. In this book we need a few of these quantities, and we will explain some of the cast of characters here. Others we identify in the text as they are needed. A simple and fundamental example is that of speed or velocity. A speed measures how rapidly distance changes in time. So its dimension is distance/time—that is, $[v] = [L]/[T]$, or alternatively $[v] = [L] [T]^{-1}$. When we drive a car or ride a bicycle, we usually measure this speed in miles per hour. But in metric units we would measure it in meters per second or possibly kilometers per hour. One kilometer is roughly six-tenths of a mile, so a speed-limit sign in Europe that says "100" corresponds to about sixty miles an hour.

A speed we will encounter later in the book is the speed of light in vacuum. It is usually denoted by c and is very special to physicists. It is special because Einstein showed in his theory of relativity that this was the greatest speed any material object can attain. Nothing can go faster than light in a vacuum. There is no law preventing material objects from exceeding the speed of light in a physical medium, such as water, in which light travels at only about $0.75\ c$. In fact, fast electrons exceeding the speed of light in water produce so-called Cerenkov radiation, which is very useful for tracking such particles. In metric units, c, which can be measured to a huge accuracy, is exactly 299,792,458 meters per second. This is about 3×10^8 meters per second or 3×10^5 kilometers per second. It is this speed that enters Einstein's equation $E = Mc^2$.

If we multiply a speed by a mass we have a quantity of dimensions $[M] [L] [T]^{-1}$—gram-centimeters/second—which is known as the momentum. Isaac Newton showed that it requires a force (and a time) to change a momentum. Indeed, the rate of change of momentum in time is this force. The dimensions of a force are evidently $[\text{momentum}] \times [T^{-1}]$. Since this book is about energy, the units in which energy is to be measured are of great importance to us. Because one kind of energy (kinetic energy) is found to be proportional to mass and velocity-squared—for velocities small compared with that of light— the dimensions of energy are

$$[E] = [M] \times ([L]/[T])^2 \text{ or } [E] = [M] \times [L]^2 \times [T]^{-2},$$
and all energies have the same dimensions.

We see also from Einstein's equation, in which E is an energy, that its dimensions are $[E] = [M] [L]^2 [T]^{-2}$. This quantity occurs so often in physics that the unit of 1 gram $\times$ 1 centimeter^2/1 second^2 is given a name—the erg. We shall often have to deal with much bigger and much smaller amounts of energy; hence it is convenient to introduce names for these as well.

If we take ten million ergs—10^7 ergs—we have a unit that is called a joule. James Prescott Joule was the nineteenth-century British owner of a brewery who did physics on the side. He investigated how various forms of energy can be interconverted. For example, to raise the temperature of water requires energy, typically a certain number of joules; so the joule is a useful unit for common tasks. On the other hand, even the erg is a monstrously large unit of energy for individual nuclear processes. This may seem ironic considering what damage a nuclear explosion can cause. But these explosions consist of adding up the effects of an almost incomprehensibly large collection of nuclear events such as fission. These individual events are conveniently measured for energy in a unit that is called the electron volt—abbreviated eV. This is the energy that an electron—the lightest unit of matter that carries electric charge—would acquire in dropping through a battery-sized potential of one volt. One electron volt is 1.6×10^{-19} joules—close to nothing. A nuclear process produces some millions of electron volts (MeV)—still almost nothing. But they add up.

When a scientist gives you a formula and claims that it explains something, it is a good idea to "check the dimensions." One side of an equation cannot be equal to the other side unless the dimensions are the same on the two sides. If not, the formula will not be valid if the lengths are measured in kilometers instead of meters or even in yards; for instance, the perimeter of a square, $P[L] = 4 \ S[L]$ (where the dimensions of perimeter and side are written in square brackets next to the symbols), and the area, $A[L^2] = S[L] \times S[L]$, but it is clearly an error to imagine that $A[L^2] = 4S[L]$.

DEFINITIONS, UNITS, AND DIMENSIONS
FOR INDUSTRIAL ENERGY

Despite the authors' preference for metric units, in this section we present the common units of the energy industry, among which is the barrel ("bbl") of crude oil, defined as 42 gallons of 231 cubic inches each. The volume of a barrel can readily be calculated in metric units:

$$1 \ bbl = 1 \ bbl \times (42 \ gal/bbl) \times (231 \ in^3/gal) \times (2.54 \ cm/in)^3$$
$$= 1.590 \times 10^5 \ cm^3 = 0.1590 \ m^3.$$

The chemical energy content of a barrel of crude oil to be burned with air is the "barrel of oil equivalent," BOE: 1 BOE $\equiv$ 5.8 million British Thermal Units (BTU). Fine, but what is a BTU? It is a unit of energy, defined as that amount of energy that will warm one pound of water by one degree Fahrenheit. Since a pound is 0.4536 kg, and there are 180 (212 − 32) Fahrenheit degrees between the boiling point of water and the freezing point (compared with 100 degrees Celsius), 1 BTU would raise the heat of 453.6 g of water by 100/180°C.

The "calorie" is defined as the energy required to warm 1 g of water by 1°C, so 1 BTU = 453.6 × (100/180) = 252 calories. Since measurements of the heat produced by a certain amount of mechanical work yield 4.186 J/cal, we have 1 BTU = 252 cal × 4.186 J/cal = 1055 J = 1.055 kJ.

Thus 1 BOE = 5.8 × 10^6 BTU × (1.055 kJ/BTU) = 6.12 × 10^9 J = 6120 MJ (megajoule), or 6.12 GJ (gigajoule).

To relate the barrel of oil equivalent (BOE) to a familiar quantity of energy, we note that it is also the energy equivalent of 5614 cubic feet of typical natural gas, or about 0.22 ton of typical bituminous coal.

Large modern power plants, whether nuclear- or fossil-fueled, typically produce electrical energy at a rate of a billion watts — a million kilowatts — of electrical power. This could alternately be stated as 1000 MW (megawatts) of electrical power or 1 GW (gigawatt). To eliminate confusion between the energy input to the plant and its energy output, it is conventional to refer to the "megawatt thermal" — MWt or MW(t) — as distinguished from the "megawatt electric" — MWe or MW(e) — in referring to an electricity-generating plant. A nuclear plant generating 1000 MWe or 1 GWe at a "thermal efficiency" of 30% must produce some 3.33 GW of thermal energy; the plant is a 3.33 GWt plant.

Amounts of energy needed for a major country, one that has on the order of 100 million people, require a large unit of measure. Since Americans use electrical energy at a rate of some 1.3 kW per head (13 hundred-watt lightbulbs), each nominal 1-Gwe power plant will supply a population of 800,000 using energy at the rate typical of Americans.

We recall that 1 watt used, as in a lightbulb, for 1 second is 1 joule, so that 1 kilowatt-second is (1 kilojoule). The number of seconds in a year is obtained in multiplying 60 × 60 × 24 × 365.25 to give 31.56 million seconds or 3.156 × 10^7 s in a year. So the million-kW plant generates 3.156 × 10^{16} J/yr. A typical 100 W lightbulb is built to burn about a thousand hours, so in its lifetime it burns 1000 × 60 × 60 × 100 = 360 megajoules or about 0.2 BOE (used at 30% efficiency).

The "quad" is a residue of the Anglo-Saxon heritage.

One quad ≡ 1 quadrillion BTU; as the trillion is a thousand billion, the quadrillion is a thousand trillion and is thus 10^{15} BTU. The quad as 1 quadrillion (10^{15}) BTU or 1055 × 10^{15} J is almost identical to the metric unit the exajoule (EJ), which is 10^{18} J. Thus, in terms of GW-yr of electrical energy, 1 quad = 1055 × 10^{15}/3.156 × 10^7 = 33.4 GW-yr. As for the measurement and supply of primary energy (converted at about 30% efficiency into the electrical product, with 70% of the primary energy converted into waste heat), 1 quad of primary energy will yield about 10.0 GW-yr of electrical output.

A unit frequently encountered is the "ton of oil equivalent" or "toe." Since oil yields about 10,000 calories per gram on combustion, the ton of oil equivalent is defined as 10 million kcal, or 41.87 gigajoules. One million tons of oil equivalent (1 Mtoe) is thus 41.87×10^{15} J, or 0.042 EJ. While not quite so vague as the weight of a quart of milk (what butterfat content?), the differences in quality of various oils or coals lead to definitions that are different by a few percentage points.

Access to the Web makes it easy to convert one unit to another. A useful site is http://www.digitaldutch.com/unitconverter/, where one confirms, for instance, that one food calorie equals 4186 joules. And at a site more specialized in fuels, http://www.processassociates.com/cgi-bin/convert.exe, one can easily convert one equivalent ton of crude oil to 42.41 GJ.

The most authoritative source on units of measurement is the U.S. Government's National Institute of Standards and Technology (NIST), at http://physics.nist.gov/cuu/. The reader will have no difficulty with the "Système Internationale" (SI units)—the modern metric system, for which the basic units are the meter, the kilogram, and the second.

Megawatts and Megatons

Introduction

IF IT IS TO benefit humanity, concern for our planet and the future of our civilization needs to be matched with an understanding of the facts. High among such apprehensions should be the potential of destruction of populations and the entire fabric of organized society by the use of the existing arsenals of nuclear weapons in Russia, the United States, and other countries. A second worry is the plundering of the planet for carbon-based energy, with the accompanying increase in atmospheric carbon dioxide and the resulting rise in world temperature. Our purpose in this book is to provide sound information for resolving the conundrum of control of nuclear weaponry and for providing acceptable energy for the future, whether or not society uses nuclear power.

Nuclear energy burst on the world in August 1945 with the explosion of atomic bombs over the Japanese cities of Hiroshima and Nagasaki. Never since employed in warfare, nuclear weapons have nevertheless played an enormous role in U.S. national security. With an overall cost of about $4 trillion for the weapons, their command and control, and their delivery systems, they are regarded by some as having prevented a third world war for more than 50 years. Others believe that they have created an unparalleled hazard for the United States and for the world. Perhaps both views are correct.

Soon after the Second World War, nuclear energy was applied to less destructive purposes, in the propulsion of submarines and naval surface ships, as well as icebreakers. For the first time, the United States (and soon thereafter the Soviet Union and then Britain and France) had a true submarine—one that could travel submerged for months, capable of destroying shipping or preventing its passage, threatening opposing naval forces, or launching strategic ballistic missiles to destroy cities or military targets.

In parallel with the mastery of nuclear energy for warfare arose the nuclear power industry. Under normal circumstances, more than 430 large commer-

3

cial reactors now provide almost 20% of the world's electrical power—the equivalent of 340 plants of a million kilowatts of electricity, each of which could serve an American city of nearly one million. In France, 80% of the electrical power comes from nuclear facilities.

This book first provides the background of the mastery of the release of nuclear energy over the past sixty years—explosively in weapons, and more gradually in nuclear reactors producing heat and electricity. Next there is a section on the technology and evolution of nuclear weaponry—simpler in principle than power reactors—followed by a chapter on the fundamentals and implications of the impact of nuclear radiation on health, which is of importance both in nuclear weaponry and in nuclear power. Then comes a section on nuclear reactors, their common features and their variants, as well as on experience in the supply of nuclear electricity. Following a discussion of the long-term energy future, the book ends with a view of the policy options for managing the world's nuclear weaponry and for obtaining the benefits of nuclear power at tolerable cost and risk.

Nuclear Weapons. The first nuclear weapons had an energy yield equivalent to about 20,000 tons of a powerful explosive (TNT)—20 "kilotons," or 0.02 "megaton." A typical strategic weapon now has a yield of 0.5 megaton. The number of U.S. nuclear devices grew to a peak of 33,000 in 1967 and is now about 12,000. The Soviet Union had some 45,000 nuclear weapons available in 1986, and Russia now has perhaps 18,000. For years, the push of a button in Moscow would have ensured the destruction of the United States and of Western Europe by a Soviet Union that was in turn threatened with devastation by the more accurate U.S. weapons.

A decade after the end of the Cold War, U.S. and Russian nuclear arsenals remain ready for use. The use of a mere 20 weapons would kill 25 million people in the United States or Russia. In the United States, the political debate focuses on absolute security against the explosion on American soil of even a single weapon from some new ("rogue") nuclear state—perhaps North Korea, Iraq, or Iran. With the evolution and diffusion of industrial technology over the last half century, accelerated by the Internet revolution, the barriers to the acquisition of nuclear weapons are primarily political; but there also remains the requirement of obtaining materials like plutonium or enriched uranium.

Nuclear Power. The generation of electricity from nuclear reactors is a mature industry with potential for further technological innovation. The importance of nuclear power, however, has a dimension beyond the question of price and operating risk. Over the last 20 years, the scientific understanding of climate has led to the conclusion that the continued increase in carbon dioxide in the atmosphere (largely from the burning of fossil fuels—coal, oil, and gas) can transform the earth's climate, causing great human and economic losses. Whether this "enhanced greenhouse effect" is already reflected in the fact that the 1990s were the warmest decade (and 1998 the warmest year) in recorded human history is still debated by some. Nevertheless, the nations of the world, meeting at the Conference of the Parties of the United Nations Framework Convention on Climate Change in Kyoto in 1997, agreed that by the year 2010 or 2012 the industrialized countries should reduce their overall carbon emissions—more precisely, their carbon-dioxide-equivalent emissions—by at least 5% below the 1990 level, despite the continued growth in population and industrial output. Even if these goals are met, atmospheric carbon dioxide will still increase in absolute terms.

If the level of carbon dioxide in the atmosphere (and its equivalent from other "greenhouse gases") were to be held only to a doubling of the preindustrial level—i.e., that of 1850—more than half the world's energy would need to come from noncarbon sources[1] by the year 2050.

Nuclear energy can fill part of this gap, but only if there are assured

- availability of nuclear fuel,
- safety against catastrophic accident,
- management and disposal of nuclear waste, and
- affordable cost.

In contrast to the preponderance of nuclear power in France (and the many reactors built in Japan and Korea in recent decades), not a single reactor built in the United States was ordered after 1973, Sweden is committed to the elimination of nuclear power, and Germany agreed in June 2000 with its nuclear industry to close reactors over the next 20 years—more specifically, after producing an amount of electrical energy equivalent to an average working life of 32 years for each nuclear plant.

The Link Between Nuclear Weapons and Nuclear Power. The materials that produce heat in nuclear power plants—uranium and plutonium—can be used more and more readily to make nuclear weapons. The number of nations and groups in the world capable of fabricating nuclear devices has increased greatly with the diffusion of technology, so that the real barrier—once the decision to build nuclear weapons has been made—is the availability of plutonium or enriched uranium. Since the 1950s, the expansion of nuclear power in states not possessing nuclear weapons has been undertaken with an emphasis on strict controls preventing the diversion of civilian materials into military armories, but these controls consist mainly in accounting and do not of themselves constitute a true security system.

Something New: Megatons to Megawatts. Even before the end of the Cold War and the dissolution of the Soviet Union, that nation and the United States (with more than a hundred times as many nuclear weapons as the rest of the world combined) had undertaken bilateral agreements to limit and reduce their nuclear weaponry. In 1993 the United States contracted to buy 500 tons of weapon uranium from Russia, its 90% enrichment in U-235 so reduced (to about 5%) that it is useful in nuclear reactors and impossible to use directly in weapons. Over 20 years, $12 billion is to be transferred for the acquisition of material from more than 10,000 disassembled weapons. In 1998, the United States and Russia agreed that each would declare 50 tons of weapon plutonium excess and transfer it out from the military sector, either for disposal as waste or for burning in reactors. This quantity corresponds to plutonium from about 20,000 weapons. The Russian uranium purchased by the U.S. side is handled by the recently privatized (July 1998) United States Enrichment Corporation; the program goes by the name "Megatons to Megawatts." In fact, a megaton of explosive energy release corresponds to the heat produced in about 20 days of operation of a plant supplying 1000 megawatts of electricity—thus 20,000 megawatt-days.

One of the major controversies concerning the future of nuclear power is the near-term employment of plutonium, with one extreme position consisting in rejecting the use of nuclear reactors to burn plutonium from excess nuclear weapons (a "Megatons to Megawatts" approach), while another insists that plutonium produced in the normal operation of civilian power plants should always be separated and recycled in those reactors rather than disposed of as "waste."

In this book we introduce the tools for understanding both nuclear weaponry and nuclear power, highlighting the options available, those under investigation, and the decisions that must be taken to reduce the threat of nuclear weapons. We analyze the prerequisites for the expansion of nuclear power that would have to be met if it is to make a substantial contribution to limiting the warming of the earth by enhanced greenhouse effect.

Making the wrong decisions in nuclear weaponry and nuclear power can greatly imperil the security of the United States and, indeed, of the entire world. Right decisions may provide the tools that can supply clean energy for millennia. Although we do not hesitate to draw conclusions from our analysis, we encourage interested readers to reach their own. We begin with a clarification of the nature of energy and of nuclear fission—the genie that poses both such peril and such promise.

A Fable for Young Readers
(and whimsical elders)

—In this bit of chocolate candy, there are about one hundred thousand billion billion atoms, which can be written with a 1 followed by 23 zeros:

100,000,000,000,000,000,000,000, phew!

—Tell me, Grandpa, do you scribble zeros all over the page when you do your sums?

—No. We nuclear physicists like to save ourselves trouble and paper, because beautiful trees are destroyed to make paper, and we have to stop the destruction of trees. We write simply: 10^{23}. We put, above and to the right, in small characters, the number of zeros after the 1. We save paper and valuable time. So 10^2 is 100; 10^1 is 10; and 10^0 is 1—no zeros after the 1.

—And what is the candy atom? A tiny bit of candy?

—No, it is a mixture of three little grains stuck together, atoms of hydrogen, carbon, and oxygen—the same carbon that we burn as charcoal to grill hot dogs, and the same oxygen that is in the air you breathe and animals and plants breathe, and that is absolutely necessary to life, and the same hydrogen that amounts to one-ninth the weight of every body of water.

—There are only three kinds of atoms?

—No, there are about a hundred, and they can mix together in countless ways, either by chance during the billions of years of the history of our universe, or because they are rearranged by people, as, for example, when your mom makes an apple pie from her own secret recipe.

—And an atom—tell me how it's made. Can it be cut, like candy?

—Yes, it can. Think of the atom as a little hollow ball. It is surrounded by a very thin cloud of electricity, in the middle of which there is a little seed, the nucleus, which is 100,000 times smaller.

—You mean 10^5 times smaller.

—You've got it. You have understood how to keep zeros from monopolizing our paper.

—And the little seed, the nucleus, does it have parts?

—Yes, it does. Looking at the parts is what nuclear physicists do. The different types of atoms have different nuclei. Nuclei are made up of two types of little seeds—different, but with very nearly the same mass—that we call the proton and the neutron. The nucleus of the lightest atom, hydrogen, has only one proton, and its cloud of electricity is made up of exactly one electron. For the heaviest atom in nature, uranium, the total number of protons and neutrons is, for the most part, 238, but there are exactly 92 protons, and its cloud of electricity is made up of 92 electrons.

—And the protons and neutrons—can they be cut?

—Yes. That's the work of particle physicists. In a proton or a neutron, there are quarks, which are 1000 times smaller than the nuclei.

—You mean 10^3 times smaller.

—That's it. The quarks are stuck together by gluons.

—And these quarks—can they be cut?

—No.

—Why?

—Because.

—Later, when I am big, I will cut the quarks.

—It's not possible.

—Why?

—You will understand that when you grow up, if you stick to your studies.

—It must be interesting to be old. I would like to grow old much faster.

—Since you are no longer afraid of zeros, let's take a closer look at some of the things that make up our universe. If, in a tiny piece of your candy, there were only 1000 atoms, it could be made of 10 layers of 100 atoms each—a cube 10 atoms high, by 10 atoms wide, by 10 atoms deep. And the size of the candy would be ten times the distance between neighboring atoms. If there were 10^{21} atoms, the size of the chocolate would be 10^7 times larger than the distance between atoms.

—Why?

—Because $7 \times 3 = 21$. Think of the preceding calculation where there were 1000 atoms, that is to say 10^3 atoms, and the rule will become obvious. Multiplying by ten means another little zero, so 10^7 atoms are just that—10^7; and 10^7 rows, each of 10^7 atoms, make 10^{14}; and 10^7 layers of 10^{14} atoms make 10^{21}.

—I see.

—If your candy is a bit more than a tenth of an inch, or 0.3 centimeter, across, the distance between its atoms is 0.000,000,03 centimeter, which can be written 3×10^{-8} to save paper. There are 7 zeros between the decimal point and the "3." That gives you an idea of the size of an atom, which is quite close to the distance between atoms in a candy. We can put together ten million of them side by side to reach across this bit of candy. Ten million is a figure easy to imagine, since there are about that many people in a large city, like New York, Los Angeles, London, or Paris. In a large country there are 100 million people—

—You mean 10^8 people.

—Yes, 270 million people in the United States, 60 million in France, 150 million in Russia, and more than 10^9—a thousand million or a billion—in China and in India. And the richest man in the world has about $80 billion— almost $300 for each of the people in the United States.

—How did he make that much money?

—He is a wizard at telling computers what to do.

—What does he do with it all?

—He has been working hard at using it to help people in the United States and the world—so that children are protected against disease, and people who use libraries can easily read rare books or see marvelous paintings at the art museums of the world.

—See pictures at the museum! I wouldn't like that at all. And besides, I want to go far away. Far! Far! Far away to the end of the universe, to explore new worlds.

—That is very difficult. Because you only live about 100 years.

—You mean 10^2 years.

—That's it, and the end of the universe is 10^{10} years away if you travel at the speed of light, which is a billion kilometers per hour. And even the nearest stars with planets possibly inhabited by living creatures are too far away for us to be able to even dream about going there. It would be marvelous enough if we could one day receive intelligent messages sent by radio. Explore our planet Earth, which you will find beautiful, very beautiful, and still full of wonders to explore.

—One planet—you mean 10^0 planets?

—You've got it. Now you can become a nuclear physicist or an astronomer.

—Bah! I would rather be a firefighter or an explorer.

—Don't be too hasty. During the last fifty years, exploring the nucleus of the atom has led to the production of the electricity you read by, but it has also brought about the danger of humanity's extinction. That is a fire to fight! You and your friends should learn to use it wisely to contribute to the well-being of everyone, and not only to the unlimited riches and power of a few madmen whose heads and guts have remained in the stone age.

You will be convinced by what you learn when you are able to read this book. And give it to your parents. A few chapters may seem too difficult even for them. Tell them to skip those. It is not a textbook, where you start with the simple to arrive at the more complicated. It is an overview of vast domains, very different, connected by their relation to nuclear energy, and some parts are more difficult than others.

When you grow up, you will explain even the hard parts to your parents. They will be delighted.

CHAPTER 1

All Energy Stems From the Same Source

ATOMS, ELECTRONS, AND NUCLEI

ALL MATTER is an assembly of atoms. A liter of water, for example, contains about 10^{26} atoms of hydrogen and oxygen. Each atom has at its center a tiny nucleus, in which is concentrated nearly all of the mass and which occupies about a millionth of a billionth (10^{-15}) of the volume of the atom, which in turn is of the order of 10^{-24} cm³. The nucleus consists of neutrons and protons—particles having about the same mass. The proton carries a positive electric charge; the neutron, on the other hand, has no charge at all—it is electrically neutral. Neutrons and protons are usually referred to as nucleons, since nuclei are composed of them.

The nucleus is surrounded by orbits or "shells" of electrons; an orbit is some 10^{-8} cm in diameter. The mass of the electron is about 2000 times smaller than the mass of the proton or neutron. The atomic electrons are bound to the nucleus by electrical attraction. Each electron has a negative charge, and when all these charges are added up they exactly compensate for the positive charge of the nucleus. An atom either with one or more extra electrons or lacking one or more of its normal electrons is an "ion."

Common sense is not much help in trying to visualize an atom. The electrons and the nuclei that make them up obey laws peculiar to the infinitesimally small. This is the realm described by quantum theory and to some extent by the theory of relativity. These theories are among the great achievements of the twentieth century. They were created by geniuses who were capable of turning common sense on its head. To understand them in their full depth requires years of study. The discussion that follows aims to acquaint the reader with the basic elements of these discoveries.

The electrons, which are far away from the positive nucleus but bound to it by the attraction of their negative electric charges, have well-defined energies as determined by the quantum theory of the atom, published in 1913 by the Danish physicist Niels Bohr. The familiar "periodic table" of high school chemistry was introduced by the Russian chemist Dmitri Mendeleev in 1891, arranging the elements roughly in order of atomic mass but—it turns out— precisely in order of "Z"—the "atomic number," the number of protons in the nucleus (see below). Of course, Mendeleev and his contemporaries knew nothing of protons or atomic number. The "periods" of the table seem at first sight to be rather arbitrary, but they are really of a length that ensures that elements in the same relative position in a period or row have similar chemical properties. These periods are of length, 2, 8, 18, 32, 50, 72, and 98, and were first explained as an early triumph of quantum mechanics as applied to the atomic structure—they correspond to concentric shells, in which the electrons are located at various distances from the nucleus.

Chemically, an element is defined by the number of protons in the nuclei of its atoms. Each atom of a given element contains the same definite number of protons in its nucleus; it is this number that determines the element's chemical properties, and that is known as the atomic number (Z) of that element. Thus, there are seven protons in the nucleus of every atom of nitrogen, and so nitrogen's atomic number is 7.

But while for any element the number of protons in the atomic nucleus is by definition fixed, the number of neutrons in the nucleus may vary. Atoms of an element that have certain but different numbers of neutrons in their respective nuclei are called isotopes of that element. For example, the simplest element, hydrogen, with a single proton in its nucleus ($Z = 1$) exists in three forms: the single proton by itself as the nucleus, the proton with one neutron, and the proton with two neutrons. The two heavier forms, or isotopes, of hydrogen are known as deuterium and tritium, respectively.

Most elements have several isotopes. Two important isotopes of uranium ($Z = 92$)—an essential element used to produce nuclear energy—are uranium-235, with 92 protons and 143 neutrons, and uranium-238, which also, of course, contains 92 protons but has 146 neutrons. Thus $Z = 92$ for both isotopes, but uranium-238 has three more neutrons.

INTERACTIONS

To extract an electron from the atom requires an investment of energy, enough to overcome the electrical energy that binds that electron to the nucleus. Chemical reactions, combustion, and everything that affects our senses are related (directly or indirectly) to exchanges of energy among the electrons that surround the nuclei. These reactions sometimes involve some transfer of electrons among atoms. The nuclei are indifferent to these flirtations or marriages between electrons in the distant shells. Such liaisons can affect several atoms, which attach themselves to each other in a stable way to form a molecule, the smallest unit, or quantity, of a chemical compound; or they can involve hundreds, thousands, or millions, combined to form crystals and all of the various forms of matter that make up our universe.

As we have said, a single proton forms the nucleus of the simplest chemical element in nature—hydrogen—the most abundant element in the universe, making up about three-quarters of its observed mass. That is to say, the mass that emits starlight. There is now proof that 95% of the mass of the universe has not yet been directly observed, and one of the most active and fundamental current questions of science is to identify this "missing mass." Protons and neutrons are bound together by a special short-range force. This nuclear force is much stronger than the electric force, which tends to repel electric charges of the same sign and hence if not opposed would render nuclei unstable. The nuclear force is felt only when the nucleons are closer to each other than about 10^{-13} cm. It is necessary to expend a million (10^6) times more energy to extract a nucleon from the nucleus than to extract an electron from an atomic shell; and since the distance is 10^5 times shorter, the force is therefore 10^{11} times greater. The nuclear force acting on a single proton is approximately what is needed to support a mass of a hundred kilograms, while the atomic force is about what is required to hold up a microgram; even some of our scientist friends are unlikely to have figured this out.

ENERGY

The temperature of a medium is a reflection of the random motion of its atoms. Raising the temperature corresponds to increasing their agitation. The average speed of the atoms or molecules increases—they are animated in a kind of incessant motion. They collide with their neighbors or with the atoms of the walls that contain them. The burning sensation that is felt when a finger is stuck into a flame reflects the atoms dancing in the flame and entering into collisions with those of our flesh. They tear out the outermost orbital electrons and thereby provoke the dissociation of the groups of atoms of which our tissues are made. (At modest temperatures, specialized nerves in humans and other animals warn of potential damage.) From a microscopic point of view, the energy of these atoms is of the same nature as that of a moving automobile, or of a thrown baseball or any projectile. It is called "kinetic energy," i.e., energy of motion. The sum of all the kinetic energies of these agitated atoms constitutes the thermal energy of the medium. In solids or liquids, because the atoms or molecules exert forces on one another, raising the temperature increases the energy associated with these forces, and this increase in energy is similar in amount to the increase in kinetic energy.

Energy comes in a variety of forms that appear very different from one another: kinetic, thermal, electric, magnetic, chemical, and many more. They can be transformed, one into another. But this transformation obeys a rule that has no exceptions: the total energy is conserved. Any energy that appears does so at the expense of another form of energy. At the end of the transaction the accounts must balance. The other forms of energy are usefully termed "potential energy"; they have the potential to be converted into kinetic energy.

The formulation of the law of conservation of energy was a great theoretical and practical breakthrough. It applies just as well to collisions between infinitesimally small particles as to complicated systems like a battery that heats the filament in a flashlight bulb, a motor that operates a crane, a bird that swallows a fly. In short, it applies to everything.

The law of conservation of energy was put to a severe test with the discovery of radioactivity in 1896 by the French physicist Henri Becquerel. In 1898, Pierre and Marie Curie, working in Paris, isolated two intensely radioactive elements, which they named polonium and radium, and Marie Curie introduced the term "radioactivity" for the process of spontaneous emission of radiation from such substances. Where does the energy arise which, for example, ejects helium nuclei—"alpha particles"—from atoms of radium? An alpha particle has kinetic energy some 200 million times greater than the agitation energy of a radium atom at room temperature. How does radium manage to glow and remain warmer than the bodies that surround it—for weeks, months, even centuries?

The law of conservation of energy is a very reliable guide for scientists. It allows them to greet discoverers of perpetual motion with appropriate skepticism: the problem always boils down to the presence of some hidden form of energy that had not been taken into account. If the Curies did not wish to believe in miracles, there had to be a vast source of energy that was being tapped by radioactivity. Explaining this, as will become clear, was one of the triumphs of the theory of relativity, and of nuclear physics.

The light from the sun conveys energy, ultimately absorbed by terrestrial objects which transform it into heat or chemical energy. Electrical energy can be transported by electrons circulating in metallic conductors, set in motion by electromagnetic forces generated by a variety of machines. Chemical energy comes from the change in the binding energy of the orbital atomic electrons which combine in chemical reactions. The energy of radioactivity is something different; it manifests itself not only in the form of energetic alpha particles but also in the emission of electrons (known as beta particles) with a million times the energy of an ordinary atomic electron. Many radioactive materials emit gamma rays—the equivalent of super-energetic X-rays. To begin to understand radioactivity required a revolution—in this case Albert Einstein's "special theory of relativity." One of Einstein's many claims to fame is to have shown that all variations of energy E are associated with a change of mass M, by the simple universal law

$$E = M \times c \times c \text{ or, more commonly, } E = Mc^2$$

where c is the speed of light in vacuum.

This extraordinary law—which is so simple and universal and relates three different quantities, mass, velocity of light, and energy—is easy to use. According to Einstein, all that is needed is to write M in kg, E in joules (J), and c in meters per second (m/s): E (joules) = $M(\text{kg})c^2$. Since $c = 3 \times 10^8$ m/s and its square is 9×10^{16}, E (joules) = 9×10^{16} (joules) for the energy equivalent of a single kilogram of mass. Our task as authors is made much easier by the discussion of "units" and "dimensions" in the "Optional Review" section in the front matter, which will now come in handy.

In everyday life, a 100-watt lamp is a common item around the house; it provides a pretty bright light. The watt—named after Scottish physicist James Watt (1736–1819)—is a unit of "power," which is the rate—the per-second amount—at which energy is created or absorbed. One watt is one joule per second, or $1\,W = 1\,J/s$. Ten such bulbs consume 1000 W, or one kilowatt (1 kW) or 1000 J/s, as does also a household toaster. In one hour or 3600 s, 1 kW accumulates to a total of 1000 J/s $\times$ 3600 s = 3.6×10^6 J or 3.6 megajoules = 1 kilowatt-hour = 1 kWh.

Using Einstein's equation, one can easily deduce that one kilogram of mass disappearing would liberate 9×10^{16} joules, equivalent to 25 billion kWh of electrical energy. Thus, a typical million-kilowatt power plant—be it nuclear or fossil fuel—that supplies a city of 800,000 people would require 25,000 hours, or about three years of operation at full power, to produce electricity corresponding to the transformation of a mass of one kilogram. Over those three years, in fact, about 75 tons of nuclear fuel would be removed from the reactor, 3 kilograms less than went in—for three kilograms of mass would have been converted to heat at 100% efficiency. Only 30% of this heat would be converted to electrical energy by means of steam turbines and electrical generators of the usual kind.

To consider an earlier unit of energy, it is known from advertising that eating an egg gives your body 90 Calories' worth of energy, that a one-pound steak supplies 800 Calories, and that one shouldn't consume more than 2000 Calo-

ries per day if one doesn't want to put on weight. (Note that these "Calories" are "large calories," used as a convenient unit for measuring the energy value of food.) A Calorie of heat will raise the temperature of 1 kg of water by 1°C; it corresponds to 1000 calories. An ordinary calorie, correspondingly, will raise the temperature of 1 g of water 1°C. (A Calorie is 4186 joules.)

These Calories are produced by a change in mass. When an atom of carbon ultimately combines with two atoms of oxygen to yield a molecule of carbon dioxide, a reaction that takes place when carbon burns in air, or with the carbon that is ubiquitous in our diets, the molecule of resultant carbon dioxide has a greater kinetic energy (of vibration, rotation, and overall motion) than that of the initial carbon and oxygen atoms. This energy is supplied by the energy that was stored in the atomic orbital electron shells. It corresponds to a change of mass: the atoms to be burned weigh more than the cooled combustion products. The change in mass represents about one five-billionth of the mass of the atoms to be burned. Thus, the well-known law of chemistry, which chemists up until Einstein's time accepted, that the mass of the products of a reaction is equal to that of the reacting products is false—although by an almost imperceptible amount. The heat produced in the reaction comes from an inevitable but tiny loss of mass.

In the sun, four protons ultimately combine to form a nucleus of helium, after a few intermediate stages, and the difference in mass between the four nucleons and the nucleus of helium is close to 1% of the initial mass, large enough to be easily measurable in the laboratory. One consequence of this mass difference is experienced every time a person gets a sunburn or sees our lovely earth, or uses the fossil fuels created by sunlight millions of years ago.

The difference in scale between the mass changes associated with nuclear and chemical reactions is colossal: 1% instead of a fraction of a billionth. For chemistry, the measurement of heat release is far more feasible, accurate, and convenient for predicting potential reactions than is the measurement of mass differences. But when particles annihilate entirely, their entire mass is available as energy. For instance, a positron (an antielectron with a positive electric charge, emitted by certain radioactive nuclei such as sodium-22 or copper-64) coming to rest in matter combines with an ordinary atomic electron to form two high-energy X-rays, each with an energy close to the rest mass of the electron or positron—510,000 eV, or 510 kilo-electron-volts. This annihilation process is the basis of positron-emission tomography (PET), now widely used in medical research and diagnosis.

The luminous energy radiated by the sun and all of the stars, which are suns in their own right but are farther away from us, is emitted at the price of a loss in their mass. Stars can last for billions of years because they tap the store of nuclear energy—a million times greater than the thermal energy in the interior of the hot star.

Kinetic energy (the energy of motion) is readily transformed into heat, as is evident from the warming of the brakes on an automobile or bicycle as they stop the vehicle. When a lead ball travels at the speed of sound in air, three hundred meters per second, it has a kinetic energy such that, if stopped by a steel armor plate, the transformation of the ball's kinetic energy into heat would raise the temperature of the lead by 300°C. The speed of light in vacuum is a million times greater than that of sound in air. The delay due to the finite velocity of propagation of light is easily observed in the course of a telephone conversation with someone in a distant country if the call is carried via a relay placed on a satellite at forty thousand kilometers above the earth. The delay in the voice carried by a radio wave traveling at the speed of light is perceptible in the conversation when its echo arrives at its destination, although the delay is a million times shorter than it would be if the signal were traveling at the speed of sound.

This enormous speed of light, which appears squared in the formula $E = Mc^2$, means that the total energy associated with the mass of the lead ball at rest (its "rest mass") is some trillion times—a million squared—greater than the kinetic energy the ball has when it moves at the speed of sound in air. Moving at a speed equivalent to 2% of the speed of light, a lead ball of six kilograms mass would liberate as much energy in a collision as was produced by nuclear fission in the six kilograms of plutonium of the bomb that razed the city of Nagasaki.

THE DISCOVERY OF NUCLEAR ENERGY

Until 1945, nuclear energy manifested itself either on a very large scale appropriate to the evolution of the universe, in the sun and the other stars, or on the microscopic scale of reactions that could be produced in the laboratory between isolated nuclei. Many, even among the greatest physicists, believed that it would never be possible to tap this immense reserve of energy contained in the mass of atomic nuclei. Einstein once said that it was as likely as a blind man hunting a bird in a country where there were very few birds. Nuclear physics was long considered simply a fundamental science without any prospect of serious application. But as sometimes happens in science, a basic discovery, by researchers who had no inkling of the consequences, turned everything topsy-turvy. The fate of our civilization was suddenly put at risk.

In 1932, the British physicist James Chadwick discovered the neutron, emitted when beryllium was bombarded with alpha particles from radioactive materials. He proposed that the alpha particle combines with the beryllium nucleus of 9 atomic mass units to form carbon with 12 atomic mass units, plus a free neutron. The term "neutron" had been applied long before its discovery to a hypothetical particle that would solve some of the paradoxes inherent in imagining nuclei to be made up of protons and of nuclear electrons that compensated the charge of about half of the protons. Chadwick himself thought he had discovered a proton-electron compound much smaller than the hydrogen atom. It took a little while before he and other physicists became convinced that what he had actually found was a new fundamental particle that accounts for more than half of the mass of the objects around us. With the neutron one had a projectile to induce nuclear reactions which, in contrast to the proton, was not inhibited by the electric charge of the nucleus. A proton approaching a nucleus is repelled by the positive charge of the nuclear protons; with neutrons it is possible to penetrate heavy nuclei which, like uranium, contain many protons (uranium has 92) and hence strongly repel positively charged projectiles such as protons.

HOW TO START A NUCLEAR REACTION: THE NEUTRON AS A PROJECTILE

Alchemists had tried for centuries to transmute lead into gold, but by Chadwick's time it was well understood that the nucleus was unaffected by chemical manipulations. Up to that time the projectiles that had been used to probe the interiors of atoms were positively charged protons, or alpha particles emitted by radioactive sources. Before the First World War, Pierre and Marie Curie were pioneers in the fabrication of powerful radioactive sources. (Their long-term exposure to intense radioactivity eventually ruined the health of both.) In the infant science of nuclear physics, these sources played the role that particle accelerators were later to play. Their use led to many fundamental discoveries, in particular to the discovery of "artificial radioactivity" by Irène and Frédéric Joliot-Curie in 1934: radioactive substances are produced when one bombards a normally nonradioactive (stable) element with a nuclear projectile (in their case, boron with alpha particles). These sources were soon replaced by protons brought to high speeds in the cyclotron, a particle accelerator invented and built at Berkeley, California, by the American physicist Ernest O. Lawrence in 1932 (Leo Szilard, who appears later in this saga, independently invented the cyclotron). Following the discovery of artificial radioactivity, Enrico Fermi in Rome had the idea of using neutrons as electrically neutral projectiles to achieve nuclear transmutations, that is, the transmutation of the atomic nuclei of one element into those of another.

Although great energy is required for charged particles to penetrate the electrical barrier that surrounds a nucleus, neutrons can slip into a nucleus while being practically at rest. In March 1934, Fermi submitted a paper demonstrating the production of artificial radioactivity by neutron bombardment of aluminum and of fluorine. By December 1934, Fermi had discovered that neutrons whose energy had been dissipated by collisions in paraffin were much more effective than neutrons that had not been slowed down. This was an unexpected result, explainable only by quantum mechanics; naively one might have thought that more energetic neutrons would be more efficient at producing nuclear reactions. Despite the fact that the neutron sources available at the time were extraordinarily weak, the use of slow neutrons opened the door to a flood of discoveries.

Physicists were particularly tempted to irradiate uranium, because it was the heaviest element known to exist in nature: its nucleus contains (as already noted) 92 protons. If a neutron was captured and added to the already existing stock of neutrons, it could create an unstable nucleus that might disintegrate by emitting an electron, so that the nuclear charge would increase by a unit as a neutron transformed itself into a proton plus an electron.

In this process, it turned out that another particle was also emitted, called the neutrino, which has no electrical charge, has a mass small even compared with the mass of an electron, can cross the entire earth without interacting, and is therefore difficult to detect. But it has been detected, and even before it was actually observed, its existence was postulated to maintain the conservation of energy and momentum in these decays.

The capture of a neutron is an inexpensive way to add a proton to a nucleus and thus to produce atoms that are not found in nature because they are unstable. Normally, however, the result of such transmutation is an element of familiar chemistry, since it will have the same number of protons (hence will bind the same number of electrons) as the next higher element in the periodic table. In the case of uranium, there was no known chemical element with one more (a 93rd) proton. Uranium would then have to be transmuted into a completely novel chemical element. By June 1934, Fermi's group in Rome published the results of experiments using neutrons to produce many new types of artificially radioactive material. Among these, the team provided evidence for the creation of a new element, which they ultimately called "ausonium," one whose nucleus was supposed to contain 93 protons, one more than uranium. Their evidence misled them; they had not discovered element 93, but something much more important. The actual discovery of element 93 did not take place until 1940, when Edwin M. McMillan and Philip H. Abelson at Berkeley created and named "neptunium," with a mass number of 239, the result of the decay of uranium-239, whose half-life is 23 minutes. Neptunium-239 decays with a half-life of 2.3 days to plutonium-239; plutonium is element 94.

Over the years 1934 to 1938, the best German, French, and Italian teams attempted to clarify the nature of the radioactive entities produced from uranium by the capture of neutrons. No one except Ida Noddack — a respected German radiochemist, co-discoverer of the element rhenium — thought of the idea of fission — that is, the splitting of a heavy nucleus into two lighter ones, in contrast to the shedding of one or two protons or neutrons.[1] But when, at the end of 1938, fission was at last discovered and identified, it was soon realized that it had the potential to revolutionize warfare and could determine the balance of power among nations.

This had been recognized by a Hungarian physicist, Leo Szilard, who figures importantly in this history as a man of vision, vigor, and influence.[2] Living in London, Szilard had filed a patent on March 12, 1934, on the concept of a nuclear chain reaction in which a neutron bombarding a mass of material (he cited beryllium, uranium, or thorium) would produce two neutrons of high energy, which would in turn produce 4, which would produce 8, which would produce sixteen, and so on. Szilard at this point had no clear idea of transmutation caused by neutrons and certainly none of fission. He was later to play a key role in the United States in realizing the first nuclear chain reaction.

The recognition of fission itself is a scientific detective story. In his Nobel

Prize acceptance speech in Stockholm on December 10, 1938, Enrico Fermi related that the neutron bombardment of uranium led to one or more elements of atomic number larger than 92 and specifically mentioned elements 93 and 94, now called neptunium and plutonium. His prize was awarded "for his demonstration of the existence of new radioactive elements produced by neutron irradiation, and for his related discovery of nuclear reactions brought about by slow neutrons." He had no idea that his group in Rome had produced fission in uranium for the preceding four years; they did not detect the fission products as they were emitted, because a thin aluminum foil that was intended to shield the detector from the alpha particles from uranium stopped the fission fragments from entering the detector. They had also produced elements 93 and 94, but their experiment provided evidence for the fission products and not for elements of higher atomic number.

Within a month, Otto Hahn and Fritz Strassmann in Berlin published their work of 1938 identifying some of the products of neutron bombardment of uranium as the element barium, which has 56 protons. As chemists, they were sure of their results; as "nuclear chemists," they were reluctant to bring themselves, as they wrote, "to take such a drastic step which goes against all previous experience in nuclear physics."[3] They could not conceive of a physical mechanism that would burst a uranium nucleus to yield two of about half its mass.

At Christmas 1938, Lise Meitner—a colleague of Hahn and Strassmann's who as a Jew had been forced into exile in Sweden—and her nephew Otto Frisch first thought through the implications of the barium discovery and explained it by applying the "liquid drop" model of the nucleus that had just been invented by Niels Bohr; they published their work in *Nature* on February 6, 1939. Capture of a neutron by a uranium nucleus would set the liquid drop into oscillation violent enough so that it would split in two. Frisch termed the process "fission" by analogy to the division of biological cells. The fission products, made up of lighter nuclei, were created with considerable kinetic energy whose value, expressed in mass units, represented about a thousandth of the initial mass of the uranium nucleus.

In December 1938, Fermi had left Italy with his family to receive the Nobel Prize in Sweden, intending not to return. Arriving in New York on January 2, 1939, he was welcomed to the physics department at Columbia University. Two weeks later, Fermi and his wife, Laura, greeted Bohr as he came from Denmark with the as-yet-unpublished news that Meitner and Frisch had confirmed the process of nuclear fission by "radiochemical" experiments (i.e., using chemistry to characterize the radioactive materials resulting from neutron-induced fission of uranium). Fermi was interested in fission as a new physical phenomenon, but Szilard, who had moved to New York in November

1938, saw in fission the near certainty of nuclear explosives—the realization of his five-year fixation on the chain reaction. In 1932, Szilard had read the H. G. Wells novel *The World Set Free*, published in 1914, in which the major cities of the world are destroyed by atomic bombs in 1956.

On February 19, 1939, Frisch (in Copenhagen) published an experimental verification of the Hahn-Strassmann results by observing large signals from fission fragments in an ionization chamber. The ionization chamber is a basic tool of physics used at that time to detect alpha particles. After being emitted by a nucleus, such a particle travels a few centimeters in air, losing its energy by stripping electrons from hundreds of thousands of atoms. A sensitive electronic amplifier detects the electric charge from all these ions drawn to a negative electrode (or the electrons drawn to a positive metal plate). Comparison with the signal from a typical 6 MeV (million-electron-volt) alpha particle made it clear that the fission process liberated nearly 200 MeV.

In February 1939, at Princeton, Bohr conjectured that fission by slow neutrons occurs in one special isotope of uranium—uranium-235. That month, using the liquid-drop model, Bohr and John Wheeler calculated neutron energies required to induce fission. A neutron captured by a heavy nucleus provokes the nucleus to vibrate with an energy equal to the neutron binding energy plus any kinetic energy the neutron may have had; with sufficiently large neutron energy, the excited nucleus rapidly splits apart, while for lesser neutron energies the excitation is eventually emitted as gamma rays. They found that a neutron energy of 0.6 MeV would be needed to cause fission in uranium-238 and about zero for uranium-235 (slightly negative—i.e., the energy given to the uranium-235 nucleus by the 8-MeV binding energy of an additional neutron is above the threshold for fission, without the need for any kinetic energy of the neutron), thus confirming Bohr's conjecture that fission by slow neutrons in natural uranium was due to the rare (0.71% abundant) uranium-235. This theoretical prediction was proved in March by Alfred O. Nier and his colleagues in Minnesota. They observed a higher fission yield from a sample of uranium that they had managed to enrich in uranium-235 content.

In 1939, more than a hundred articles devoted to fission were published in scientific journals. Work had begun in the United States, Germany, and Japan, among other countries. In France, Frédéric Joliot (who had married Irène Curie and taken the name Joliot-Curie) and his collaborators had applied for a patent on a "device for producing energy" and "improvements to explosive charges." Of utmost importance, the fission was accompanied by the emission of several neutrons—a possibility mentioned by Fermi in a January 1939 speech in Washington and established by March 1939. In April 1939, Frédéric

Uranium-235 fission scenario

1) Initial state: the neutron and the uranium-235 nucleus are almost at rest, at room temperature.

2) Intermediate state: the neutron has been incorporated into the nucleus, which vibrates like a drop of water before breaking up. The reader can fill a balloon with water, tap on it, and see how it vibrates.

3) The nuclear "droplet" has split, giving way to two lighter radioactive nuclei whose kinetic energy is 150 MeV, and to two or three neutrons whose energy is 2 MeV each.

The electron volt (eV) corresponds to the energy acquired by an electron accelerated in vacuum by a potential of one volt, about the voltage of the familiar dry cell. It requires about ten electron volts (10 eV) to extract an electron from an atom, and about ten million electron volts (10 MeV) to extract a neutron or a proton from a nucleus.

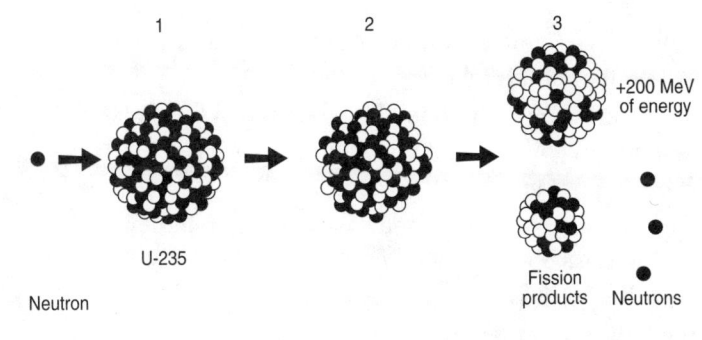

Fig. 1.1. Chain reaction scenario.

The energy of the fission products corresponds to the energy of the atoms in a medium raised to a temperature of many billions of degrees. The sum of the kinetic energies of the two fission-product nuclei is almost equal to the energy $E = Mc^2$, where M is the difference in mass between the initial nucleus and the sum of the final nuclei.

Joliot-Curie, Lew Kowarski, and Hans von Halban in Paris found an average of 3.5 ± 0.7 neutrons per fission, while Szilard and Walter H. Zinn at Columbia University found about two. For uranium-235 fission by slow neutrons—i.e., neutrons that had only the thermal energy of the environment—the number was later determined more accurately as 2.4. This made a chain reaction possible, a neutron from each fission causing an additional fission. Even in natural uranium, a chain reaction could take place if somehow the large amount of uranium-238 could be prevented from gobbling up too many of the emitted neutrons; uranium-238 does not fission with slow neutrons and has only about one-fifth the probability of fission with fast neutrons that uranium-235 has. The chain reaction, if it can be achieved, allows the passage from the infinitesimally small individual nuclear reaction to a reaction involving millions of billions of billions of nuclei.

Such a chain reaction using neutrons of fission energy is equivalent to compound interest that doubles capital every hundred-millionth of a second. It was evident to the physicists of the day that if one could produce fission on a large scale, an energy equivalent to that from the explosion of a ton of a powerful explosive like TNT could be extracted from sixty thousandths of a gram (60 milligrams) of uranium. It follows from Frisch's discovery that uranium fission liberates as energy about a thousandth of the mass of the material: from 60 milligrams of uranium could be obtained 60 micrograms of energy, and $E = Mc^2$ gives us $6 \times 10^{-8} \times 9 \times 10^{16} = 5.4 \times 10^9$ joules. The energy release of one gram of high explosive is about one thousand calories or 4186 joules, so that one ton of high explosive is just about 4.2×10^9 joules.

At Columbia, as in laboratories the world over, physicists scrambled to explore the new phenomenon of fission, while Fermi, with Szilard's urging, worked toward the goal of a self-sustaining fission reaction. In July 1939, Szilard prodded Fermi to start a large-scale experimental pile with "perhaps 50 tons of graphite and 5 tons of uranium," and by November 1, a new "Advisory Committee on Uranium" recommended purchases on that scale.

The Second World War completely changed the conditions under which work on fission was conducted. At the strong suggestion of Albert Einstein—who had been alerted to the possible uses of fission (and to the fact that the Germans were already working on it) by the Hungarian refugee physicists Eugene Wigner and Leo Szilard—President Franklin Roosevelt decided to launch research to avoid being caught short by the Nazis. His conviction, and that of his scientific advisors, was reinforced by a secret report by the physicists Rudolph E. Peierls and Otto Frisch, two German Jewish emigrants who had taken refuge in Birmingham, England. They calculated that a relatively small amount of uranium-235—something like ten pounds as opposed to tons—

would suffice to retain enough of the fission neutrons to cause an expanding number of fissions and so to make a bomb. They communicated this to the British authorities, who in 1941 passed the information on to Washington. Specifically, in July 1941 the "Maud Committee" established by the British government concluded that it would take some three years to make a nuclear weapon and would require a few kilograms of uranium-235.

In the United States, the Manhattan Project, as the nuclear weapons program was called, benefited not only from the excellent community of American physicists and the American capacity for organization and production but also from the help of European physicists who had fled Nazism. With the Japanese attack on Pearl Harbor on December 7, 1941, and the immediate U.S. entry thereupon into the war against Japan and Germany, all scientific and technical resources in the United States were mobilized in the service of the war effort for the duration of the conflict.

Szilard's efforts bore fruit. On December 2, 1942, Fermi put into operation at the University of Chicago the first atomic "pile"—a "nuclear reactor"—using the controlled propagation of fission in uranium-235 in a stack (or pile) of natural uranium lumps (metal and oxide) distributed in a pile of graphite blocks. Although at Columbia Fermi had made experimental piles of graphite and uranium oxide, it was only at Chicago that he had sufficiently pure material and enough of it for a self-sustaining fission reaction.

By the beginning of the Second World War, physicists had concluded that it was possible to exploit the formidable binding energy of atomic nuclei to make either explosives of unequaled power or fuel, a fuel whose mass would be millions of times smaller than that of traditional fuels, whose energy production is based on the interactions between atomic orbital electrons. Fermi had the good fortune to be able to accomplish this work in the United States, far from the ravages of war and, above all, with the immense resources that the United States would then contribute to drive the Nazis and the Japanese from conquered territories.

TWO METHODS OF GENERATING NUCLEAR ENERGY

As a result of the enormous investment made in it during the Second World War, our knowledge of the nuclear field grew rapidly. There is now a much better understanding of the laws governing nuclear structures made up of protons and neutrons (jointly called nucleons) than when fission was first discovered. Nucleons in a nucleus are, as we have mentioned, bound together by the strong nuclear force. It takes a certain amount of energy to extract a nucleon from the nucleus. A nucleus is lighter than the total of the protons and neutrons that constitute it by an amount that is the "nuclear binding energy." The mass of a nucleus is obtained to sufficient accuracy by the use of a precision mass spectrometer—a kind of one-line television tube in which ions of the particular material (atoms with a single electron removed) are accelerated and then deflected in a magnetic field. The resulting spot position is precisely measured and is inversely proportional to the mass. Such measurements were an active field in the 1930s and 1940s. The amount by which the mass of the atom—the mass of the nucleus plus the mass of the accompanying atomic electrons—falls below the mass of the same number of hydrogen atoms and neutrons is the nuclear binding energy. It is more useful to divide this by the total number of protons and neutrons to obtain the "binding energy per nucleon." These average binding energies are plotted over the full range of nuclei from the lightest to the heaviest in Figure 1.2, giving a graph that is called the "curve of binding energy." Studying it suggests two methods of generating energy in nuclear reactions. The first is the long-unsuspected process of nuclear fission by neutrons, only discovered in February 1939. The second is nuclear fusion.

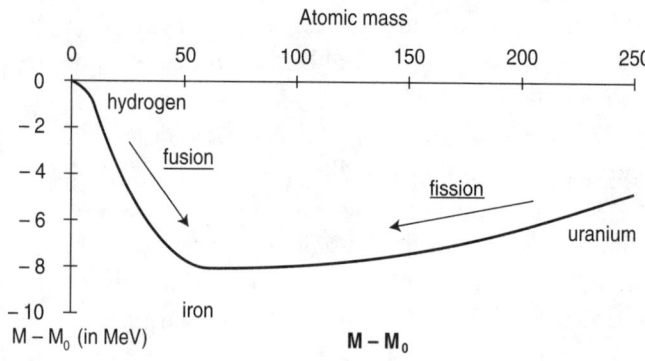

Fig. 1.2. Binding energy (BE) per nucleon as a function of mass number.

The curve of Fig. 1.2 shows the general trend of the variation of binding energy per nucleon as a function of the mass number. It can be seen that

• if the lightest nuclei combine to yield a heavier nucleus (fusion), energy is liberated because the mass of the collection of nucleons decreases; and
• if the very heavy nuclei break apart (fission) into nuclei of intermediate mass, energy is released.

The situation is analogous to that of a billiard ball rolling on a surface with a vertical profile similar to that of the curve. Initially at rest at the left or right extremity, the billiard ball will pick up speed and reach the bottom with usable kinetic energy. The curve is more than a sketch. The most stable elements are those around iron, with a binding energy per nucleon of about 8 MeV. If 56 hydrogen atoms could somehow be induced to form a single iron-56 atom (26 protons remaining as such, and 30 paying the price of combining with their electrons to form neutrons in the iron with atomic number $Z = 26$), $56 \times 8 = 448$ MeV would be liberated according to Einstein's formula $E = Mc^2$. This is almost half of the mass of a single proton (931 MeV). More precisely, the atomic mass of hydrogen is 1.00783 atomic mass units (u), on a scale based on carbon-12, defined as having an atomic mass of 12.000 u, and the atomic mass of iron-56 is 55.93494 u; the "mass deficit" or increase in binding energy is 0.504 u or 469 MeV, since 1 u = 931.49 MeV. Using the precise atomic mass of U-238 (238.05078) and that already stated for iron, it can be seen that if uranium could be provoked to break up into four nuclei of mass similar to that of iron, there would be liberated energy equivalent to 0.334 u, or 311 MeV.

The problem for physicists is to produce the fission or fusion to liberate the energy promised by the differences in mass—to get the ball to move, despite the equivalent of pebbles blocking it.

The process known as fusion uses light nuclei like those of hydrogen and helium. The idea is to combine two of these nuclei into a larger nucleus that is less massive than the total mass of the light nuclei that have been combined to produce it. The same number of nucleons are present in the combined nucleus as in the two lighter ones; the billiard ball starting in the region of the light nuclei would like to end up farther to the right—at the position of the heavier combined nucleus as represented in the curve above. But how to bring about the nuclear reaction that it represents?

The difficulty comes from the fact that in order to obtain the fusion of two light nuclei, they have to be knocked against each other with a good deal of energy, because these two nuclei both have a positive electrical charge and hence mutually repel each other—more and more strongly as they approach. To get them to come together closely enough for the very short-range nuclear forces to take over and overwhelm the electric repulsion, the fusing nuclei have to be accelerated. It is easy to produce fusion reactions on a small scale with instruments that accelerate individual nuclei. It is vastly more difficult if one wants to ignite a substantial mass of billions of billions of nuclei.

One method consists in heating this mass to a temperature at which the average kinetic energy of agitation of the atoms in the medium approaches the energy of a particle in a high-energy accelerator's beam. The energy threshold for these reactions is such that one has to reach hundreds of millions of degrees in order for them to take place. This is what happens in the sun, where gravitation has compressed the mass and heated it to a temperature high enough to stop the gravitational compression. As a result of the high temperature, fusion takes place—so slowly that it will take billions of years to consume a typical proton. Chapter 3 shows how nuclear fusion has been exploited on a massive scale in thermonuclear weaponry.

We have already discussed the other method of generating nuclear energy, fission, and introduced the chain reaction in uranium. To relate our discussion of fission to the curve of binding energy: In very heavy nuclei such as lead or uranium, the nucleons are less and less tightly bound; the binding energy per nucleon is less than it is in the lighter structures into which they are transformed by fission. These fission fragments have masses between 70 and 170 times the mass of hydrogen. Even a very low-energy, nearly stationary neutron, when it is captured by the uranium-235 nucleus, destabilizes it and makes it explode into lighter fragments: the uranium-235 is called a "fissile" element. As noted, uranium-235 contains 92 protons and 143 neutrons, and uranium-238 has 92 protons and 146 neutrons. The curve of binding energy shows only the available energy; it remains for human ingenuity to provide the path by which the nuclei will split or merge. It is not necessary to initiate the fission reaction by raising the temperature. All that is required is to know how to combine the fissile element, uranium-235, with a medium that allows enough of the two or three neutrons liberated in the fission process to remain unabsorbed so as to be able, in turn, to produce other fissions, thus giving rise to a chain reaction. Uranium-235 and uranium-238 have quite different probabilities of reaction with an incident neutron, although the energy released by fission of one or the other nucleus is about the same.

CHAPTER 2

The Nuclear Chain Reaction

THE PURPOSE of this chapter is to provide the basic understanding of what is needed to exploit fission and to produce fissionable materials for use in nuclear weapons. Nuclear reactors are discussed only for the production of plutonium-239; reactors for electrical power are treated in Chapter 5.

By the end of the Second World War, physicists understood how to produce both explosives and electricity by exploiting the energy liberated from atomic nuclei. Much development has ensued in both fields.

As discussed in the previous chapter, the two processes for creating nuclear energy are fission and fusion—the splitting of heavy nuclei into lighter ones, and the creation of heavier nuclei by fusing together lighter ones. In both cases, the energy exploitable is equal to Mc^2, where M is the difference in mass between the initial and final nuclei involved in the process and c is the speed of light in vacuum. It is millions of times greater per atom than the energy resulting from chemical reactions.

The discovery of fission and the fact that each fission releases two or more neutrons gives the possibility of a neutron chain reaction that would allow a change of scale from the few neutrons created with natural radioactive sources or by cosmic rays to numbers of neutrons (and fissions) that would enable large power plants and dwarf the largest chemical explosive releases.

For an explosive chain reaction to take place, the mass of fissionable material must exceed a threshold value called the "critical mass," or else too many neutrons escape without collision and the chain reaction dies out after a few successive fissions. It was this critical mass of 60 kilograms of uranium-235 that Frisch and Peierls were the first to predict by a correct approach, although inaccurate input data made their estimate low by about a factor of 10.

The power produced by Fermi's first chain-reacting pile at the University of Chicago (two watts) was that of a small flashlight, but the essential feature of nuclear reactors was there—the possibility of making a large-scale controlled neutron chain reaction. This stunning achievement of December 1942 was followed by the construction, at Hanford, Washington, of a reactor 100 million times more powerful than Fermi's first demonstration reactor. Here the objective was not to produce energy but to make in sufficient quantity an element—plutonium-239—that could replace the fissile element uranium-235 in the fabrication of nuclear-explosive weapons. This was an alternative to the separation, thus far never attempted, of U-235 isotopes on an enormous scale.

Plutonium-239 (with $Z = 94$, i.e., there are 94 protons) is produced by the reaction uranium-238 + neutron => uranium-239 –> neptunium-239 –> plutonium-239. (The symbol => means "produces" and the symbol –> means "decays to," by the emission of an electron—a beta ray.)

The "capture" of a neutron by uranium-238 (U-238) liberates immediately several gamma rays of total energy equal to the binding energy of a neutron, transforming U-238 to U-239. The ensuing decays of U-239 and neptunium 239 (Np-239) each involve the emission of an electron and a neutrino. The emission of one negative electron (identical to the usual atomic electrons) increases the charge of the nucleus by one unit. Thus the nucleus of neptunium has 93 positive charges while that of plutonium has 94. Uranium-239 has a half-life of 23 minutes, neptunium-239 2.4 days; they are fleeting intermediaries in the production of plutonium-239 from uranium-238, which is 140 times more abundant than uranium-235 in uranium ore. Plutonium-239 (Pu-239) has a half-life of 24,000 years.

Some Definitions

Lifetime (or, more technically, "mean life"): the time after which, on average, a freshly produced radioactive nucleus decays. Some individual nuclei will decay sooner and some later. One cannot predict at what time a specific nucleus will decay but can only predict when it is most likely to do so.

Half-life: the time after which half of any given sample will have decayed. For radioactive decay, the half-life is precisely 0.697 as long as the mean life. Because the individual decays occur at random, this time is the same no matter how large the sample. The time it takes for half of a trillion nuclei to decay is the same as the time it takes for half of a billion nuclei. And the time for decay of half of the remaining 500 billion or 500 million nuclei is again the same.

Decay rate: the rate at which an average sample of a species of radioactive nuclei will decay. It is the inverse of the lifetime; the longer the lifetime the slower the rate.

"Fertile" materials are those that do not produce energy through fission, but are transformed by neutron capture into material that can be fissioned. That is the case for uranium-238, which can be transformed into plutonium-239. The absorption of a neutron by uranium-238 is, therefore, important because plutonium-239 is easy to fission and can make a good explosive. In the operation of a reactor, the capture of neutrons by uranium-238 does not lead to a total loss of the neutrons in the long run—i.e., in a nuclear reactor where the Pu-239 thus produced might be fissioned in a few months or years. But excessive capture by U-238 can so degrade the neutron economy that no chain reaction is possible. Several other isotopes can undergo fission by the absorption of slow neutrons, i.e., are fissile: the most important ones are uranium-233 and plutonium-241.

Like Fermi's first reactor, the Hanford reactors used pure carbon to slow (or "moderate") the fission neutrons from uranium-235 so that they would produce new fissions, while minimizing the chance for them to be captured by the much more abundant uranium-238 that would otherwise quench the chain reaction. But their enormous heat production meant that the Hanford plutonium reactors needed to be cooled by the waters of the Columbia River.

The first pile at Hanford generated 250 million watts—250 megawatts or MW—of thermal power and produced each year almost a hundred kilograms of plutonium. A rule of thumb is that a megawatt of fission heat in a natural-uranium reactor accompanies the production of about a gram of plutonium-239 per day. About six kilograms were sufficient to make a bomb. The first plutonium bomb was tested on July 16, 1945, in the New Mexico desert, and the second was dropped three weeks later on the Japanese city of Nagasaki, three days after a uranium fission bomb had destroyed the city of Hiroshima.

PROPERTIES OF THE NEUTRON

In view of the vital role played by neutrons for producing nuclear energy, controlled or explosive, we should dwell a bit on their properties.

A free neutron in vacuum has a half-life of about a thousand seconds (about 16 minutes). It disintegrates spontaneously into a proton, an electron, and a neutrino. A neutron bound to protons can live eternally (in a stable nucleus such as deuterium) or disintegrate into a proton, an electron, and a neutrino (in a radioactive nucleus) in a time that varies between a fraction of a second and a billion years or more, depending upon the type of nucleus.

But a neutron's life is typically far more complicated. A free neutron traveling through matter undergoes collisions with nuclei through which it is slowed down or captured. When capture occurs, the nucleus is converted into another somewhat heavier nucleus, which can be either stable or radioactive. The interactions of neutrons with nuclei are astonishingly varied—ranging from those

Specifics about the Neutron Chain Reaction in Nuclear Weapons and Nuclear Reactors

Each neutron released in fission can either be captured by a nucleus, escape from the nuclear fuel, or give rise to a fission which, in turn, produces two or three neutrons. If this latter process dominates and more than one neutron from each fission provokes a new fission, then there is an exponential multiplication in an infinite medium. That is, if one fission provokes two more, and if each of these two similarly provokes two, the number of fissions in each successive interval of ten billionths of a second is 1, 2, . . . 32, . . . 1024, 2048 . . . Thus if the mass M is sufficient, the system makes the transition from the infinitely small scale of the initial fission reaction to that of the immense number of nuclei contained in M — in less than a microsecond.

For an explosive chain reaction to take place, the mass M must be greater than a threshold value called the "critical mass," so that enough neutrons from the average fission remain in the region and so can carry on the chain reaction. This critical mass M_c is about 10 kilograms for plutonium-239 and about 60 kilograms for uranium-235, when these materials are in the form of metal spheres at their normal density. In nuclear weapons, these values might be reduced by a factor of 4 if the force of a powerful explosive is sufficient to double the density of the plutonium by compression,[2] and by a further factor of 2 by having the sphere of fissionable material surrounded by matter that can "reflect" some of the escaping fission neutrons back into the sphere. This possibly surprising fact is of great practical importance. In a just-critical mass, there is a certain probability that a neutron will collide and cause a fission in going from the center of the sphere to the edge. That probability will remain the same if the density is increased by a factor F, while the radius is reduced by a similar factor. Since the mass of a sphere is proportional to the density and to the cube of the radius, the mass of a just-critical sphere — Mc — is inversely proportional to the square of the density (that is, $F \times F^{-3} = F^{-2}$).

The play *Copenhagen* by Michael Frayn, produced in both London and New York, alludes to some of these matters in the context of a brief but historic meeting in Nazi-occupied Copenhagen in September 1941 between the leader of the German nuclear program, Werner Heisenberg, and Niels Bohr, the inventor of the Bohr atom and of the liquid-drop model of the nucleus that in 1939 had been accepted as explaining neutron-induced fission.

When the fission propagates, the mass M is raised to the enormous temperature of tens of millions of degrees; the internal pressure corresponding to metallic density at this temperature (and to the pressure of the X-rays that correspond to the light from such superhot materials) drives the ball apart. The nuclei of M, therefore, don't all have time to fission; the system becomes subcritical and the neutron chain reaction dies out before it involves most of the U-235. In the Hiroshima bomb, only about 2% of the uranium fissioned.

In a nuclear reactor, the goal is to maintain a chain reaction in which each fission leads to precisely one subsequent fission (and so on) to prevent a runaway chain reaction from taking place. Just one neutron coming from the fission reaction, or from the fission products, is able to produce a succeeding fission; the rest are lost to absorbers in the reactor, and some to neutron-absorbing control rods used specifically to maintain the reactor just critical. In reactors producing heat to be used to generate electrical power, one takes advantage of the fact that the fission of each nucleus produces 25 million times more energy than the combustion of an atom of carbon. In such a reactor of 1000 MW, a thousand times more fissionable material is present than exists in a bomb. This produces in a continuous, controlled fashion a great deal of energy over many years at a rate equivalent to the energy released by one Hiroshima bomb every eight hours. The essence of reactor design and operation is to prevent a divergent chain reaction, that is to say, one in which the multiplication coefficient (number of daughter fissions divided by number of mother fissions) is greater than 1.

The primary fission products are, in general, very unstable: they and their radioactive decay products make up the radioactive waste whose management is one of the most vexing problems of the nuclear industry.

causing fission or other types of nuclear alchemy to elastic collisions in which only momentum is exchanged without any nuclear reaction occurring at all.

When a nuclear projectile, such as a neutron, moves through matter, the chance that it produces a reaction is determined by what is called the "cross section," defined, in this case, as the apparent target area offered by a nucleus. The common unit of measure is the "barn," equal to 10^{-24} cm^2, the area of a little square 10^{-12} cm on a side—a unit used by the Los Alamos physicists, who thought that this cross section on a nuclear scale was so large that it was, figuratively speaking, as big as the side of a barn. The cross section takes into account not only the actual physical size of the nucleus but also the likelihood of a collision of a particular kind. A homey analogy might be to assign an "area" to a window at which a child throws a rubber ball. This "area" would measure the likelihood that the child will break the window. Clearly that likelihood will depend not only on the actual area of the window but on the kind of glass it is made of and how hard the ball is thrown. The "area" will increase as the ball is thrown harder and harder.

Similarly, the cross section for nuclear collisions varies with the energy of the neutrons, but by no means in such a simple fashion; a larger set of possibilities exist. These include elastic collisions (i.e., a "bounce" with all the energy remaining in the kinetic energy of the colliding particles) and inelastic collisions, in which the target nucleus is excited by the collision, with later decay by the emission of a gamma ray—the neutron loses energy in the inelastic collision but remains free. Then there is "absorption," in which a neutron is totally absorbed into a "compound nucleus" that later decays by emission of a gamma ray; or even an "n-p" (neutron-proton) reaction, in which a neutron striking a target is absorbed but a proton is emitted. So there is a whole array of cross sections for the collision of a neutron with a nucleus like that of uranium-235. In the example of the window, at low speed, the "elastic cross section" of the window equals the "geometrical cross section"; as speed increases the "fracture cross section" increases while the elastic cross section falls, because for a window (but not for a nucleus), the "total cross section" is equal to the geometric cross section. See Fig. 2.1 for an example.

It is remarkable that this capture cross section can change by a factor of 1000 or more as one goes from a given nucleus to a neighboring nucleus that has only one proton or neutron more. The situation is analogous to that in chemistry when there is an enormous difference in chemical properties between one atom and its next-higher neighbor in the periodic table, which has only one electron more. For example, chlorine, with 17 orbital electrons, is highly reactive, and argon, the next element, with 18 electrons, is chemically inert.

Absorption of Neutrons

Great ingenuity was necessary to construct a reactor with natural uranium. No matter how much uranium with its natural content of 0.72% U-235 is amassed, the system will be subcritical, because a fission neutron has only a small chance of causing fission in U-238 (and less in the small amount of U-235) before it loses enough energy by inelastic scattering so that only the scarce U-235 is susceptible to fission. The problem is evident from the four rather complicated graphs of Figs. 2.2, 2.3, 2.4, and 2.5.

In a nuclear weapon, a fission chain reaction takes place in nearly pure uranium-235 using directly neutrons of fission with an energy of 2 MeV. In a typical nuclear reactor, the neutrons are slowed down from the initial fission-neutron energy of 2 MeV to thermal energy of 0.025 eV; a light-element moderator avoids capture in the uranium-238 resonances while slowing to thermal energy.

The genius of the nuclear pioneers was the observation that a light-element moderator could bring the fission neutrons safely through the complex "reso-

Fig. 2.1. A neutron, represented by a frog, escapes from a nucleus
that has undergone fission, represented by an exploding airplane.

The frog falls toward the sea and sees two islands of about the same size: Uranium-235 and Uranium-238. The lucky frog manages to open his parachute, which greatly slows his fall. But when he looks down again at the two islands, he is astonished to see an important difference:
Uranium-235 now looks big, and Uranium-238 seems tiny by comparison.
"It's weird," says the stupefied frog to himself. "These neighboring islands have almost the same name, they were about the same size, but one has grown to immense proportions as I slowed."
That is because this apparent area has nothing to do with a geometric area. It reflects the appetite of the uranium-235 nucleus for the approaching neutron. This appetite is measured by the effective area or cross section of the nucleus.

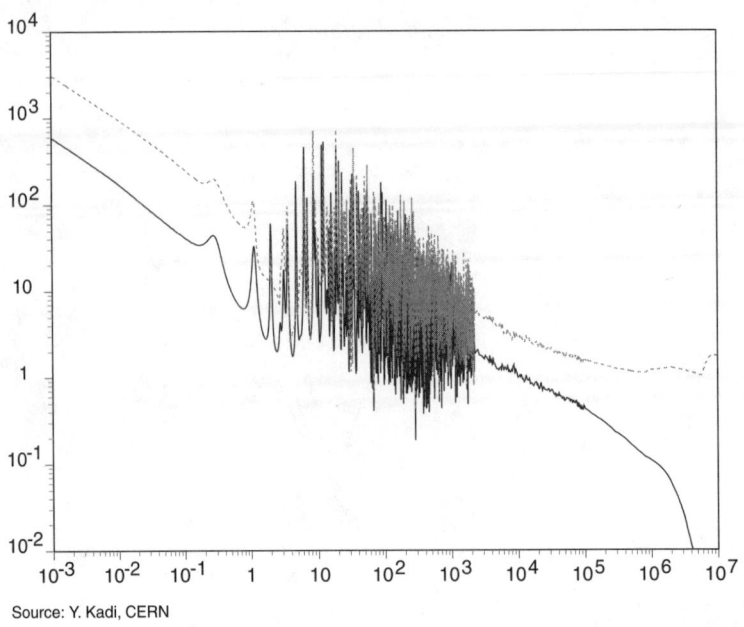

*Fig. 2.2. Plot of the capture (solid line) and fission (dotted line) cross sections of U-235.
Cross sections are in barn and neutron energies in eV.*

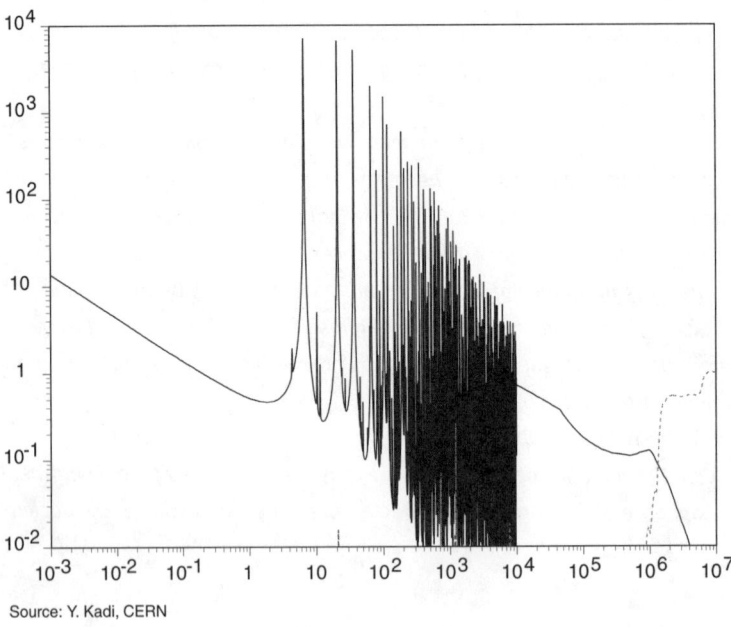

*Fig. 2.3. Plot of the capture (solid line) and fission (dotted line) cross sections of U-238.
Cross sections are in barn and neutron energies in eV.*

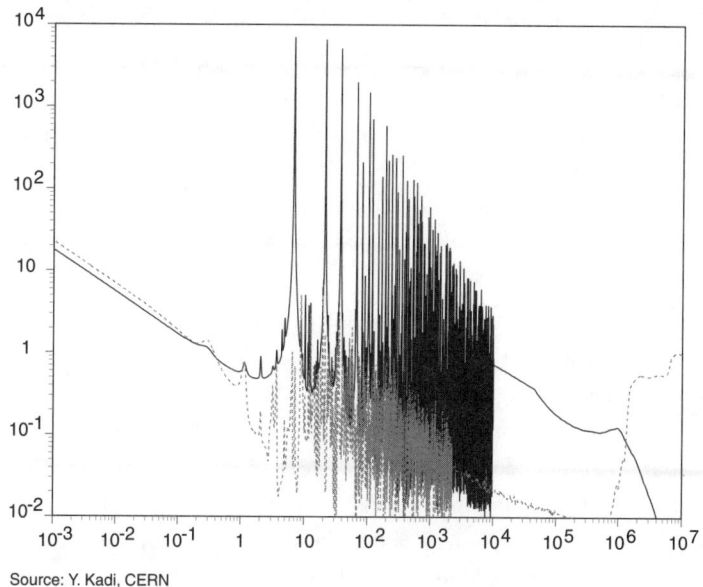

Source: Y. Kadi, CERN

Fig. 2.4. Plot of the capture (solid line) and fission (dotted line) cross sections of U-natural. Cross sections are in barn and obtained by multiplying the U-238 curve by 0.9928 and adding the U-235 curve multiplied by 0.0072. Neutron energies are in eV.

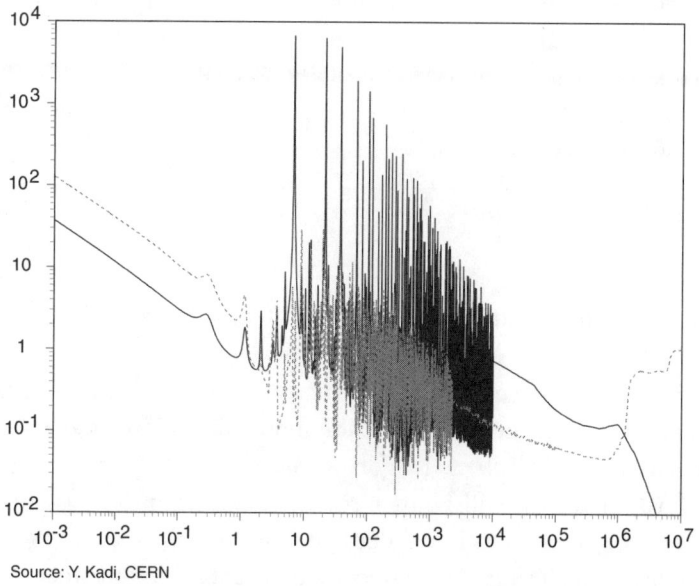

Source: Y. Kadi, CERN

Fig. 2.5. Plot of the capture (solid line) and fission (dotted line) cross sections of U-enriched. Cross sections are in barn and obtained by multiplying the U-238 curve by 0.959 and adding the U-235 curve multiplied by 0.041 (typical of the fuel in a pressur-ized-water power reactor enriched to 4.1% U-235). Neutron energies are in eV.

The fission cross-section and capture cross-sections for U-238 (Fig. 2.3) show that once the neutron gets below a fraction of an MeV, the probability of capture is 100,000 times as high as the probability of fission. A chain reaction requires that at least one of the two or so neutrons produced in fission cause another fission, but in U-238 below an MeV only about one neutron in 100,000 would cause fission; so the neutrons would die out.

The situation for a neutron chain proton is very favorable with pure U-235 (Fig. 2.2). Inelastic scattering is not shown on these curves, but it is clear that the fission cross section for U-235 is almost ten times its capture cross-section at almost any energy, so that only 10% of the neutrons will be lost to capture, making a chain reaction in pure U-235 of sufficient mass impossible to avoid.

The fission and capture cross sections for natural uranium are obtained by multiplying the curves for U-235 by 0.7% and adding them to the corresponding curves for U-238, multiplied by 0.993. When this is done (Fig. 2.4), the fission curve except at the highest energy is entirely due to U-235, and the capture curve at all energies is essentially identical with that for U-238. It is clear that the situation for a chain reaction in natural uranium is unfavorable for neutron energies of 10 eV or more, but is quite favorable for thermal energies, where the capture cross section is 4.1 barn, and the fission cross section of natural uranium (due entirely to U-235) is about 3.4 barn.

Fig. 2.5 plots the cross sections for uranium containing 4.1% U-235. A neutron chain reaction can occur only below the resonance absorption region of U-238; but at energies below about 1 eV, the fission cross section is large enough to permit substantial structure in the core of a power reactor.

nance region" in the middle of the graph to enter the favorable region between thermal energies and 10 eV or more. Let's see why this is so.

Neutrons can be slowed down by elastic collisions similar to those between billiard balls: they lose their energy all the more rapidly if the target nuclei are light—that is, of mass similar to that of one or a few neutrons, not one hundred or more. A billiard ball will bounce several times from the cushions of the table before coming to rest. It will do the same if it strikes a large wood block placed on the table, losing a small fraction of its energy in the process. But if it strikes another ball at rest head on, it comes to rest and all the energy is taken by the target ball. When a cue ball strikes a target ball in a glancing encounter, it loses hardly any energy, and it can lose at most 100% in a direct, head-on collision. On the average the cue ball loses 50% of its energy when it strikes another ball at random. The proton, which is the nucleus of the hydrogen atom, is the lightest nucleus: the neutron thus loses, on the average, half of its energy in each collision with a proton, and occasionally all of its energy in a single collision, while it loses a much smaller fraction of its energy when the target is made up of heavy nuclei.

A neutron emitted in the fission of a heavy nucleus has kinetic energy of 2 MeV—it travels at a speed of 20,000 km/s (about 7% of the speed of light) and is called a "fast" neutron. A thermal, or slow, neutron, which has an energy of 0.025 eV, travels at a speed of about 2 km/s. Twenty-six collisions are enough, on the average, for a neutron of 2 MeV to be gradually slowed down to the average energy of the atoms that make up a hydrogen-rich medium at room temperature. It takes 31 collisions in deuterium (heavy hydrogen, with one neutron and one proton in its nucleus), 120 collisions in carbon, and 2200 in uranium.

The medium that slows down the neutrons is called the "moderator." This medium varies from one type of reactor to another. In most reactors, the moderator is the hydrogen contained in ordinary water, but natural uranium will not produce a self-sustaining chain reaction with ordinary water as moderator, because of the hunger of protons to capture a neutron (with the emission of a gamma ray); hence all water-moderated reactors use slightly enriched uranium. Fig. 2.5 shows the capture and fission cross sections of uranium containing 4.1% uranium-235.

In his 1934 experiments, Fermi and his team stumbled onto the importance of moderated neutrons accidentally. After observing peculiar results that turned out to be due to a wooden table supporting the experiment, the experimenters put a piece of paraffin in front of the target, and they noticed that for the creation of radioactivity this greatly enhanced the effectiveness of the neutrons that passed through it. In paraffin, two out of three nuclei are protons, the nuclei of hydrogen atoms, which slow the neutrons so effectively. Thus was invented the science of slow-neutron physics, for which Fermi received the Nobel Prize. The citation for Fermi's 1938 Nobel Prize reads: ". . . for his demonstrations of the existence of new radioactive elements produced by neutron irradiation, and for his related discovery of nuclear reactions brought about by slow neutrons."

As a neutron is being slowed down to thermal energy by collisions, it passes through intermediate energies and it can be lost in a collision or absorbed without producing fission. The probability of absorption (i.e., the absorption cross section) of slow neutrons varies considerably from one element to another: boron-10 (an isotope with 5 protons and 5 neutrons—which constitutes one of every 5 atoms of natural boron) has a considerable cross section of

3800 barns; the 80%-abundant boron-11 (5 protons plus 6 neutrons), 5 milli-barns (i.e., 0.005 barn); cadmium, 2000 barns; carbon, only 3 millibarns; and iron, 10 barns. Such differences in the characteristics of these materials allow their judicious use in the control of nuclear reactors: by inserting them in the right combination, one can finely control the fate of neutrons emitted in the reactor core.

The extraordinary differences in the properties of U-235 and U-238 nuclei can be seen in Figs. 2.2 and 2.3: the fission cross section for uranium-235 is about 583 barns for neutrons with an energy of 0.025 eV—an energy compara-ble to the thermal energy of an atom at the temperature of a typical nuclear fuel—and falls to 2 barns for neutrons of 2 MeV, which is the average energy of neutrons emitted by fission. For uranium-238, it is effectively zero (10 micro-barns) for thermal neutrons, while it is only 0.5 barn for 2 MeV neutrons. That is why a chain reaction cannot take place in uranium-238: a neutron liberated by fission has an energy close to 2 MeV; the likelihood that a single neutron will produce another fission with a probability greater than $\frac{1}{3}$ (so that a fission reproduces itself) is slight, because it is too quickly slowed down by a few inelastic shocks and drops below the energy threshold for fission. In the lowest energy range, all the cross sections simply rise in inverse proportion to the velocity of the neutron—that is, as 1 divided by the square root of the kinetic energy.

In the energy region above a few electron volts one sees also that a very small variation in the neutron energy changes enormously the probability of capture of these neutrons. These narrow peaks, called "resonances," illustrate the fact that these nuclei, combinations of neutrons and protons, are systems capable of vibration, like musical instruments.

Quantum mechanics tells us that the neutron projectiles have a wave nature as well as a particle nature. These waves can resonate in the nuclei they enter. The large absorption cross section for neutrons of these uranium-238 res-onances would result in the capture of so many neutrons as they are slowing down in the moderator (considering the 141-fold larger abundance of uranium-238 than uranium-235 in natural uranium) that the uranium must be arranged in lumps within the moderator; thus the slowing takes place outside the lumps, and the neutrons diffuse into the uranium only after they reach thermal energy. Even so, with a uranium-235 absorption cross section of 98 barns at thermal energy, and a uranium-238 absorption cross section of 2.7 barns (multi-plied by 141, this adds an effective 381 barns to the uranium-235 absorption cross section), the proportion of neutrons reaching thermal energy that can cause fission is only 55%, which is the ratio of the number of fissions divided by the total—the number of neutrons that cause fission, the number captured by uranium 235, and the number captured by uranium 238, i.e., 583/(583 + 98 + 381) = 55%. With 2.4 neutrons emitted per fission of U-235, the situation is marginal for a reactor using natural uranium, even with a perfect moderator, since only 2.4 × 55% = 1.32 fission neutrons are produced even if every fission neutron is brought without loss to the thermal range. Nevertheless, with sufficiently pure graphite or with heavy water, natural uranium reactors work very well.

THE DOPPLER EFFECT

There is the perhaps apocryphal but instructive story of a troop of soldiers singing merrily while marching in step on a suspension bridge. The bridge collapsed when the rhythm of the march corresponded exactly to that of the natural swinging motion of the bridge. There was a "resonance" between the frequency of the marching and the frequency of the vibration of the bridge. Troops are required to break step under such circumstances.

A commonplace phenomenon familiar in acoustics and optics, the Doppler effect, plays a critical role in reactor design and operation. It is now routinely used in noninvasive monitoring of blood flow in arteries. When a whistle on a moving vehicle emits a sound, the pitch appears higher if the vehicle is approaching the listener. In the nineteenth century, the Doppler shift was demonstrated for skeptical audiences in Europe by having trumpet players on an approaching railway car play a given note; musicians with perfect pitch stationed at trackside would hear a higher note. Even those of us without perfect pitch notice a drop in pitch of a train whistle as the train passes by.

When a star recedes from us, the frequency of the emitted light decreases and it becomes redder—the result of the wave nature of light. The same happens in nuclear physics. If a neutron has an energy that corresponds to a resonance of a compound nucleus (the nucleus composed of the original nucleus plus the neutron), it has a great probability of being absorbed. If the nucleus is in motion at a significant speed, because the atoms are agitated by an increase in temperature of the medium, then the absorption probability for this neutron decreases, but can rise for another neutron that initially did not have the right energy for resonance.

When a neutron is slowed down by collisions, it can lose too much energy in a collision to fall into the very narrow region of a resonance. With the Doppler effect, because if the agitation of the atoms in a heated medium, it is as if the strong resonances were wider so that the energy range of the neutron for which it can be absorbed is also wider; the phenomenon is important for

Fig. 2.6. The Chain Reaction in a Nuclear Reactor,
Tragi-comedy in five acts
With, in order of appearance:
THE URANIUM-235 NUCLEUS, THE MASTER OF CEREMONIES,
played by the big top hat
THE NEUTRONS, played by the frogs
THE FISSION PRODUCTS, played by the little top hats
THE PROTONS, played by the birds
THE URANIUM-238 NUCLEI, played by the snakes
Authors: Joliot−Halban−Kowarski−Perrin−Fermi
Stage effects: Sempé

Fig. 2.7. Act I: A slow neutron is swallowed by a uranium-235 nucleus.

Fig. 2.8. Act II: The nucleus bursts and liberates three new neutrons and two lighter nuclei, the fission products.

reactor safety. If the core of a reactor starts to heat up because of a lack of cooling liquid resulting from a sudden leak, the large amount of uranium-238 exhibiting Doppler broadening allows the neutrons to be more readily absorbed, and the chain reaction is spontaneously smothered, without the need of external intervention. The uranium-238 does not fission, so it acts like a neutron absorber that is the more effective the warmer the fuel is.

Figures 2.6 to 2.11, in which the neutron is disguised as a frog, illustrate the problems to be resolved in order for the neutron to lose its energy before being captured, to enable a chain reaction to occur in natural uranium or in uranium slightly enriched with uranium-235.

In Act IV, the frog, which is at the head of the stairs, has to jump down the steps until it reaches one of the lower ones. Then it has to dive into the top hat, which is full to the brim and which breaks apart (fissions) because it is very fragile. The pool then releases three little frogs that it had kept prisoner, to carry on the game.

But there are traps lying in wait. The snakes on each step represent the possibility of capture in uranium-238 without fission.

To help the frog get down the stairs, that is to say, to slow down the neu-

Fig. 2.9. Act III: The neutrons gush forth from the fission at high speed. They are slowed down by collisions with the protons (the birds), which sometimes absorb them.

trons, the medium where the fission takes place is structured in such a way that the neutrons collide elastically with the light nuclei of the moderator and give up a part of their energy: the lighter the moderator nucleus, the greater the fraction of its energy given up by the neutron at each collision, thus allowing it to avoid as many traps as possible at intermediate energies. The proton, the nucleus of the hydrogen atom, allows the frog to jump many energy-steps with a single bound, rapidly losing its energy, thus avoiding the danger of being trapped by a snake. The traps are the most numerous and the most treacherous on the middle steps. They illustrate the action of uranium-238 (Fig. 2.3). The art of the nuclear physicist is to help the frog get to the bottom of the stairs with a minimum of losses.

But life wouldn't be interesting if it weren't complicated. A proton also has a rather high probability of capturing a neutron. There are enormous differences in the voracity of different nuclei for the neutron-frogs. For example, deuterium (D), heavy hydrogen made up of a proton already bound to a neutron, is less voracious than protons. And that is why heavy water (D_2O) is very useful as a moderator in certain reactors. As we have noted, the number of collisions necessary for the neutron to reach thermal energy is a bit higher for deuterium than for collision with protons (31 vs. 26), but the capture probability per collision is much lower.

The physicist still has many more obstacles to overcome. To use uranium-

Fig. 2.10. Act IV: The uranium-235 is mixed with uranium-238. The uranium-238 (the snakes) captures the neutrons while they are slowing down and gradually changes into plutonium-239.

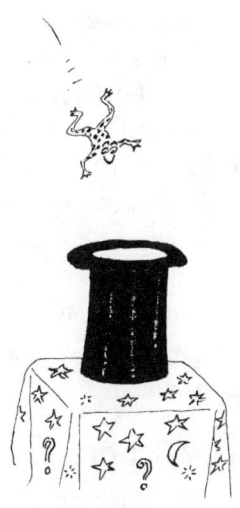

Fig. 2.11. Act V: The neutron moderator, the uranium, and a few traps are laid out so that only one neutron, of the three that burst forth from the fission, is available to be absorbed by the uranium-235, to restart and continue the operation indefinitely.

235 in natural uranium, the fission neutrons must be kept out of the traps set by the uranium-238 with which it is mixed. That makes it impossible to arrange a chain reaction without a moderator (and therefore impossible to trigger a bomb) when the fuel is natural uranium, and that's fortunate for civilization. To make a uranium bomb, the natural uranium has to be enriched in uranium-235 (usually to a concentration of 80% or more) so that the reaction can proceed entirely with fast neutrons. That is how the bomb that destroyed Hiroshima was made.

Fundamentally, the rate at which a neutron is absorbed (whether in a bomb or in a reactor) is the product of an average cross section, the neutron speed, and the number of nuclei present per cubic centimeter; and the time for the absorption is 1 divided by this product. Checking dimensions, we state that $[T] = 1/([L^2] \times [LT^{-1}] \times [L^{-3}])$. True. (In technical notation, $\tau^{-1} = <\Sigma \times v> = <\sigma v N>$, where Σ is the so-called "macroscopic cross section" in cm^2 per cubic centimeter, and the $< \ldots >$ is a particular average that takes into account the different cross sections for different neutron velocities at a given temperature.) Whereas a bomb explodes in a time of about 10 nanoseconds and has an explosive power corresponding to the few kilograms of material moving at a speed of a corresponding 10^9 cm/s, a thermal neutron in a reactor takes about 40 microseconds to cause fission, so the maximum explosive yield one could obtain from a thermal reactor is far less—because it will blow itself apart before much fission has a chance to happen.

A vast amount of radioactivity is stored in a power reactor that has been operating for some months or years, and the reactor must be carefully designed and operated to contain this material, but damage to the public from blast from a thermal reactor is not possible in view of the small kinetic energy of the expanding reactor, even if driven by explosives into a region in which the chain reaction goes at its maximum possible rate. If there were an attempt to make a nuclear explosive of natural uranium arranged in a graphite-moderated reactor, the explosive yield of the reactor would be something on the order of a millionth that of the simple nuclear explosive, and the radioactive materials produced in the actual explosion similarly less.

ENRICHMENT OF URANIUM

A nuclear weapon requires plutonium, or uranium highly enriched in uranium-235. We have explained why enrichment is also useful in providing more flexibility in the design of power reactors, enabling them to use ordinary water, instead of heavy water, as moderator and coolant. Uranium enrichment to 3% to 5% (i.e., by a factor of 4 to 7 from the natural 0.71%) is enough to permit close-spaced fuel rods and hence a more compact reactor than with natural uranium, where the moderation must occur in regions free of uranium. The higher concentration of U-235 also allows the fission cross section of U-235 to compete favorably with the capture cross section of the water moderator.

In early years after the Second World War, enrichment procedures needed gigantic installations which cost hundreds of millions of dollars, consumed enormous amounts of electrical power, and were impossible to camouflage. That is no longer true. Nowadays there are simpler techniques, based on centrifugation, that are employed on a large scale to produce uranium enriched to a modest extent (4% uranium-235) for normal reactors that produce electrical power. The centrifuges are also more suited to the clandestine manufacture of enriched uranium, because they are less visible and use less electrical power.

Separating uranium-235 from uranium-238 is a process based on the 1.5% mass difference between the atoms. Two techniques were used in the Manhattan Project for enriching the uranium for the Hiroshima bomb. The first was an "electromagnetic separation" process in which uranium atoms have an electron stripped off, and these ions are accelerated to form a beam like the electron beam in a television tube or computer monitor. Uranium ions are 400,000 times heavier than electrons, and some of the ions are 1.5% lighter than the others. A TV tube using uranium instead of electrons would therefore give a double image, one about 0.75% bigger than the other; the larger image due to uranium-235 would be 140 times less intense than the image formed by uranium-238. The electromagnetic separation process used such vast amounts of energy that it was soon dropped in favor of the "gaseous diffusion process," which takes advantage of the fact that uranium-235 diffuses through small pores slightly faster than uranium-238. Thousands of "stages" of porous "barrier" materials and pumps and piping were required to enrich uranium by this method.

Gaseous diffusion makes use of uranium in combination with six fluorine atoms (uranium hexafluoride, UF_6), which is a stable substance, although highly reactive chemically. It has the virtue that fluorine occurs in nature as a single isotope (9 electrons and hence 9 protons in the nucleus, and with 10 neutrons, so that it is F-19). So the difference in the mass of the various UF_6 molecules in the gas is due entirely to the difference in mass between the uranium-238 atoms and the uranium-235 atoms. If fluorine were not "monoisotopic," these enrichment processes would work far less well. The UF_6 molecules penetrate through a porous barrier at slightly different rates because of the three-unit mass difference (out of 352 or 349 atomic mass units) between the molecules containing uranium-238 and the ones containing uranium-235.

The uranium-235 molecules penetrate through the barrier about 0.43% faster than the molecules containing uranium-238—just the difference in the average speed of the molecules at the same temperature.

More recently, the use of highly precise lasers has been demonstrated to be an effective means of exciting and inducing chemical reactions or ionization in uranium-235 atoms without affecting uranium-238. It remains to be seen whether this technique can compete economically to enrich reactor fuel; in actual fact, in 1999 the United States Enrichment Corporation abandoned the atomic-vapor laser isotope separation process, for which it had expressed high hopes at the time of its stock sale to the public in 1998. The "noncompetitive" electromagnetic enrichment process will appear again in Chapter 12 on nuclear weapon proliferation and terrorism.

Most U.S. and French commercial enrichment is still done by gaseous diffusion, but the rest of Europe and Russia have moved to the gas centrifuge, which requires larger capital investment but uses much less electrical power. The modern high-performance gas-centrifuge isotope enrichment plant consists of many tall cylinders each spinning in a vacuum at some 1500 m/s surface speed; this enormous rotational rate produces high equivalent gravitational fields (evident in a mild form in amusement-park rides), and separation factors per stage are larger than in the gaseous diffusion plant. Even so, and despite the fortyfold larger power requirement of gaseous diffusion for a given amount of "separative work," it remains cheaper for the United States to use its old gaseous diffusion plants, which are supplied with low-cost power under a long-term contract. A nuclear power plant producing 7000 GWh/yr of electrical output needs 250 GWh of electrical energy each year for enrichment in a gaseous diffusion plant, in comparison with 6.3 GWh/yr in a gas-centrifuge plant.

NUCLEAR REACTORS

Nuclear reactors use nuclear fission to produce heat. This heat is extracted from the fissionable fuel by fluids called "coolants," which, in addition to preventing overheating, transport the fission heat to generate steam for the turbines that drive the electric generators. From the point of view of producing energy, a reactor is an incredibly complicated steam kettle. We concentrate in this chapter on the fission aspect of a reactor and shall explain in Chapter 5 those aspects related to the electricity they generate.

Three reactor types are in common use for supplying electrical power. The first uses natural uranium, in which the proportion of uranium-235 is only 0.71%. In such a reactor, the fast fission neutrons must lose their energy to arrive at almost zero energy (0.025 eV), where their probability of fissioning the uranium-235 nuclei is maximum. The tiny proportion of uranium-235 makes this operation difficult and requires that the loss of neutrons by absorption in the moderator be minimal. As was previously discussed, this objective can be attained with a heavy-water moderator, or with pure carbon in the form of graphite, these two materials being light and also having a small probability to capture neutrons. The heavy-water moderator is at atmospheric pressure and room temperature, although the fuel in the double-walled pressure tubes is so hot that the heavy-water coolant operates at a temperature near 300°C and produces steam of 150 bar (atmospheres) pressure to drive the steam turbine. These reactors can be built without the thick steel pressure vessels that must be used with light-water reactors, fundamentally because the neutrons diffuse much farther in heavy water or pure carbon and so can diffuse back from the moderator into the fuel that is enclosed in relatively thin-walled pressure tubes, widely spaced in the moderator.

In the second and most common type of reactor, the fuel is composed of uranium-oxide ceramic fuel pellets containing an enhanced proportion, of the order of 4%, of uranium-235, called "low-enriched uranium"; ordinary water — "light water" — is both coolant and moderator. A variant in use in about 20 of the most modern French reactors, as well as elsewhere in the world, replaces the uranium-235, in a third of the fuel elements, by plutonium diluted in uranium-238. This mixture is called "MOX," for Mixed OXide; the MOX fuel is almost interchangeable with normal low-enriched uranium fuel in producing fission heat in the reactor.

In these water-moderated reactors, the fuel mixtures are sufficiently dilute so that the reactor fuel cannot be used directly to make bombs, the heat produced is spread through a large amount of material to make it easier to remove, and the later partial fission *in situ* of the plutonium produced by transformation of the uranium-238 contributes also to the useful heat. In these reactors the neutrons are consumed before they are entirely "thermalized" — the fissions are caused primarily by neutrons of energy slightly higher than the 0.05 eV that corresponds to the operating temperature of the water moderator and the fuel (not room-temperature thermal energy of 0.025 eV, but twice that because of the elevated temperature).

In reactors using low-enriched uranium, the nuclear fuel is shaped in pellets or thin rods, to enable removal of heat while respecting a maximum operating temperature in the center of the fuel element. The pellets or rods are immersed in moderators that are as little absorbing as possible, made up of light elements (the frog can thus jump an enormous number of steps in a single bound), so that when the neutron encounters another fuel element, it has the low energy necessary to be swallowed with high probability and provoke another fission. It is barely possible to have a "homogeneous" natural uranium reactor, in which the uranium is well mixed, even with a perfect moderator that absorbs no neutrons. However, two billion years ago, not only was it possible but it actually happened, as can be seen in the accompanying insert. The discovery of this event gave rise to wild speculation on the part of some who chose to believe the fantastic story that our planet was once inhabited by beings who killed themselves off in a nuclear war.

OKLO: FOSSIL NUCLEAR REACTORS

In June 1972, French authorities observed a tiny anomaly in the course of analysis of the concentration of uranium-235 in an ore sample: a concentration of 0.7171% was measured instead of the normal value of 0.7202%. This little variation of 4 parts per thousand recurred in subsequent control measurements on the same sample. When it was observed in material destined for the Soviet Union, where it was to be enriched commercially, the possibility of cheating was raised: the Soviets would have suspected France of sending them uranium from which part of the uranium-235 had already been extracted.

At the end of what virtually amounted to a criminal investigation to trace the origin of this depleted ore, it had to be admitted that no criminal hand had slipped artificially depleted material into a natural ore sample. It turned out that the abnormal sample came from the northern extremity of the Oklo deposit in Gabon, Africa. In some soundings, samples whose concentration of uranium-235 was only about half normal were found.

At first it was thought that there might be as-yet-unknown isotope separation phenomena, which would have had enormous economic importance. In that case, however, there would have been enriched samples next to these depleted samples. Such samples were indeed "found," but this turned out to be the result of invalid measurements conditioned by a preconceived notion of some natural process of uranium enrichment.

All physicists have come across aberrant measurements in the course of their careers. That is of no importance if they know how to keep their heads. As Einstein said: "Anyone who has never made a mistake has never tried anything new."

In the Oklo case, a French Atomic Energy Commission team explained the mystery. Analysis of the ore showed that the uranium had been the site of 100,000 times more fissions than would be expected from the accumulation of spontaneous fissions over two billion years. Only a long-sustained chain reaction could have caused these fissions that consumed up to half of the uranium-235 in the ore deposit.

The half-life of uranium-235 is 700 million years. Two billion years ago, the fraction of uranium-235 in natural uranium ore was 7.26 times as much as it is now (because U-238 has a much longer half-life, the fraction of uranium-235 is divided by two every 0.7 billion years, so two billion years ago it was $2^{2/0.7} = 7.26$ times greater than it is now). This amounted to about 5% uranium-235, as in our modern nuclear power reactors. The formation of the natural reactor is also linked with the appearance of life on Earth, because the great concentrations of natural uranium in the ore at Oklo resulted from successive chemical fractionation in which oxygen played an essential role. It is generally accepted that oxygen first appeared in the terrestrial atmosphere when living organisms became capable of photosynthesis. That played a major role in eliminating the once-dominant carbon dioxide from the earth.

One had to wait, therefore, for favorable conditions to be attained, two billion years ago, for the natural uranium to be concentrated in the ore to the value necessary for a chain reaction, and these favorable conditions lasted only a billion years, because of the decay of the uranium-235. In the sequence of events leading to a reactor, as deduced by the engineers, organic matter—acting in the sea on sedimentary layers containing uranium—played a preponderant role in precipitating uranium in the form of pitchblende in geological traps. Oklo was a reactor made not by men, but by plants.

When the necessary favorable conditions came together and the ore found itself in contact with water from the swamps acting as moderator, the reactor was able to operate for almost a billion years without a violent dispersion of the ore body—totally different from the disastrous 1986 Chernobyl accident in Ukraine, about which we shall have a lot more to say.

There is much to be learned from the study of these fossil reactors about the geological evolution of nuclear waste over long periods of time.

Very few of the uranium deposits of the planet could have exhibited the spontaneous arising of natural reactors, and the traces of any that did arise have been blotted out by geological upheavals—except at Oklo. It was an extraordinary discovery. Some writers and pseudo-thinkers have seen in it proof of a visit to our planet by extraterrestrials, much more advanced than we, for whose existence they claim that this is only one among many indisputable signs. Experience shows that one cannot convince cranks by making use of scientific proof; but if one can't persuade the public new to the controversy, then there is reason to worry.

Our old planet has already encountered, in the course of its long history, the same radioactive entities, artificial and fleeting (a lifetime of only a few million years), that it is now proposed should be entombed as nuclear waste. It makes one wonder whether this buried waste will one day surprise a future generation as much as this one has been surprised by natural reactors like Oklo. Engineers don't yet know if they have enough information to choose burial sites as safe as that of Oklo.

CONTROLLING THE FISSION REACTION

A nuclear power plant engineer's job is somewhat like that of a tightrope walker. With a good balancing pole, it is easy, particularly if there is also a safety net. In the role of the balancing pole, various design features and devices permit control of fission within the mass of the fuel, the process being stabilized at the desired power level by the extraction of the generated energy. For the net, safety procedures are designed to stand up to any foreseeable eventuality that might lead to an uncontrolled chain reaction.

The evolution of a chain reaction is characterized by a multiplication factor K, which is the ratio between the rate of fissions at a given time and the rate one "generation" later—that is, when the neutrons produced in these fissions are absorbed. The time between generations is about 40 millionths of a second for water-moderated reactors, because of the great distance a neutron travels in slowing and its slow speed in the later portions of this process. This time is only several billionths of a second for a nuclear weapon and less than a millionth of a second for a fast-neutron reactor which uses neutrons that have not been slowed down by a moderator.

If this ratio K is greater than 1, the number of fissions increases with time and the system is supercritical.

If this ratio is equal to 1, the system is critical, and the number of fissions per second remains constant in time.

If this ratio is less than 1, the system is subcritical, and the fissions die away.

The "criticality excess" is a measure of the amount by which the factor K differs from 1. So the criticality excess is K-1.

In nuclear weapons, one tries to make K as large as possible, by bringing together in a short enough time, on the order of some microseconds, more than the minimal mass necessary for an explosive chain reaction to take place. This mass is on the order of a few kilograms or tens of kilograms, depending on the density and nature of the material, and for relatively slow assembly it is important that there be no neutron to start the reaction between the time the system is just critical until it is finally assembled.

In nuclear reactors, the factor K must be equal to 1 to permit producing and extracting the energy resulting from fission in several tons of fuel nuclei in a stable fashion. The trick to operating nuclear reactors is to maintain the multiplication factor strictly equal to 1, by continually compensating for the exhaustion of fissionable material over time and by preventing any disruption from this balance that would allow this factor to exceed the critical value of 1.

But a reactor is not like a wild mustang that a fearless rider is trying to stay astride as it snorts and paws the ground.

Unlike a nuclear weapon, a reactor cannot release an enormous amount of fission energy in a fraction of a microsecond. In a reactor, the multiplication factor due to those neutrons instantly emitted in fission, called "prompt" neutrons, is always maintained at slightly less than 1. Nature has very kindly introduced a tiny gimmick which slows or delays the neutron multiplication. Providentially, certain fission products emit neutrons delayed by about ten seconds when they disintegrate in a cascade that finally results in stable nuclei.

The delayed neutron fraction is small—0.65% for uranium-235, 0.21% for plutonium-239—but it is sufficient to maintain the reactor in equilibrium in conditions where the multiplication factor including prompt neutrons is less than 1, and where an uncontrolled chain reaction cannot take place. The ten or so seconds of breathing space given by the delayed neutrons allow the use of an essential safety element: control and safety rods that one can introduce rapidly into the fissionable core of the reactor and that contain nuclei that absorb a sufficiently large fraction of the neutrons to maintain the multiplication factor below 1. When they are partially withdrawn from the reactor, they absorb fewer neutrons; the "neutron flux" increases—that is to say, the number of neutrons that, each second, cross a square centimeter in the center of the reactor. The power of the reactor then increases continually at a rate that is a function of the delayed neutron lifetime and of the excess in criticality. The time available for the control of a normal reactor is typically many minutes, since the multiplication factor including delayed neutrons is only slightly greater than 1.

PARAMETERS OF CRITICALITY

In the course of time, because of transmutations and disintegrations, the chemical composition of the core evolves. Uranium-238, which acts as a neutron trap, changes into plutonium-239, which is as active in the production of energy as uranium-235 and winds up contributing considerably to the energy supplied by the reactor. Uranium-235 is consumed.

But certain fission products, whose quantity in the reactor varies over time, can be ferocious poisons for the reactor. That is the case for xenon-135, whose neutron capture cross section is colossal. It is not produced directly but comes from the disintegration of a direct fission product, iodine-135. If a high-power reactor is shut down, the quantity of xenon-135 increases to such an extent after a few hours that the reactor cannot be restarted, even if the control rods are completely withdrawn. In fact, this was the one surprise in scaling from Fermi's two-watt pile of December 2, 1942, to the 250-megawatt Hanford reactor. In a low-power reactor, the amount of fission product present at any given time is not sufficient to produce poisoning. Even with a high-power reactor, however, it is enough to wait a day or so, because, just by chance, the half-life of xenon-135—the time in which half of it decays—is only eight hours. Under normal operation at high power, the xenon-135 is burned as rapidly as it is created, because it is transmuted by absorption of neutrons and is thus present only in small quantities. Nevertheless, in a large reactor, one can have "waves" of xenon that, from time to time, poison portions of the reactor. The Xe-135 concentration can be locally high, depressing the local fission rate and heat production (and also the production of additional Xe-135 in that locality); this excess of poison can then decay, only to appear elsewhere in the reactor.

In addition, physical parameters (cleanliness and purity of materials, for example) are not uniform throughout the reactor, and the criticality depends on the temperature.

Let's suppose, for example, that there is a large region in the core of the reactor where the multiplication factor rises to 1.01. Keep in mind the essential character of a chain reaction; seventy generations are enough for a 1% criticality excess to double the number of neutrons, and this is reached in less than a tenth of a second in a thermal reactor that, we have supposed, has been driven to criticality by prompt neutrons alone. We recall the Oriental tale in which a prince wants to thank a subject and allows him to choose his own reward. The subject points to a chessboard and asks for one grain of rice on the first square, two on the second, and so on, doubling each day until the sixty-fourth square. The sovereign, astonished and enchanted by so modest a request, accepts and discovers a little late that on the sixty-fourth square he will have to give the subject 10^{19} grains of rice, or about 30 billion tons, more than his kingdom has produced during its entire existence. The chagrin of the ruler would be matched by that of the reactor operator and supplier if a reactor were so large and poorly controlled that one portion could be "prompt critical" while the rest of the reactor was critical only with the aid of delayed neutrons. However, for reactors containing U-238, the Doppler effect reduces the reactivity in the regions of high power and hence high fuel temperature.

The pressurized water reactors used in both the United States and France are designed so as to be undermoderated: when the temperature increases, the expansion of the moderator (which is also the coolant) reduces the criticality and stabilizes the power. This is all the more true if there is a sudden loss of moderator because of a leak or a catastrophic failure of the reactor vessel. But an inundation of cold water coming from the emergency cooling system can cause a sudden increase in criticality. These safety considerations have led to the almost universal adoption of pressurized light-water reactors or of a variant, the boiling-water reactor, which differs little from the pressurized water reactor in its nuclear properties.

Having provided the background for a full understanding of nuclear reactors, we shall return later to the engineering features of the nuclear plants producing 17% of the world's electrical power, with their problems and promise. But let's first attack what is, technically speaking at least, an "easier" problem, although it is at once both a great threat and a great responsibility for American society—our nuclear weapons and those of the rest of the world.

CHAPTER 3

Nuclear Weapons

HAVING SUPPLIED some basic information on nuclear physics and fission, we can now discuss important applications—nuclear weapons and reactors for producing electrical power. Although some details remain officially secret and hence cannot be published in this book, the main design features of nuclear weapons are much simpler to understand, in principle, than those of power reactors, because there is no need to provide for transfer of heat from the fission chain reaction, and the mechanical aspects of a bomb are of interest only to the moment of detonation. Furthermore, the analysis of possible accidents is easier than in the case of nuclear reactors. For these reasons, we first consider nuclear weaponry.

In discussing the functioning of nuclear weapons, we will restrict ourselves to facts and principles abundantly documented in the literature in the public domain. Weapons proliferators will find nothing new in our book; we are not writing a text for nuclear terrorists. We shall explain the general principles of the different kinds of nuclear weapons and describe their effects, as well as the planetary dangers that will exist until the absurdly large existing stock of weapons is liquidated. We shall finally discuss some of the aspects of nuclear weapon explosion tests, and the costs incurred in the weapons program.

THE FISSION WEAPON

The basic problem in designing a fission bomb is to bring together a sufficient amount of fissionable material to sustain a neutron chain reaction, and to do it in a short enough time. The requirement is that the material be "subcritical" in its initial configuration (i.e., that fission neutrons produced in the plutonium or in the uranium-235 should have enough likelihood of escape that each produces, on average, less than one other fission neutron). It is also desirable in the assembled supercritical configuration that the fissionable material be dense so that the time required for a neutron to provoke fission will be short; for this reason, plutonium or uranium in the form of metal is the material preferred for nuclear weapons.

The Los Alamos New Mexico Laboratory was created in April 1943 (as "Site Y" of the secret Manhattan Project) to design and build nuclear weapons with enriched uranium that would be shipped from Oak Ridge, Tennessee, and with plutonium produced in the reactors at Hanford, Washington.

The first approach used for assembling a nuclear weapon was the "gun," so called because the pieces were placed in a steel gun barrel and brought together by a propellant such as gunpowder or nitrocellulose. A neutron injected at the instant the uranium mass is fully assembled "initiates" the chain reaction, of which some 80 generations—i.e., 80 doublings—are completed in a millionth of a second.

A typical fission weapon consumes about 1 kg of its fissionable uranium or plutonium; the complete fissioning of this much material provides an explosive yield equivalent to that of 17 million kg, or seventeen kilotons (17 kt), of high explosive such as dynamite or TNT. The fission also releases 8 g of neutrons—4.8×10^{24} neutrons. When captured by the surrounding materials, these neutrons produce hundreds of grams of radioactive substances, which add their radioactivity to that of the fission products and supplement the enormous quantity of gamma rays emitted instantly during the fission process. The nuclear weapon that destroyed Hiroshima, which contained about 60 kg of almost pure uranium-235, had an explosive yield of 13 kt; thus only 13/17 kg or about 0.8 kg was actually fissioned, and the efficiency of the weapon was 0.8 kg/60 kg, or about 1.3%. The rest of the uranium was dispersed into the atmosphere, together with the highly radioactive fission products.

The "gun" technique is no longer much used by the nuclear weapon states, because a different approach employs less fissile material and does so more efficiently—at the price of increased complexity. Such was the confidence in the design of the gun-type bomb that the design was never verified by a test explosion before the bomb was dropped on Hiroshima. Other nations might confidently build and stockpile gun-type nuclear weapons without test.

The other assembly mechanism developed at Los Alamos was the "implosion bomb," of the type used against Nagasaki, in which high explosive is used to assemble a supercritical mass more rapidly than is possible with propellant. The gun-assembled weapon that works with uranium-235 is incompatible with a particular property of plutonium: when plutonium is produced in reactors, the isotope Pu-240 is created along with the more common isotope, Pu-239, and has a high rate of spontaneous fission, with the emission of neutrons; one of these can cause the material to chain-react almost as soon as it becomes critical in the assembly process and before it is fully compressed, thus leading to premature disassembly and producing a "fizzle." When this high spontaneous neutron emission rate was discovered after the first samples of plutonium produced in reactors were delivered to Los Alamos, almost the entire laboratory had to be refocused to solve the problem. The "plutonium gun" was no longer an option. An approach was needed that could complete the assembly before one of the troublesome spontaneous neutrons would arrive. The key was to use implosion.

In the implosion design, a sphere or hollow shell of fissionable material is compressed by powerful chemical explosives distributed over its external surface. Because the explosives compress the core of the bomb, the implosion concept also yields a more efficient bomb—that is to say, one that produces the same amount of energy with a smaller mass of fissionable material. The Nagasaki bomb contained about 6 kg of plutonium and had an explosive yield of about 22 kt, corresponding to the fission of about 1.3 kg of plutonium; its efficiency was thus about 1.3/6, or about 20%—in contrast to the 1.3% efficiency of the uranium gun. Thus, the implosion method, once mastered, is doubly attractive—it allows better efficiency and higher yield using either plutonium (as we will show in Chapter 12, even plutonium from civil power reactors with a very high concentration of Pu-240) or uranium-235 in a smaller quantity than in the case of the gun-type design.

The feasibility and efficiency of an implosion weapon depend on the precision of the shock wave producing the compression. If one tries to use a number of "detonation points" on the surface of a sphere of explosives to create the converging detonation wave that is eventually to compress a central plutonium sphere or shell, the wave starts out as expanding from each detonator. The plutonium shell is therefore distorted or broken, because it is not compressed uniformly. The problem for the engineer is to distribute a number of detonators in such a way that the propagating wave is everywhere aiming for the center of the plutonium shell. In the same manner that a glass lens in air converts the expanding sphere of light from a flashlight bulb into a parallel beam or even focuses it on a point, systems of "lenses" made with high explosives allow the conversion of detonation waves diverging from each detonator into one spherically convergent wave. Such lenses were initially made with two kinds of explosive—a "fast component" in which the detonation wave moves at high speed (analogous to the way light moves in the air in an optical lens system), and a "slow component" in which detonation moves more slowly (analogous to the way glass slows the motion of light).

The injection of the neutrons to initiate the chain reaction must be precisely synchronized with the implosion, or else the bomb will function at reduced power, or may not function at all if the material "bounces" and becomes subcritical before a neutron arrives.

The bomb's efficiency can be improved by adding an additional shell of beryllium or uranium-238, which reflects escaping neutrons back to the fissionable core, and by shaping the plutonium or uranium-235 into a hollow shell rather than a solid sphere.

The early Los Alamos implosion bombs employed an "internal initiator" in which a small amount of beryllium, together with the alpha-particle-emitting radioactive element polonium (Po-210), was used to produce neutrons (in the same fashion that Chadwick produced neutrons in 1932). But a simple neutron source would not do the job, because in this design of weapon, neutrons that are injected too early result in an explosion yield reduced by a factor of 10 to 20 from the design yield of 20 kt. Accordingly, the polonium and beryllium are separated by a thin layer of material (just a bit thicker than the range of the alpha particles from polonium), which is disrupted by the shock from the conventional explosives that travels through the plutonium core to the initiator at its center. The shock mixes the polonium and the beryllium so that the alpha particles can suddenly begin to produce neutrons. To be sure of having a neutron available in a fraction of a microsecond, many curies of polonium are required, and the polonium (usually produced by neutron irradiation of large amounts of bismuth in a nuclear reactor) has a half-life of four months and must be replaced every six months or so.

Boosted Fission Weapons

The fission energy released can be further increased by incorporating hydrogen gas composed of deuterium and tritium in the fissioning core, where they undergo a fusion reaction—which releases a large number of neutrons—within the fissioning mass. This procedure is called "boosting." Boosted fission weapons were tested for the first time in 1951. U.S. weapons now contain a shell of plutonium, into which a deuterium-tritium gas mixture is introduced. When the plutonium is imploded by the conventional explosive that surrounds it, and produces a nuclear explosion, the temperature becomes high enough to provoke fusion of the deuterium-tritium, with the sudden emission of the order of a gram of neutrons. These, in turn, cause additional fissions in the surrounding fissionable metal. Only a little of the energy released by a boosted fission weapon comes from the deuterium-tritium reaction: each fission supplies about 150 MeV, whereas each deuterium-tritium reaction itself delivers a total of 17.6 MeV (the reaction is $D + T \Rightarrow He^4 + n + 17.59$ MeV). The helium nucleus carries away 3.5 MeV, the neutron 14 MeV, in inverse proportion to their masses. The neutrons from the fusion induce additional fissions in the surrounding plutonium, and it is this "boost" that gives a major enhancement to the explosive yield. Tritium has a half-life of 12 years, and so the supply must be refreshed—typically every couple of years or perhaps at intervals as great as 10 years.

Pure fission weapons have been used to reach a maximum power of the order of 500 kilotons of dynamite (forty times the yield of the Hiroshima bomb) and also for much lower yields, of the order of 20 t. After developing the first implosion weapon in 1945, the United States then designed fission weapons with the fissionable material split between a central ball and a hollow shell, making possible much smaller and lighter implosion weapons. Boosting helped to increase the yield. Further advances were made in safety against unintended detonation and in protecting the weapon against tampering or theft or unauthorized use. To increase safety, the most important advances have been the Enhanced Electrical Detonation Safety System, which prevents the weapon from being set off by lightning or accident; the use in some weapons of "insensitive high explosive," which cannot be detonated even by direct impact from a rifle bullet; and the development of fire-resistant "pits" (that is, the metal-encased fissile-material shell that is mounted inside the assembly of high explosives), which will prevent the dispersion of plutonium in a fire fed by aircraft fuel and the combustion of the high explosive. For prevention of unauthorized use, Permissive Action Links (PALs, which were initially electromechanical combination locks) were first introduced on U.S. weapons deployed in Europe in 1962 and have expanded in deployment and capability ever since. The PAL prevents the firing circuit in the warhead from detonating the high explosive unless the warhead receives the preset code. Modern PALs allow only a limited number of attempts before disabling the warhead. The PAL decouples physical possession of the warhead from the ability to use it and thus both improves control over the nuclear force and allows its broader deployment without fear of unauthorized use.

THERMONUCLEAR WEAPONS

In 1952 the United States detonated an explosive device based on a new principle that enabled the creation of weapons of unlimited yield—10 megatons, or 100, or even 1000. This was the thermonuclear weapon, or hydrogen bomb.

In this test, a fission device such as the one discussed above was used to provide the energy necessary to trigger a nuclear fusion reaction in a large amount of material containing deuterium. Fusion on a large scale can occur only at temperature, pressure, and density greater than found at the center of the sun. It is this nuclear fusion reaction that is at the core of thermonuclear weapons. The fission device is the "primary," while the system that

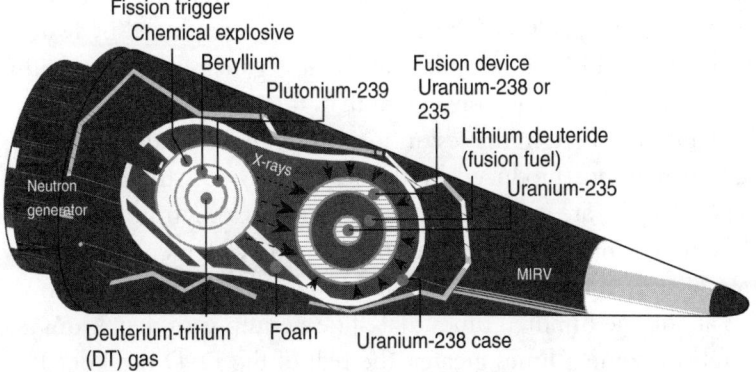

A modern thermonuclear explosive
This W87 thermonuclear warhead is launched on an MX intercontinental missile. Packed into a multiple independently targeted re-entry vehicle (MIRV, shown below), it splits off from the missile to strike its target.

Fission trigger
Chemical explosive
Beryllium
Plutonium-239
Fusion device
Uranium-238 or 235
Lithium deuteride (fusion fuel)
Uranium-235
X-rays
Neutron generator
MIRV
Deuterium-tritium (DT) gas
Foam
Uranium-238 case

MIRV length: 5.7 feet   MIRV base diameter: 1.8 feet
Explosive power: 300,000 tons of TNT

Explosion process The compression of plutonium with a chemical explosive (above, left) starts a fission explosion that, in turn, is boosted by the fusion of DT gas. X-rays then compress the second component, causing a larger fission/fusion.

Fig. 3.1. A cartoon (not a blueprint) of one of the most modern U.S. thermonuclear weapons appears in a May 1999 report of a congressional committee— The Cox Report—and is reproduced here.

provides the bulk of the fusion energy is the "secondary" which contains deuterium. Fig. 3.1 is a diagram of a modern thermonuclear warhead.

For fusion to take place efficiently, the secondary is compressed and heated not by high explosive but by the enormous pressure of soft X-rays from the primary explosion—as prescribed by the "radiation implosion" approach invented at the Los Alamos Laboratory by the mathematician Stanislaw Ulam and the physicist Edward Teller in March 1951. Radiation implosion allows the fission energy produced by the primary to be transferred to the secondary capsule by the X-rays produced by the hot fissile core of the primary, in just the same way that a hot tungsten filament in a hundred-watt bulb produces a bright light—except that within a hundredth of a microsecond the kilograms of plutonium transfer most of their thermal energy to this bath of electromagnetic radiation, which carries with it not only most of the energy of the fission explosion but also enormous pressure.

Various materials and configurations may be used in the secondary of the thermonuclear weapon. The neutrons coming from the thermonuclear charge can induce fission in adjacent enriched or natural uranium (the fission probability for a 14-MeV neutron on uranium-238 being substantial), so that typically 50% of the energy of a "thermonuclear" weapon is produced by fission. The thermonuclear fuel of the secondary element is generally made of lithium-6 and deuterium; this is more convenient than the ultracold liquid that was the form of deuterium used in the test in 1952, and it has the additional following feature: the neutrons produced by the reaction D+D—the collision of two deuterium nuclei—are captured in the lithium-6 to form tritium, which accelerates the thermonuclear reaction in the secondary element. Since the reaction rate of tritium with deuterium at temperatures reached in a hydrogen bomb is about one hundred times that of deuterium with deuterium and the energy release some 4 times greater, the half of the D+D reactions that produce tritium (plus a proton—the other half produce helium-3 plus a neutron) more than double the energy released per billionth of a second by the deuterium fuel. And the lithium-6 both generates tritium from the other half of the D+D reaction and regenerates whatever tritium is used by reacting with D.

The first thermonuclear bomb tested by the United States at Eniwetok Atoll in the Pacific Ocean on November 1, 1952, was a 10-megaton explosive called Mike. Designed as a means to obtain much more powerful weapons without the necessity of investing large amounts of uranium-235 or plutonium, this two-stage approach (radiation implosion) to the release of nuclear energy has been favored even in the power range accessible to fission weapons, because of several qualities: safety, lightness, and a minimum consumption of fissionable material that was expensive and difficult to obtain. A yield of many megatons might be obtained in this way with the investment of only 6 kg of plutonium in the primary. Even if a 10-megaton yield could somehow be obtained from a fission weapon, 600 kg of fissionable material would have to be consumed; therefore, at 30% efficiency, it would have been necessary to invest 600/30% = 2000 kg of plutonium. On the other hand, with the two-stage weapon, the same 2 tons of plutonium might suffice for 333 such weapons. The two-stage weapon is safer than a comparable fission weapon because the fissionable material is present in a far smaller amount, and hence can be maintained much farther from criticality until the weapon is detonated.

The most powerful thermonuclear bomb ever exploded was tested in the atmosphere by the Soviet Union in 1961. Its power reached nearly 60 million tons of classical explosive, that is to say, the equivalent of 4600 Hiroshima bombs. Had it been built as planned with uranium surrounding the thermonuclear fuel in the secondary (rather than with a lead surround, to reduce its yield so as to avoid damage to inhabited regions during the test), it would have had a yield of one hundred megatons.

Since the 1950s, rather than fundamentally new principles for nuclear weapons, there has been a continuous refinement of procedures. Major improvements have involved the miniaturization of nuclear warheads, the addition of safety systems as noted for fission weapons, the introduction of a less hazardous chemical explosive, the reduction of the weight of the housing of the weapon and of the casing that prevents the soft X-rays from escaping so that they can efficiently compress the secondary, and the development of weapons with selectable explosive power. A weapon of selectable yield could evidently be derived from a normal two-stage weapon provided with an optional feature to sever the secondary from the primary or prevent the boost gas from entering the pit of the primary; such an approach would give a choice of three explosive yields—the full yield of the two-stage weapon, the yield of the boosted primary, or the yield of the unboosted primary explosive. So-called "dial-a-yield" weapons in the U.S. inventory may make use of these techniques.

With these improvements in weapons systems, the military had available an arsenal covering all imaginable, and even unimaginable, needs. The lightest bombs can be launched by a hand-carried device. Many of the U.S. strategic weapons—those mounted on long-range ballistic missiles or carried by the nuclear-capable B-52 bombers—have destructive power of a half megaton or so, close to forty times that of the Hiroshima bomb. All U.S. nuclear weapons have been developed at Los Alamos, or at the Lawrence Livermore National Laboratory created in 1952 at the urging of Edward Teller, who was dissatisfied with the progress Los Alamos was making on the hydrogen bomb.

NEUTRON BOMBS

For a very small nuclear explosion of 100 tons, the effect of the blast and direct heat—which destroy all buildings in proximity—can be negligible if the bomb explodes at an altitude of some 300 meters, but the radiation, the neutrons in particular, can be fatal at this distance from the explosion. That gave military thinkers the idea of ordering bombs in which the production of neutrons is increased by using the heat of the fission to trigger fusion reactions between the two types of heavy hydrogen, deuterium and tritium, each reaction releasing, as has been discussed, a particularly energetic neutron of 14 MeV, with great ability to penetrate steel, concrete, and (most important here) air. That's the principle of the neutron bomb. It would, the claim was, kill soldiers in their tanks and civilians in buildings without destroying libraries and churches. In reality, it seems that a neutron bomb should more properly have been called a "reduced-blast weapon" and one that therefore was less effective than its predecessors; the ordinary 20-to-50-kt nuclear weapon would have killed more soldiers by radiation and many more by blast. The neutron bomb does not instantly disarm a tank—even radiation of twenty times the lethal dose will take at least thirty minutes, perhaps more, to disable the members of a tank crew, although they are sure to die in a week or two.

Advocates of the neutron bomb argued that it would make U.S. use of nuclear weapons against advancing Warsaw Pact armies more credible—cities could have been evacuated, then enemy troops killed by radiation, and the buildings housing them would survive the neutron bomb rather than be destroyed by blast as they would be by an ordinary high-yield bomb.

The neutron bomb troubled relations among NATO partners during the 1960s, because it became possible to imagine a "clean" nuclear war on the territory of the European allies, as an alternative to the use of nuclear weapons only inside the borders of the Soviet Union or its allies. This weapon was favored by some American strategists, who saw in it the best way to pin down the Soviet tank regiments stationed in East Germany if they were to head west.

Fig. 3.2.

West Germans obviously were afraid that this first use of nuclear weapons on the ground would be followed by a salvo of dirty tactical Soviet nuclear weapons against NATO targets in Germany. They were no doubt right, although fortunately this hypothesis was never tested.

When the French army procured short-range nuclear weapon launchers, President de Gaulle insisted that they must never be used in France but only on Soviet troops operating on German territory and threatening France. Regiments of French Pluton missiles and then of Hadès rockets were installed for this mission. That enraged France's German allies, who succeeded in obtaining the commitment that these missiles would not become operational. It was only in 1996 that it was decided to disband the Hadès rocket regiments altogether.

In the opinion of the authors of this book, the neutron bomb does not strengthen deterrence of attack by an adversary possessing nuclear weapons. If you have made the effort to acquire these "clean" weapons and the adversary has only "dirty" weapons, which do not distinguish between a library and its librarian, you would be quite reluctant to order the firing of nuclear weapons on allied territory unless, of course, you had made the chivalrous effort to offer to your enemy, free of charge, a stock of neutron bombs to replace his dirty weapons. The situation is different if the goal is not deterrence but rather the routine use of nuclear weapons. Could one thus imagine the bombing of individual leaders such as Saddam Hussein or Slobodan Milosevic, while leaving intact the ancient structures dear to the heart of civilized humanity? All such a despot need do to protect himself against this weapon is to build a swimming pool 3 meters deep over his head; he will then be perfectly protected from the neutrons. An underground room at a depth of 3 meters or more would also do the job.

Today it is widely appreciated that there exist nonnuclear antitank weapons of such precision that they can transform tanks into rolling death traps, taking the place of such nuclear weapons. The saga of the neutron bomb and the persistence of thousands of nuclear weapons, beyond the time when guided conventional weapons could do the job better, testify to a bureaucratic momentum and a reluctance to give up most of the nuclear weapons even when the threat they pose to civilization far exceeds their benefit to national security.

EFFECTS OF NUCLEAR WEAPONS

The effects of a nuclear weapon depend on its yield, the altitude at which it explodes, and meteorological conditions. A nuclear explosion within the atmosphere produces blast and high-speed winds and debris that are similar to those from a similar energy release from conventional explosives. Of course, a megaton of high explosive would be almost 10^6 cubic meters—a cube of one hundred meters on a side—stacked on two football fields side by side, to a height equal to the length of a football field. In addition, there is an enormous amount of heat radiated by the explosion, capable of burning people, vegetation, and structures at distances of 20 kilometers, in the case of a megaton explosion on a clear day (or night). These major effects of a nuclear weapon were predicted before the first test in an analysis performed by British fluid-dynamicist Geoffrey I. Taylor, who during the Second World War was asked to predict the effect of liberating in the atmosphere an amount of pure energy similar to that from one thousand tons of explosive, but without any significant mass.

As the bomb explodes, most of its energy is in the emerging thermal radiation (soft X-rays) and the thermal and kinetic energy of the fissile core. This is transferred to the surrounding atmosphere, in which the radiation initially moves rapidly to create the "fireball." The fireball grows and cools, and the visible portion of the thermal radiation from the larger but still enormously bright fireball surface chars exposed surfaces at distances of kilometers—or tens of kilometers, in the case of multimegaton explosions. Soon the expansion of the fireball slows, as the cooler thermal radiation from the fireball eats into the air more slowly; fireball expansion is overtaken by a "shock wave" as the principal route for energy to leave the explosion. As long as the pressure within the shock wave is much greater than normal atmospheric pressure, the strong air shock moves much faster than the speed of sound in normal air, which is one thousand feet per second or 300 m/s.

As a colleague who witnessed two aboveground atomic bomb explosions in the Nevada desert in 1957 reports, the sequence of events one observes is first the almost unbearable light, then a clicking, somewhat painful sensation in one's ears as the shock wave passes, and finally the rolling thunder of the explosion; meanwhile the fireball turns an ominous black as it picks up soil from the ground. After the explosion the desert surface had been changed into an eerie glass.

In a nuclear explosion, the neutrons and gamma rays from fission and fusion are also an important cause of death. Within a microsecond of a nuclear explosion in the atmosphere, the gamma rays and neutrons from the fission reaction emerge from the exploding fissile core. They expose any people within hundreds of meters to a lethal dose of radiation. Most individuals at that range will be killed by the heat and blast from the bomb. In the bombing of Hiroshima and Nagasaki about 15% of the deaths are believed to have occurred from radiation. Only for the smallest yields does the radiation account for a large fraction of the deaths from a nuclear explosion; it would be the case for a low-yield neutron bomb, intentionally detonated at an altitude to reduce destruction by blast. In the case of the explosion of a megaton thermonuclear weapon in contact with the ground (a "ground burst"), radioactivity can be spread over large areas by the fallout of particles of soil and can continue to have an effect for decades after the explosion, but for densely populated areas near the impact point, the delayed effects are minor compared to the effect of the immediate radiation.

Imagine that one of the thousands of megaton-class nuclear warheads explodes at a typical height of 6500 feet (1.98 km) over Times Square in New York City (see Fig. 3.3). By 1.8 seconds after detonation, the shock wave has raced roughly 0.9 km ahead of the fireball. Until the shock strikes the ground, only the nuclear radiation and the devastating thermal radiation have been the cause of damage. Now buildings are crushed and swept away, objects and people turned into projectiles as the pressure and the winds behind the shock front take their toll. At 11 seconds the shock will have reached 5.5 km from the epicenter—the point on the surface just below the detonation—with overpressure (in comparison with the normal atmospheric pressure of 14.7 pounds per square inch, or psi) of 6 psi—864 pounds of force on each square foot of the surface. The wind speed corresponding to that shock pressure is 312 km/h. At 37 seconds the front of the shock wave will have reached 16 km; the overpressure will have dropped to 1 psi and the wind behind the front to a mere 61 km/h. Glass windows will be stripped from the buildings down to about 0.5 psi, but half the buildings will be left standing at overpressures of 4 to 5 psi.

Now the intense heat from the bomb will have dissipated, although fires and even a firestorm may have been ignited. The prompt nuclear radiation from the explosion is followed by the gamma rays from fission products in the expanding cloud of debris. The hot fireball entrains debris from the ground as it rises, cooling so that the radioactive materials largely condense on debris particles. The fine particles begin to fall to the ground, but unless there is rain they do so over a period of hours and days in regions to which the by-then-invisible radioactive cloud is carried by high-altitude winds. By 110 seconds the cloud is at an altitude of seven miles; it attains its maximum height of 24 km at about 10 minutes.

The photographs taken of Hiroshima after the explosion of a bomb only one-seventieth as powerful as our megaton example show the damage to the city (Fig. 3.4). As was explained above, the instantaneous destruction is due to the blast effect and the fires set by the intense heat radiated in the form of light by the fireball at the heart of the explosion.

Let us add to this rather dry picture some excerpts from *A Path Where No Man Thought* by Carl Sagan and Richard Turco:[1]

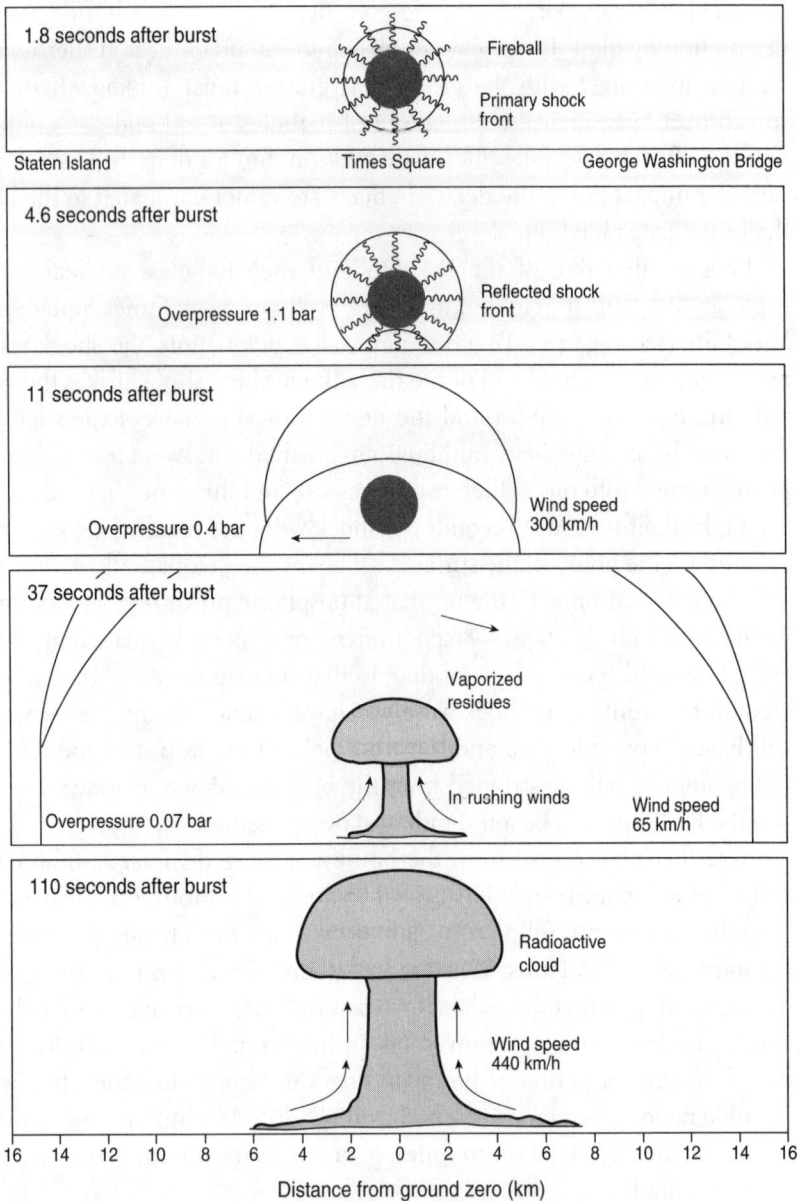

Fig. 3.3. The effects of a one-megaton nuclear warhead exploding at 2 km over the heart of New York.

Fig. 3.4. Photographs of Hiroshima after the explosion.

The Japanese city of Hiroshima was wiped out on August 6, 1945, by an approximately 13-kiloton-yield nuclear weapon. Some 200,000 men, women, and children were killed, many from lingering deaths. Col. Paul W. Tibbet, Jr., was the pilot of the *Enola Gay*, the B-29 that, for the first time in human history, dropped an atomic bomb on a city. He had named the airplane for his mother. Here is his description of what he saw:

"What had been Hiroshima was going up in a mountain of smoke. . . . First I could see a mushroom of boiling dust—apparently with some debris in it—up to 20,000 feet. The boiling continued three or four minutes as I watched. Then a white cloud plumed upward from the center to some 40,000 feet. An angry dust cloud spread all around the city. There were fires on the fringes of the city, apparently burning as buildings crumbled and the gas mains broke."[2]

An eyewitness description of the early effects of the Nagasaki explosion two days later comes from a young Japanese physician, T. Akizuki:

The sky was dark as pitch, covered with dense clouds of smoke; under that blackness, over the earth, hung a yellow-brown fog. Gradually the veiled ground became visible and the view beyond rooted me to the spot with horror. All the buildings I could see were on fire. . . . Trees on the nearby hills were smoking, as were the leaves of sweet potatoes in the fields. To say that everything burned is not enough. The sky was dark, the ground was scarlet, and in between hung clouds of yellowish smoke. Three kinds of color— black, yellow, and scarlet—loomed ominously over the people, who ran about like so many ants seeking to escape. . . . That ocean of fire, that sky of smoke! It seemed like the end of the world.[3]

One study of the effects of the two bombs noted that: "Radioactive black rain fell on both Hiroshima and Nagasaki; in Hiroshima the rain was accompanied by a sudden chill, many survivors shivering in midsummer."[4]

THE ULTIMATE VICTIM: LIFE ON EARTH

Shocking as they are, the above descriptions and the photos of Hiroshima in Fig. 3.4 fall far short of providing a full appreciation of the effects of a nuclear weapons exchange using even 10% of the current U.S. or Russian weapons stock—even if a thousand explosions are put side by side, and with each covering an area sixteen times larger than the area destroyed in Hiroshima or Nagasaki, as would be the case with a bomb of one-megaton yield. In addition to the local damage in Hiroshima, replicated 16,000 times, a new effect on a planetary scale would have to be considered.

Not until the 1980s was it suspected that the effects of many simultaneous nuclear explosions could influence the climate by obscuring the sun, bringing temperatures down perhaps far below the level of the last ice age. Scientists concluded that one thousand simultaneous one-megaton explosions would be enough to produce an ecological catastrophe without precedent in historic times: nuclear winter. And yet, at that time, as is the case now, many times that explosive power was available. It is useful to discuss this phenomenon, which, whatever the numerical conclusions drawn, forces us to consider a dramatic reduction of the stockpile of nuclear weapons—too large for any reasonable military strategy that can be imagined—as a first priority for mankind.

In 1982, in the journal *Ambio*, specializing in human environmental issues and published by the Swedish Royal Academy of Science, two chemists, Paul J. Crutzen, 1995 Nobel laureate in chemistry, and John W. Birks, published an article entitled "Twilight at Noon: The Atmosphere After a Nuclear War." They drew attention to a previously neglected aspect of a nuclear war and the fires that it would cause: the emission of great quantities of dust and soot capable of absorbing a large fraction of the sun's rays for several months, which could change the physical parameters of the atmosphere for a long time, with grave consequences for the earth's ecosystems. "Under these conditions," they wrote, "it is likely that the agricultural production of the Northern Hemisphere would be almost entirely eliminated, so that those who might have survived the immediate effects of the war would have nothing to eat."

Scientists then proceeded to make detailed calculations on various nuclear war scenarios, taking into account fires in forests, oil and gas installations, and urban areas. These were published and compared.

The results: cold and darkness would descend on the earth. Detailed studies catalog the associated effects: the formation of heavy clouds of toxic gas at ground level, caused by the destruction of cities; the spreading of radioactive fallout; the thinning of the protective ozone layer which prevents the sun's ultraviolet radiation from reaching the ground.[5]

According to the book by the late Carl Sagan and Richard Turco, the conclusions of the studies make it possible to envisage, if not the total extinction of the human species, at least something approaching the Apocalypse.

In 1983, Sagan and Turco organized a meeting in Washington of important advisors and high-level members of the preceding government—the administration of President Jimmy Carter. "This was at a time," they write in their book, "in the Reagan years when 'fighting' and 'winning' a nuclear war was considered feasible, and merely describing the dangers of nuclear war—nuclear winter aside—was judged, if not unpatriotic, then at least eroding the will of the American people to oppose Soviet tyranny and therefore naive and foolish." Using slides and a handout, they presented their scientific findings to this informed but nontechnical audience. They argued that nuclear winter was so serious that it had implications for nuclear strategy, policy, and doctrine and for attitudes that seemed to be shared by almost all American and Soviet officials toward the Cold War.

The authors go on:

As you might imagine—because our findings were so unexpected and because their implications ran so much at cross-purposes to what then passed for prevailing wisdom—there was a fairly spirited discussion. The remark that we found most memorable, as well as most useful, was uttered by one senior practitioner of dark arts: "Look," he said, "if you believe that the mere threat of the end of the world is enough to change thinking in Washington and Moscow, you haven't spent much time in those cities."

Since then, we've spent considerable time in Washington and Moscow and many other places where nuclear war is planned and weighed. The remark was particularly helpful because it reminded us of how abstracted many officials and strategists are from the horrors they plan for, and how resistant to fundamental change the principal political and military establishments and the weapons laboratories had become. If nuclear winter were to inform, much less change, national policies, it would take time.

A COMPARISON OF NUCLEAR AND NONNUCLEAR CATASTROPHES

Table 3.1 lists some of the catastrophes that have stricken humanity. It is apparent that the use of the nuclear arsenal would be a major increase in magnitude. The table also allows us to compare the relative importance of catastrophes that have nothing to do with nuclear energy, civilian or military.

The figure given for the eruption of the Tambora volcano includes immediate deaths (10,000), those caused locally by epidemics and famine (80,000), and an estimate of those brought on by the climatic consequences the world over. The Black Plague may have had far more victims, worldwide, than the figure shown indicates, because only deaths in Europe were counted. With respect to Chernobyl, the figure for the first year is 1000. Estimates of the number of deaths due to radiation, over tens of years, vary between 3000 and 60,000 (and between zero and 900,000 for experts firmly implanted in extreme positions, as detailed in Chapter 7 of this book); the authors' best estimate is about 30,000.

TABLE 3.1. EXAMPLES OF MAJOR CATASTROPHES

(Adapted from Sagan and Turco, A *Path Where No Man Thought*)

CAUSE	LOCATION	DATE	NUMBER OF VICTIMS
Black Plague, pandemic	Europe	1347–51	25 million
Earthquake	Shaanxi, China	1556	830,000
Volcanic eruption	Mount Tambora, Indonesia	1815	160,000
Famine	Northern China	1876–79	10 million?
First World War	Mainly in Europe	1914–18	20 million
Accidental chemical explosion	Halifax Harbor, Canada	1917	1654
Flood	Huang He Basin, China	1931	3.7 million
Second World War	Entire world	1939–45	40 million
Nuclear weapon explosion	Hiroshima, Japan	1945	200,000
Cyclone	Bangladesh	1970	300,000
AIDS	Entire world	1980–	> 3 million per year seropositive
Chemical discharge	Bhopal, India	1984	5000
Nuclear power plant accident	Chernobyl, Soviet Union	1986	30,000?
Nuclear war	Entire world	?	1 billion perhaps

NUCLEAR EXPLOSIONS FOR TESTING WEAPONS

The gun-type nuclear weapon was used in combat in August 1945 without having been the subject of a test explosion, but the implosion weapon did have a full-scale test at Alamogordo, New Mexico, on July 16, 1945, in order to give assurance that the concept actually worked, and to measure the yield, for which predictions in June 1945 (by V. F. Weisskopf's group at Los Alamos) ranged from 4000 to 13,000 tons of high explosive.[6] The height of the burst over the intended target in Japan was to be chosen on the basis of this yield measurement.

After 1945, further evolution of weapon design was performed with the aid of development tests (initially in the atmosphere, and beginning in the 1950s most frequently underground), which were used to validate and to refine the principles involved. Tests could be used also to reduce the conservatism of a specific design by cutting the amount of fissionable material or the amount of boost gas, and noting in a series of nuclear explosions the threshold below which the yield suffered as a result.

Once a nuclear weapon design was deemed final and the model was put into production, it was eventually subjected to a production verification test in order to prove that the weapons manufactured by the large production complex of the Atomic Energy Commission (now the Department of Energy) worked.

Weapons drawn at random from the armory after some years are not normally subject to stockpile verification test explosions, because the number of such tests required to demonstrate the reliability demanded of nuclear weapons would be enormous. Instead the U.S. stockpile is monitored by inspecting and disassembling 11 weapons of each type each year, to observe changes such as corrosion, and to remedy them. At the urging of Congress, however, several stockpile verification tests were conducted in the 1980s.

Initially nuclear weapon tests were explosions in the atmosphere—on the ground, on a barge or tower, tethered from a balloon, or dropped from an aircraft. This provided a convenient way to measure directly the yield and other characteristics of the nuclear weapon, and it also provided interested foreign intelligence services substantial details about the nuclear weapon design. Atmospheric nuclear explosions spread the fission products, the radioactive materials formed by neutron activation, and the residual plutonium and transuranic elements in the atmosphere. We will show that these radioactive materials from the 528 nuclear tests in the atmosphere probably caused some 300,000 deaths among the population the world over.

A totally different purpose of testing was to understand weapon effects—blast, thermal radiation, the production of electromagnetic pulses from weapons, and the like. Such nuclear explosions are called weapon effects tests. Nuclear arms to be used for defense against ballistic missiles in the vacuum of space needed to be tested for their output of neutrons (and the proportion that came from the D-T reaction, with their 14 MeV of energy, in comparison with the lower-energy fission neutrons); it was also important to understand the yield of X-rays, which make up some 90% of the energy output of a large thermonuclear weapon detonated in space. Some reentry vehicles and their nuclear warheads are more susceptible to hard X-rays, which would be produced by a defensive weapon of very small mass and substantial yield, while other reentry vehicles are more vulnerable to soft X-rays, which can blow off a thin layer of material at a greater distance than can hard X-rays. The effective destructive range against satellites of a nuclear weapon exploded in space was also of interest.

Allied to the nuclear weapon effects test were vulnerability tests, performed in order to ascertain the robustness of U.S. nuclear weapons against defensive nuclear weapons.

One type of nuclear explosion test, the so-called "one-point" safety test, is successful when it produces zero explosive yield, or a yield a million times smaller than the full-scale explosion of that same weapon. The purpose of one-point tests is to ensure that accidental detonation of the explosive of an implosion weapon (for instance, by a rifle bullet or a fire) would give no significant nuclear yield. The actual permitted one-point yield for U.S. weapons is 2 kg of TNT equivalent, which was initially set so that any hazard from radiation would be contained within the area destroyed by the weapon's high explosive itself. Existing U.S. weapons have one-point yields much less than this limit. If preliminary experiments predict a weapon to be one-point safe, the ultimate verification involves the use of a single detonator in the presence of an intense neutron source, in order to show that the yield remains less than that of 2 kg of powerful explosive. For this final test, the device resembles a real weapon, although it is triggered quite differently (with one detonator rather than all). The other safety devices of the system can be just as well tested without any explosion.

Between 1958 and 1961, the United States conducted about fifty tests with a yield less than 2 kg of explosives, to determine if implosion weapons were one-point safe.

One-point safety tests were initially carried out in shallow wells at Los Alamos. They could not be done simply by using the stockpile weapon; instead of a kilogram of explosive yield, an untested weapon might have given a kiloton, for which shallow burial would have been totally inadequate. Accordingly, a one-point test is a series of tests, beginning with an amount of fissile material that is so small that there is virtually no possibility of nuclear yield. A strong neutron source is used to ensure a measurable output if the material becomes critical. In successive shots, the amount of fissile material is increased until one either obtains a measurable yield threatening to exceed the 2-kg limit or reaches the amount of material present in the weapon to be stockpiled. All U.S. nuclear weapons have been tested and verified to be one-point safe. This is not a property that can degrade with age.

The weapon's maintenance involves "enhanced surveillance," and will eventually require remanufacture of parts that have reached an age at which degradation is possible. As noted, each year eleven weapons of each type in the enduring stockpile are withdrawn and thoroughly inspected. One of these for each type, after disassembly, is turned over to the originating nuclear weapons laboratory (Los Alamos or Livermore), where it is cut open and compared with the design specifications and expectations. We emphasize that every part of a nuclear weapon except the primary and the secondary can be thoroughly tested without a nuclear explosion. This applies to the various switches, valves, and electronics, and even to the high explosive that implodes the primary. Of course, even when nuclear testing was allowed (it is not allowed anymore, as we shall discuss in Chapter 11), one could never test the explosive of a particular weapon that was going to be used in war, since the weapon is destroyed in the test. A stockpile stewardship program for U.S. nuclear weapons includes a panoply of tools to visualize, model, compute, and inspect, with the goal of ensuring safe and reliable weapons of existing type without the necessity of nuclear explosive tests.

The U.S. stockpile contains some pits—that is, fissile-material shells—that are 30 years old, and they show no ravages of age; it is believed that pits will survive in good shape for 60 to 90 years. Nonetheless, if one assumes conservatively that a pit should be remanufactured after 30 years, then a stockpile of 3000 nuclear weapons would ultimately require a remanufacturing capacity averaging about one hundred pits per year.

A key question in ensuring reliability of remanufactured weapons is whether the nuclear weapon laboratories can restrain their desire to use the latest techniques, and instead manufacture weapons to the same specifications and preferably by the same processes as were used originally. The stockpile surveillance program can then provide high confidence that the weapons remain safe and reliable; if instead the new tools are used to design and build nuclear weapons that have never been tested, it is likely that uncertainties will eventually arise that can only be resolved with confidence by nuclear explosion testing. As will be discussed in regard to arms control and international security, this would imperil the Comprehensive Test Ban Treaty signed in 1996 by the five nuclear weapon states and almost all others—leading to increased proliferation of nuclear weapons.

Precisely 1030 tests were conducted by the United States: 215 in the atmosphere and 815 underground. There are still about twenty contaminated islands, as well as one that was entirely vaporized. Eleven bombs have been lost in accidents, and some have never been found. The Soviet Union conducted 715 tests, France 204, Britain 45, China 43, India 6, and Pakistan 6. India exploded a nuclear weapon underground in 1974, and then in May 1998 announced the completion of more underground explosions. Pakistan followed within two weeks with a declared total of six underground tests. Putting an end to nuclear testing was a long-sought goal in the control of nuclear weapons and was the purpose of the Comprehensive Test Ban Treaty of 1996—an important topic in Chapter 11.

CHAPTER 4

Natural Radiation and Living Things

In 1903 the Nobel Prize in physics was shared by Henri Becquerel and Pierre and Marie Curie for their work on what Madame Curie had named "radioactivity." This was such a new and unexpected phenomenon that it is not surprising that no one at first realized how dangerous it was. Indeed, until relatively recently radioactivity was even thought to have a "tonic" effect. As late as 1970, many European bottled mineral waters were still stating on their labels the amount of radioactivity, not as a warning, but to attract buyers.

The discovery of radioactivity played a decisive role in the evolution of science, first because it paved the way toward the discovery of the structure of the atom. Radioactive sources that are naturally occurring alpha-particle emitters were the first "particle accelerators" to allow us to transmute atoms. Ernest Rutherford used alpha particles to divine and prove the existence of the atomic nucleus, 100,000 times smaller in diameter than the atom itself (and 10^{15} times smaller in volume); and in 1934 Irène and Frédéric Joliot-Curie used alpha-particle sources to create radioactivity that had not existed in nature. Fermi then used the neutrons that Chadwick in 1932 had detected from alpha particle bombardment to produce an abundance of artificial radioactive elements. The information gained in this way revolutionized physicists' concepts of the laws governing interactions between the various components of matter.

With radioactivity, and particularly after the discovery of artificial radioactivity, scientific research has benefited from a tool that permits the detection and measurement of matter in infinitesimal quantities, a tool millions of billions times more sensitive than the classical methods of chemistry. Indeed, it is possible to detect the disintegration of a single atom.

This extraordinary sensitivity of the methods for measuring radioactivity has ironically been turned against them in some of the present debates about the potential harm from low doses of radiation. Instances of radioactive contamination that are in many cases much lower than the natural radioactivity of the human body are vehemently denounced, and presented as a particular hazard resulting from nuclear energy. Effects from other energy sources that are actually more pernicious are ignored simply because these effects are more difficult to measure. This often disserves public understanding of the relative dangers of various human activities—an understanding to which this book is intended to contribute.

In this chapter we describe the effects of nuclear radiation on people, with particular attention to low doses delivered at low rates—that is, over a long time. Experiments on large numbers of small animals, ranging from single-celled animals to mice, show that even at doses below those corresponding to clearly observable effects on each member of the experimental population, there is a probability of cancer later in life. However, the demonstration of cancer at the lowest dose rates would require unaffordably large sets of experimental animals.

We shall examine the sources of natural radiation, the effect of low doses of radiation and their influence on the incidence of cancers, and the impact of radiation resulting from the atmospheric tests of nuclear weapons. In Chapter 7, we discuss the consequences of the 1986 Chernobyl reactor accident, as well as the effects of radiation from the entire nuclear fuel cycle, from mining of uranium ore through fuel fabrication and reactor operation to the radiation from nuclear waste.

We shall explain in considerable detail why we judge that of every million people in the world about 250 will die from lethal cancers induced by the totally natural radioactivity of their own bodies (primarily from the potassium of which they are built), and about 20 times that many from diagnostic X-rays and the effects of the rocks around them.

NATURAL SOURCES OF RADIATION

Since the origins of the earth, life has been immersed in a sea of radiation. It comes from cosmic rays and the radioactive elements contained in rocks, and is now supplemented by medical radiology and certain industrial and military activities.

Cosmic rays are particles (mostly individual protons) of very high energy that approach the earth from all directions. In the upper layers of the atmosphere, they produce nuclear reactions that give rise to a great variety of particles, most of which are absorbed by the air before arriving at sea level. An important exception is the muon—a 207-times-heavier, unstable version of the electron that decays into an electron and two neutrinos. Muons were a puzzle

SOME RADIATION MEASUREMENT UNITS

In Chapter 2 we defined the mean life, the half-life, and the decay rate. We now need units that quantify the interaction between ionizing radiation and human health.

The activity—that is, the rate of decay—of a radioactive substance is measured as the number of decays per second, and is important in evaluating the hazard posed by a radioactive sample. For practical purposes, it is useful to give special names to commonly used units of activity:

The becquerel (1 Bq) is the activity of a radioactive source that yields, on average, one disintegration per second.

The curie (1 Ci)—an old unit, the activity of 1 gram of radium—is the activity of any radioactive source that disintegrates 3.72×10^{10} times per second. A 1 Ci sample therefore has an activity of 3.72×10^{10} Bq.

The physical effect of ionizing radiation is measured by the quantity of energy deposited in a kilogram of water (as a stand-in for living matter).

The gray (1 Gy)—named for the British medical physicist L. H. Gray—corresponds to 1 joule (1 J) per kilogram. Recall that 1 J is the energy needed to raise the temperature of 1 kg of water by 0.00024°C.

The roentgen (1 R)—named for Wilhelm Conrad Roentgen, who around the turn of the century discovered X-rays—is an old radiation unit related to the early use of the ionization chamber to detect and measure radiation. One R contributed one unit of charge, more specifically one electrostatic charge unit, of ionization per cubic centimeter of air. This is about 80 ergs per gram of air. In the interests of precision, the roentgen was eventually replaced both by another physical measure—the rad (1 rad = 0.01 Gy)—and by a measure of biological significance—the rem.

The biological effect of radiation is related to the physical effect by the "relative biological equivalence coefficient" Q, which is different for different types of radiation. The biological dose (the rem) is related to a dose in rads: 1 rad gives Q rem. For X-rays, $Q = 1$, so a rad of X-rays gives a dose of 1 rem; for alpha particles, $Q = 20$, so it takes only 0.05 rad to give 1 rem to the tissue.

The literature is full of doses measured in millirems or even milliroentgens (mR), but the modern international unit of biologically effective dose is the sievert, Sv, named after the Swedish radiologist Rolf Sievert. 1 Gy gives Q Sv, so that 1 rem = 10 mSv (millisieverts).

For irradiation of the whole body, the "effective dose" is obtained by averaging the biological dose over the various organs of the body with their different size and exposure—i.e., by summing the doses, multiplied by the weighting factor for the individual organs—and is also expressed in sieverts.

The "population dose" is the sum of the effective doses to the specified population. Since the effective dose of radiation to the average American is 3.6 mSv per year, the annual collective effective dose to 270 million Americans is 972,000 person-Sv, which we abbreviate p-Sv. The unit is often called the man-Sv.

The dose per time unit is called the "dose rate," e.g., "two millisieverts per year" or "2 mSv/yr," and unless otherwise specified, is the effective whole-body dose. Collective effective dose rate, then, is measured in p-Sv/yr.

The DARI is a unit proposed by the authors for measuring human exposure to ionizing radiation. A DARI amounts to 170 microsieverts. This is the dose received in a year from the potassium-40 and carbon-14 naturally occurring in one's own body. The term is an acronym of the French phrase *Dose Annuelle due aux Radiations Internes*, "annual dose from internal radioactivity."

for physicists for more than a decade, because they are produced plentifully from cosmic ray protons striking the nucleus of a molecule of air, but do not themselves interact with nuclei; the puzzle was resolved when it was discovered that muons are not produced directly but result from the decay of pions, which are the primary products of the proton collisions. The half-life of a pion is 20 nanoseconds; that of a muon is 2 microseconds, when the particles are at rest. Because of the effects of relativity in "slowing the clock" of a rapidly moving object seen from our own reference frame, a muon of a billion volts of energy has a half-life of about 20 microseconds and moves at nearly the speed of light for about 7 kilometers before decaying. Neutrinos pass through the earth with less than one chance in a billion of interacting. On the other hand, muons have an electric charge and interact with the matter in our bodies: five muons, on the average, pass through each of us every second. Because it goes entirely through the body, depositing as much energy per centimeter of path as would a lowly electron from a nuclear decay, a muon delivers to our tissues about a hundred times as much energy as an electron from an ingested radioactive substance.

Cosmic rays are largely absorbed by the atmosphere (equivalent in mass to about 10 meters of water), so that at high altitudes radiation is much more intense than at sea level. In the high mountains, at altitudes of about 4 kilometers, there are electrons, gamma rays—that is to say, energetic X-rays—and muons, and at the cruising altitude of the Concorde supersonic passenger aircraft—15 kilometers—there are, in addition, protons, neutrons, and pions.

Figure 4.1 shows what happens when a very-high-energy proton enters the atmosphere. It produces a cascade of particles with esoteric names, sporting Greek letters like μ (mu), ν (nu), and γ (gamma), in addition to the more famil-

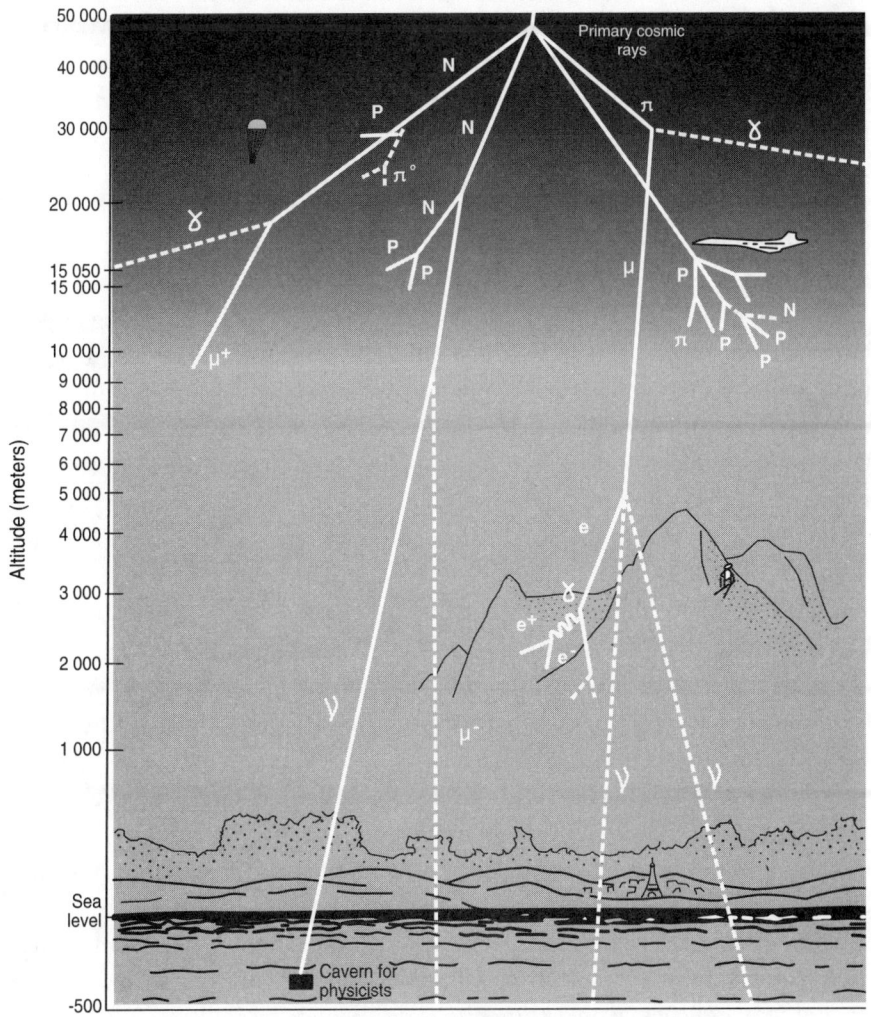

Fig. 4.1. Cosmic rays.

iar particles like electrons and neutrons. The variety of particles produced has long since exhausted the Greek alphabet, and it has been necessary to start using Hebrew letters. But for our purpose, which is to examine the effects of radiation on the human body, the names are not important.

Most of the neutrons released into the atmosphere produce carbon-14 by interaction with the nitrogen-14 in the air: by the capture of a slow neutron and the emission of a proton. Carbon-14 is a radioactive isotope of carbon (the stable form is carbon-12), with a half-life of 5730 years, and all living animals and plants inhale it. Once the carbon-14 has been incorporated into the wood of a tree or the bones of an animal and the living individual dies, the concentration of carbon-14 subsequently diminishes by radioactive decay. Carbon-14 activity is routinely used for dating objects containing carbon (organic and vegetal substances—bones and charcoal, for example) by measuring the carbon-14/carbon-12 ratio, which allows the calculation of the year in which the carbon, in the form of carbon dioxide, was inhaled and incorporated into the organism.

Natural radioactive elements, such as are present in rocks and living tissues, emit radiation. Potassium-40 (K-40), a radioactive isotope that occurs in a proportion of 1.18 in 10,000 in natural potassium, has a half-life of 1.3 billion years; it is, therefore, still present, as is uranium-235, both of which have existed since the creation of the dust from which our solar system was formed. Potassium-40 is rare compared with nonradioactive potassium-39, just as uranium-235 is rare compared with uranium-238, because the more radioactive isotopes have decayed over the billions of years since the elements were formed in ancient stars. Potassium is always present in living organisms; it is essential to life. Other mineral elements also have long-lived radioactive isotopes. The radioactive substances in the body of a seventy-kilogram human undergo 8,000 disintegrations per second, a small part of which is detectable outside the body, while most is absorbed by the tissues.

When a person is placed inside a detector designed to measure the energy of the gamma rays emitted by the body, one can identify the emitting radioelement and measure its intensity. Half of the gamma rays emerging from the body are emitted by potassium-40; for the average person, the potassium activity is about 500 gamma rays per second. In 90% of the decays, potassium-40 disintegrates by emitting an electron with no associated gamma ray, so that each second 5000 radioactive potassium nuclei disintegrate by emitting only an electron that is undetectable outside the body but perturbs the living cells. There is also in the body a carbon-14 activity of similar magnitude that cannot be detected outside the body, because it emits no gamma ray. The C-14 activity in the body amounts to 230 Bq/kg of carbon. In the 18 kg of carbon in the average person, this gives a total of 4140 Bq. Radioactive potassium from food accumulates in an adult to about 55 Bq/kg of body weight, which corresponds to 3850 Bq for someone weighing 70 kg. The total for K-40 and C-14 is thus 8000 Bq.

For potassium-40, the conversion factor from number of decays per second to dose rate is 3 µSv/yr per Bq/kg. The K-40 body burden of 3850 Bq corresponds to an equivalent dose of about 165 µSv/yr. The C-14 body burden, 4140 Bq, contributes a dose of some 12 µSv/yr. Because the electrons from C-14 decay have a maximum energy of 0.16 MeV, while those from K-40 have 1.32 MeV, the radiation exposure from the body's C-14 is ten times less than that from an equal number of Bq of K-40 decays.

Irradiation by internal sources, such as potassium-40, of sensitive parts of the body, like the bone marrow, the testicles, and the ovaries, is close to about a third of that produced by rocks and soil. With cosmic rays and the ambient radioactivity from radon, it thus constitutes a lower limit beneath which one cannot hope to measure easily any effect of an excess of external radiation.

HOW MUCH RADIATION FROM EVERYDAY ACTIVITIES?

From the natural radioactivity of the rocks and soil, the yearly average dose received by an individual amounts to about 0.46 millisievert (mSv). In adding up the various natural causes of radiation, each person in the world's population receives, on average, 2.4 mSv per year, of which about half comes from breathing radon. Radon is a radioactive gas emitted by rocks that contain uranium. It is present in many homes. Until now, no harmful effect has actually been demonstrated at residential levels—in contrast to well-documented radon-induced cancer among uranium miners, especially those who smoke. However, the U.S. government recommends that houses with high radon levels should be treated to reduce the accumulation of radon. According to the 1999 report of the National Research Council's Committee on Health Effects of Exposure to Radon, 157,000 people died of lung cancer in the United States in 1995. Of these, some 15,000 to 22,000—mostly smokers—are estimated to have succumbed to residential radon exposure. Of the 11,000 lung cancer deaths among never-smokers that year, some 2,500 are estimated to have been caused by radon exposure in the home. The committee reports that one-third of these deaths would be avoided if homes with radon levels above the Environmental Protection Agency's action guideline of 148 Bq per cubic meter of air were cut to that level by improved ventilation; this would save 5000 to 7000 lives each year among smokers and nonsmokers.

The dose measured at the surface of the body in a normal medical X-ray exposure varies over a wide range, from 0.2 to 10 mSv. But taking into account the fact that the radiation is localized and, therefore, less harmful than if it were received by every part of the body, a whole-body effective dose (i.e., an equivalent whole-body dose) amounts to 0.15 mSv for a complete pulmonary examination (X-rays from the front, back, and both sides), 0.02 mSv for a simple chest X-ray, and 1 mSv for a mammography. A CAT (computerized axial tomography) scan of the skull provides a dose of 3.5 mSv, while a CAT abdominal scan delivers some 9 mSv. Medical examinations contribute, on the average, about 1.1 mSv/yr in countries with advanced health care such as the United States—a dose half that from exposure to natural radiation, with, obviously, considerable variations among societies, and from person to person.

In contrast to these doses of radiation used for diagnosis, therapy for solid cancers delivers localized doses of gamma rays or energetic electrons of up to 100 Sv (100,000 mSv) to the cancerous tissue, over a period of five to six weeks; if this radiation were delivered uniformly to the whole body rather than only locally to a small portion of the body, death would ensue at a dose about twenty times lower—some 5 Sv. This locally cancer-killing dose of 100 Sv (of gamma rays with $Q = 1$) corresponds to an increase in the temperature of the region by 2×10^{-2} degrees. The destruction of the tissue is accomplished not by heating but by an assault on the defense and reproductive mechanisms of the cells, where very little energy is required to produce lesions. Much effort by radiologists in practice and in research is directed toward finding the conditions for effectively killing a cancer while sparing healthy neighboring tissue. Irradiation from multiple directions with X-rays or gamma rays, the use of protons that end their trajectory in the region of the tumor and deliver greater dose there (because of the "Bragg peak" in ionization near the end of their path), and division of the irradiation into multiple irradiations at intervals of days, are examples of this effort. Important advances are being made each year.

Since France obtains almost all of its electrical power from nuclear plants, radiation is of considerable interest there. The effects of the Chernobyl accident on France after the first year were officially estimated at 2% of those from natural radiation. The totality of the exposures received in occupations using ionizing radiation, if it were to be distributed over the entire French population, is, on the average, 2 microsieverts (2 µSv) per year—nearly a thousand times less than the natural radioactivity. That for the nuclear power industry (20 µSv/yr) is a hundred times less than the natural exposure. This averaging approach to evaluating the dangers of radiation, accidental or otherwise, makes sense from the point of view of public health policy if the likelihood of a cancer arising in the population is assumed to be independent of the number of persons who share the "population dose" (the sum of the effective doses to the individuals). Or whether, for example, 0.1 Sv is delivered to one person or divided among a hundred people so that each person receives 1 mSv. Of course, the risk is not equally distributed over the entire population if the dose is variable from person to person, as is most clearly the case with radon or exposure to medical X-rays, but from the point of view of public health the increased hazard to one group is precisely compensated by the diminished hazard to others.

The intensity of gamma rays emerging from a normal human body is sufficiently weak that our readers need not be concerned that they will injure a neighbor by close and even continuing proximity. A man with radioactive seed-like implants for prostate cancer is advised not to hold anyone on his lap for a few weeks. But this is far from the normal radioactivity of the body; instead, it is therapy that contributes 100 Sv locally over a few weeks, rather than 1 mSv over a year—an increase by a factor of 100,000. In seed implants, the radioactive gamma emitter is present in the body in the form of small tubes or pellets in the prostate gland; external radiation therapy with X-rays or gamma rays produces no radioactivity in the body at all.

Compared with the 120 Bq/kg radioactivity of the human body, typical masonry has some 600 Bq/kg, and granite some 1400 Bq/kg—in both cases primarily from potassium-40. A United Nations document shows granite blocks with 1200 Bq/kg of K, 90 Bq/kg of Ra, and 80 Bq/kg of Th.[1] Living in a building made of this granite would contribute a dose of 0.81 mSv/yr. For "typical masonry" the Bq/kg values are 500, 50, and 50. COGEMA (Compagnie Générale des Matières Nucléaires) has provided data on granite from specific regions of France, where a significant content of uranium and thorium (and their radioactive decay products) add substantially to the radioactivity, giving a contribution additional to the 1200 Bq/kg provided by potassium, as follows: from the Limousin district, 3440 Bq/kg; from the French island of Corsica, 3050 Bq/kg; from the Italian island of Sardinia, 760 Bq/kg; and from the northwest cape of France, Brittany, 5920 Bq/kg. A slab of granite from Brittany 4 cm thick has a radioactivity of 750,000 Bq per square meter; a person walking on this rock would receive a dose of some 1.6 mSv per year. In the United States the average annual effective dose from terrestrial gamma rays is 0.28 mSv. Low-level radiation is a fact of life.

For the past 50 years, physicists have been able to observe cesium-137 gamma rays emerging from the body. Cesium-137, a radioactive element with a half-life of thirty years, has been massively spread over the earth by nuclear explosions in the atmosphere. Part of it settles in the body. Its concentration continuously increased from 1945 to 1958, then regularly decreased beginning with the first half of 1966, as is apparent in Fig. 4.2. In 1960, it varied from 70 to 350 Bq per individual from one region of the world to another (compared with the 8,000 Bq per individual from the natural internal radiation); in the populations affected by the Chernobyl fallout, cesium-137 reappeared.

In the United States, substantial numbers of people live at relatively high altitudes, thereby increasing their radiation dose from cosmic rays. The average American receives about 2.4 mSv of radiation per year. Of this, 0.2 mSv per

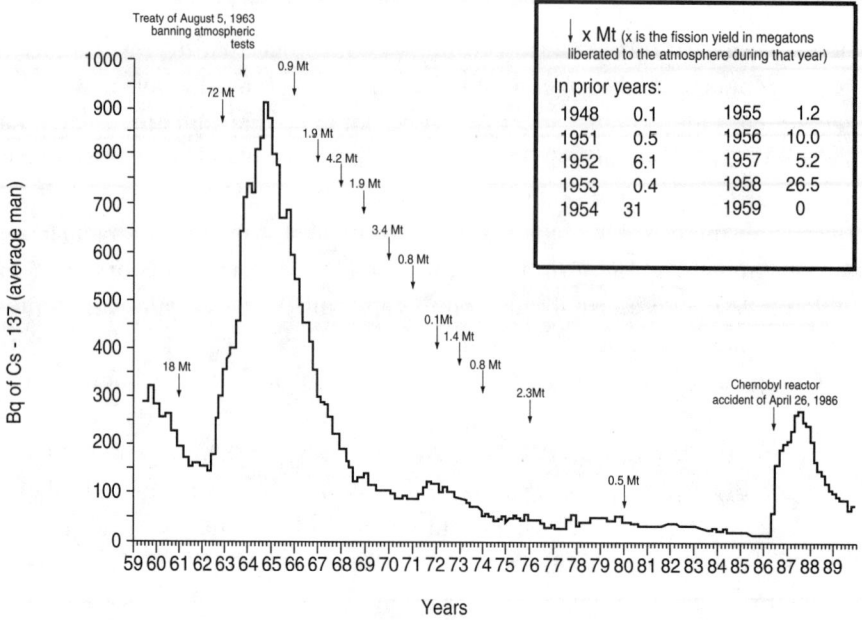

Fig. 4.2. Graph from J-L. Genicot and F. Hardeman, "A Measurement of the Ecological Half-life of Cs-137 in Belgium," Health Physics, Vol. 67, No. 6 (1994), p. 669. For each interval of 2 months, the graph shows the average amount of cesium-137 in a population of thirty to one hundred individuals — male residents of Belgium not professionally exposed to radiation. The ordinate is Bq of Cs-137 in a "standard man" containing 147 g of potassium (4300 Bq of K-40). The Cs-137 radioactivity never exceeds 25% of the K-40 radioactivity.

year comes from the body itself. Radon accounts on the average for 1.3 mSv per year for the average American — some 45% of the total radiation dose. Rocks, soils, and natural building materials contribute 0.46 mSv/yr of external radiation (we count radon exposure separately). Diagnostic X-ray examinations give an average annual dose of 0.4 mSv, while dental X-rays add 0.14 mSv. The cosmic rays contribute 0.38 mSv per year on average — 0.26 mSv/yr for a Floridian or New York City resident, 0.50 mSv/yr for a Denverite, and 1.25 mSv/yr in Leadville, Colorado, at an altitude of 3.11 km (nearly two miles). The million people in La Paz, Bolivia, at an altitude of 3900 m, receive an annual cosmic ray radiation dose of 2.0 mSv, of which 0.9 mSv comes from neutrons and 1.1 mSv from charged particles such as muons. Flying in an aircraft at the relatively modest altitude of 11,200 m (37,000 feet) gives a dose of 15 mSv per year.

In the next section we shall estimate the deaths from cancer caused by this exposure to ionizing radiation — both naturally occurring and artificial. Al-

TABLE 4.1. ORIGIN OF NATURAL RADIATION DOSE TO THE
AVERAGE U.S. INDIVIDUAL (IN mSv PER YEAR).

TYPE OF RADIATION	CONTRIBUTION		
	DETAIL	CATEGORY	SUBTOTAL OR TOTAL
Cosmic rays		0.38	
Internal radioactivity		0.19	
Potassium-40	0.17		
Carbon-14	0.01		
Uranium and heavy metals	0.01		
External exposure (rocks, soil)		0.46	
Radon and its daughters		1.3	
SUBTOTAL natural radiation			2.33
Medical X-rays	0.40		
Dental X-rays	0.14		
SUBTOTAL medical diagnosis			0.54
TOTAL (natural plus medical)			2.87

though certain cancers are eminently curable, at least in the industrialized countries, most are not, so we do not distinguish between cancer incidence and cancer death. We do not wish to discourage cancer patients or their medical teams—several types of cancer are indeed curable with appropriate treatment, and others have lower but significant rates of cure. In any case, the data we use relate cancer deaths to radiation dose.

ASSESSING THE HEALTH ISSUES FROM RADIATION

Most of the fatalities at Hiroshima and Nagasaki were due to blast and the intense thermal radiation and fire, but about 15% to 20% of the deaths (i.e., some 20,000 of the total of 120,000) were caused by the burst of nuclear radiation from the bomb, lasting much less than a second. The number of deaths years later from lethal cancers that can be attributed to these wartime explosions is uncertain. Data from the Hiroshima and Nagasaki explosions are the basis of much of our knowledge of the relative susceptibility of various organs of the human body to the initiation of cancer by irradiation of the entire body. But these data are few: among all of the people in Hiroshima and Nagasaki who were heavily exposed to radiation and survived, 3350 later died of cancer; about 350 of these deaths could be attributed to radiation exposure from the bombings. The fallout of radioactive fission products, either on the city itself or distributed over the world, contributed a small fraction of the local cancer deaths that were caused by the prompt radiation from the bomb.

The radioactive fallout that followed the Chernobyl reactor accident irradiated, in an unequal way, populations the world over. Astonishingly divergent estimates have been published for the health effects of the Chernobyl fallout. For example, one could read in a Viennese journal that "50,000 Russians have been roasted," and in a French magazine that the accident has, up to now, caused "200,000 deaths." Experts have claimed that the deaths do not exceed 3,000, while a well-known professor of medical physics at Berkeley estimated this number at 475,000.

We have tried to understand the arguments behind these estimates. In their professional lives, the authors have been intimately involved with the details of interaction of nuclear particles with matter, and they present here not only their understanding of the facts of the health risks of radiation but also statements by others with which they disagree, so that readers may be aware of the nature of the disputes in this field. In addition to his activities for the American Physical Society on a report on the safety of water-moderated reactors, the first-named author has been involved in many aspects of the American nuclear weapons program.

The authors have worked over the years with powerful accelerators without feeling that they were subjected to significant risk, so long as they respected the rules that limit the radiation doses that they are allowed to receive. But the personal experience of any single individual cannot contribute much to an understanding of the risks, because diseases brought on by moderate doses of radiation are governed by the laws of chance. If a person places a bullet in one of the six chambers of a revolver barrel, points the gun at his head, pulls the trigger, and comes out unscathed (physically), that in no way means that the "dose" can be inflicted a number of times on many people without some fatal consequences.

This analogy between Russian roulette and the effect of low radiation doses is vigorously rejected by some who estimate that there exists a radiation threshold, below which human cells have an almost absolute defense. Every official body concerned with public health, however, assumes that radiation at the lowest doses and dose rates conveys a constant risk of cancer per sievert of collective dose, although this hazard is often taken as about half that of the risk per sievert at high doses.

Radiation Effects on Living Organisms

Although the introduction of X-rays a century ago was soon accompanied by caution in their use, the immediate concern was to avoid reddening of the skin and the prompt death of tissue, just as one prudently avoids excessive heat that burns the skin. In the case of X-rays, however, it was discovered later that animals or humans exposed to large doses of radiation will show an increased incidence of cancer, an uncontrolled growth of tissue.

Only in the past four decades has the fundamental knowledge of biology been acquired even to begin to understand the origin of cancer. You can imagine the difficulties of trying to understand how radiation, sporadically passing through a few cells, can give rise to a lethal disorder in the prodigiously complicated mechanism of the cell. Radiation acts on the organism by provoking release of energetic electrons in the cells; the liberated electrons can directly attack the molecules containing the cell genetic information—the DNA molecules—or else they can, by destroying many molecules in a cell, generate what are called "free radicals," oxidizers that attack the DNA. A DNA molecule is formed of two long, intertwined helical strands, carrying identical information, and made up of large numbers of four small well-defined molecules called "nucleotides"—adenine (A), cytosine (C), guanine (G), and thymine (T).

This famous DNA "double helix" consists of two featureless backbone strings of alternating sugar and phosphate groups, bridged by pairs of nucleotides. The two strands are "complementary" in that a G on one strand is paired with a C on the other, and a T with an A. Because the G-C pair and the T-A pair are precisely the same length, the two strands coil about each other in the double helix, whatever the sequence of A, C, G, and T on one of the strands. When a cell divides into two, each strand carrying the same genetic message is the point of departure for the fabrication of new DNA molecules, which inhabit the nuclei of two new cells.

The DNA determines the structure of the animal; every schoolchild now learns that a sequence of three nucleic acids in DNA "codes for" or corresponds to each fundamental building block (amino acid) of the specific protein produced by that genetic recipe. Of the $4 \times 4 \times 4 = 64$ possible ways to select a sequence of three bases when each selection may be one of the four—A, C, G, T—only 20 of these correspond to amino acids that are used in human proteins. Many of these proteins are "enzymes"—chemical compounds of enormous specificity, with the ability to catalyze (greatly increase the speed of) particular chemical reactions. The individual stretches of DNA that correspond to one of these product proteins are the genes.

One should keep in mind the complexity of the living organism in order to appreciate the difficulty of predicting the effects of radiation. Our body contains 60 trillion cells comprising 200 different tissues. Billions of new cells are born every hour and produce, each day, half a gram of DNA, which, as an unfolded double helix, would be about 200 million kilometers long (somewhat more than the distance from the earth to the sun). Each cell contains 3 billion nucleotides, and it would require 10,000 paperback books to transcribe the information contained therein (but only one CD-ROM).

Human DNA with its 3 billion base pairs is found within the cell in the form of 23 pairs of "chromosomes"—known long before DNA was identified or understood—which altogether code for about 35,000 genes that define a human being. A typical protein may have 200 to 500 amino acids. We will mention in particular the role of the p53 protein (and the *p53* gene)—so called because it is a protein of about 53,000 atomic mass units (u). Since an amino acid has a mass of about 112 u, the p53 protein is a chain of some 473 amino acids of precisely specified sequence. The *p53* gene is about 16 times as massive as the protein, because each amino acid of the protein corresponds to three base pairs of DNA, and the average base pair plus the backbone sugar of DNA has a mass of 610 u.

Each of the proteins produced within the cell according to the DNA recipe is a single chain that is folded into a conformation specific to that protein and is usually essential to the function of the protein.

When radiation attacks a DNA molecule, it can damage a single strand or, more rarely, both strands. The damaged DNA molecule can either "die" and be eliminated or it can be repaired; and if badly repaired the cell might per-haps be transformed into a cancerous cell—i.e., a cell freed from the usual restraints on replication. Research in cell biology has made considerable progress since the 1980s. It is now known that the DNA can be repaired by a three-step process: recognition of the lesions, attack on the injured region by specialized enzymes and detachment of the injured fragment, then repair of the damaged strand by synthesis of fragments to fill in the gap, according to the information on the undamaged strand. Many experiments have been per-formed at high dose rates to determine the number of double-strand DNA lesions induced by a given radiation dose delivered in a short time or over a period of hours or days; in this way it is determined that much of the repair (or misrepair) takes place within a few hours—not in seconds, or weeks. It is not thought that a single misrepaired double-strand lesion would, in itself, induce a normal cell to become cancerous, hence escaping from the normal control of cell replication and beginning a clone of identical cells that grows out of control of the usual forces regulating cell growth. Rather, multiple hits would be required—especially to disable the vigilant DNA repair mechanisms and the mechanism that causes the cell to commit suicide (apoptosis) in many cases of faulty repair; the role of the *p53* gene in this regard is so important that it has been called "the cell's guardian angel." It is estimated that some 5 to 8 mutations must accumulate in a single cell to initiate cancer.[2]

Dose Rate and Cancer Probability

The Hiroshima and Nagasaki victims received an elevated dose of radiation in a very short time, and yet that exposure and the 350 cancers assessed as having been caused by radiation are taken as a major basis for the estimation of cancer incidence from long exposure to very low levels of radiation. Would the same dose received over a long period have the same probability of inducing cancers? At high dose delivered in short periods of time ("high dose and high dose-rate"), DNA lesions are less effectively repaired, because the same stock of enzymes has to face up to a large number of lesions in a limited time; the intracellular firefighters are overwhelmed and do sloppy work. Thus at high dose rates, the frequency of faulty repairs can increase considerably and produce cancerous cells. Is there some level of radiation below which a large enough fraction of the enzymes remain undestroyed and are thus fully capable of doing their repair job? Below this level, if there is one, the stock of enzymes should be sufficient to repair the small number of lesions produced, but they may still make an occasional error. The errors will be a constant fraction of the repairs at these low doses, just as a careful user of the computer keyboard will make about one error per thousand digits, even if unhurried.

Estimates from The International Committee for Radiation Protection

The lowest dose above which one starts to see effects on the health of any individual is 200 mSv. Above 1000 mSv, specific diseases appear. The lethal dose, if received over any time interval ranging from a second up to an hour, is 5000 mSv. Estimates of the number of lethal cancer cases induced by low doses, based on various experimental and epidemiological data, vary by a factor of 5.[3] The International Committee for Radiation Protection (ICRP), which since 1928 has been responsible for safety recommendations about radiation, calculates the coefficient for evaluating cancers induced by low doses of radiation over the entire body by extrapolating from the effects of high doses, i.e., from the observation of the cancers induced by the burst of radiation from the nuclear weapons used against the Japanese cities. If the same radiation dose per gram of tissue is given to only a small portion of the body, as in diagnostic radiology, the number of cancers is of course much lower. There is no direct statistical proof, today, of a link between the rate of cancer and natural radiation, but the ICRP used the conservative hypothesis that the incidence of cancers induced by a low dose of radiation, at very low dose rates, would be half that per Sv provoked at high dose and high dose rate. The ICRP thus takes a coefficient of 0.04 lethal cancer per Sv for exposure to low doses over a long period. We will use this coefficient each time that we have to predict the number of deaths from cancers induced by radiation.

The ICRP has modified its earlier recommendations as the estimations of the doses inflicted on the Hiroshima and Nagasaki populations have been refined. In 1990, it recommended a decrease (from 50 mSv per year to 100 mSv over five years) in the tolerable limits for nuclear workers, which led the French government to request a report from the French Academy of Sciences on the problems related to low doses of ionizing radiation. The ICRP also reduced its recommended dose limit for the public from its 1977 recommended limit of 5 mSv/yr to the 1990 value of 1 mSv/yr—applying, of course, to additional exposure to ionizing radiation above the natural background that averages 2.4 mSv/yr in the United States.

A Linear Relationship Between Dose and Cancer Probability?

The 1995 French Academy of Sciences study cites data indicating that a brief exposure of 1000 mSv produces one thousand lesions in a single strand and 40 lesions affecting both strands of DNA in each cell. Each second, there are, in addition, two spontaneous (i.e., occurring in the absence of radiation) lesions per strand in each cell, and 1.2×10^{-4} lesion affecting both strands. Radiation produces a two-strand-to-one-strand lesion ratio of 4×10^{-2}, while this ratio for spontaneous lesions is 6×10^{-5}. Therefore, for the same total damage to DNA (dominated by the single-strand events), radiation produces 666 times as many double-strand lesions as does the spontaneous process.

Recalling from Chapter 1 that one electron volt or $1 eV = 1.6 \times 10^{-19}$ J, it is easy to calculate (from the definition of the gray as 1 J/kg) that these 1000 mSv produce, in a cubic cell 10 micrometers on a side, an energy deposit of 10^{-12} J, or 6 MeV. Since each fast electron deposits about 2 MeV of energy per centimeter of travel in tissue, or about 0.002 MeV per 10 micrometers of travel, this dose would be given by 3000 particles, such as fast electrons, passing through the cell. On the assumption of a linear relation between radiation dose and DNA damage, if 3000 particles cause 40 double-strand lesions, the passage of one of these particles has only a probability of 1.3% of causing a double defect in the DNA of a single cell. While it is difficult to measure the number of cancers induced at low doses, it is entirely feasible to measure the number of double-strand lesions produced.

At a dose of 1 mSv/yr (close to half the natural dose, and one-thousandth of the 1000 mSv discussed above), there would thus be three particles per year that traverse the cell, so that they must act independently of each other, leading to an effect that is independent of dose rate. Since whatever repair to DNA is achieved is complete in a day, this suggests that 1 mSv will give the same induction of cancer whether it is given over three days or over a year. If the effects of a single ionizing particle are tiny, which they are, the mutual effects (i.e., the effects of one electron on the specific damage produced by the other) are small compared with the additive effects that the individual electrons have on the DNA. This means there is no threshold for these effects, which persist at a proportional rate down to the lowest level of radiation. That is to say, a single dose of 1 mSv is in the linear range, and so will be doses of 1 mSv per day.

What is the magnitude of this linear effect? According to this estimate, a radiation dose of 1 mSv/yr corresponds to about 4% yearly probability of radiation-induced double lesion in DNA in each cell, compared with a spontaneous double lesion every 8000 seconds (i.e., the previously mentioned 1.2×10^{-4} per second) on the average—10 per day, or 3650 per year.

The 1995 report of the French Academy of Sciences considers substantially higher doses and dose rates than those that concern us and states:

(3) . . . the process of induction of potentially carcinogenic persistent genomic lesions is significantly different at low or high doses and likewise at low and high dose rates. The differences are mainly due to DNA-lesion repair mechanisms which are not similar in the two situations.

Nevertheless, the report's authors conclude:

(4) Medical examinations are the second-largest source of exposure of populations, after natural exposure, way before other sources. Thus, to decrease significantly the radiation doses of the population, effort should be directed toward reducing the doses received during radiological examinations, especially among the young.

But earlier they note:

The absence of observable cancers or leukemias for doses inferior to 200 mSv of acute radiation or 400 mSv of chronic radiation does not allow, despite the impressive agreement of epidemiological studies, exclusion of a possible effect [i.e., cancer produced by these low doses] because of the insufficient statistical accuracy of the studies.

It is, therefore, legitimate, in the current state of our knowledge, to perform a linear extrapolation associated with a dose reduction factor. This reduction factor, however, needs further study.[4]

The dose reduction factor is the ratio between the effectiveness of a strong, brief radiation dose and the effectiveness of the same amount of radiation delivered at a low rate. The French Academy of Sciences study has taken a dose reduction factor of 2.0, in agreement with the ICRP.

To state that the "statistical accuracy of the [epidemiological] studies" is "insufficient" is equivalent to saying that these are very weak effects and the study has not been based on a sufficiently large number of individuals who have no other differences in their lives than the intensity of natural radiation to which they have been exposed. We quote this report because it presents useful and undisputed data, and because it is an up-to-date and detailed presentation that we can criticize in depth. While we don't disagree with the recommendations of the study, we do make a firm judgment that there is still at low doses of radiation, received over a long time, a linear relationship between radiation received and cancers provoked; we will explain why in the next section.

The specific charge to the study was to recommend for or against lowering the dose limits. We see little merit in the concept of dose "limits" and so we do not disagree with the French Academy of Sciences study that recommends against lowering these limits, but our reasoning is different: we consider that the amount of harm depends on the total population dose (for these relatively low individual doses) and so the overall damage would be reduced by lowering the average dose and not the dose limit; the Academy of Sciences study argues against a reduction in dose limit because the majority of the study group apparently believe it is possible that the existing dose limits are already below a threshold, and thus do not cause cancer. We note that principal conclusion (4), quoted in full above, supports our view, in that the recommended reductions would be unnecessary if there were a threshold of exposure below which there is no harm in radiation.

It is still commonly stated, though less often now than some decades ago, that "The solution to pollution is dilution." For some pollutants this is true, but if the effects of a particular pollutant are linear, dividing a given amount of pollutant dose among one hundred recipients or extending it over a long time would not reduce the overall damage. Here is an instructive analogy:

Suppose a firm produces a billion aspirin tablets a week, among which are ten containing a poison that provokes sudden and fatal heart failure. If all ten poison pills are in a batch of one hundred bottles sent to a store in a particular small town, ten people will eventually die as the pills are consumed. If the ten pills are distributed at random all over the United States, the same number of people will die. This is an example of a linear relationship, with ten guaranteed fatalities independent of the distribution of the product. (Note that the linear relationship breaks down at high doses because a person can only die once; if all ten pills are consumed by one person, there will be only one fatality, not ten.)

The hypothesis of a linear relation between radiation dose and cancer death implies that low levels of radiation (including background radiation) cause deaths, even though they can't be distinguished in any individual case from cancer deaths in the general population—almost all of which are from causes other than radiation.

A Test of the Linear Hypothesis?

Precisely because the repair of spontaneous damage takes place in a few hours, a sequence of a hundred or more repeated X-rays for following a disease, such as tuberculosis or scoliosis, can provide evidence of cancer induction by low doses of radiation and low dose rates, even though the cumulative dose to the individual may be high (3 Sv) and the individual has a 12% (3×0.04) probability of dying of cancer induced by this radiation. There are relatively few studies published in which doses have been estimated and such side effects shown to be associated with the treatment. Unfortunately, the funds for such studies have not grown, so that public discussion of this issue is not illuminated by proper information about the effects of radiation. And it is not a popular or even career-enhancing activity in the medical profession to reveal serious side effects of a beneficial treatment. In 1999 the National Research Council of the National Academy of Sciences began a three-year study on the Biological Effects of Ionizing Radiation (BEIR VII-phase 2). Its primary objective is to develop the best possible risk estimate for exposure to low-dose radiation such as that experienced from the nuclear fuel cycle. The BEIR-VII report may advance our understanding even though it may not be definitive.

Consider a hypothetical experiment to learn definitively the probability of cancer from low doses of radiation delivered at low dose rate. If 1 mSv corresponds, as indicated, to 4% probability of a double lesion, then a sudden dose of 5 mSv would give 20% probability of a double-strand break per cell—and only 4% ($0.20 \times 0.20 = 0.04$) chance of two double-stranded lesions. Since DNA repair does not appear to take place beyond a few hours or certainly a day after exposure, daily doses of 5 mSv should be independent and could be given with impunity if there were an absolute threshold that would permit a single dose of 5 mSv. Such a regime would provide a dose of 1.8 Sv/year, for which the ICRP coefficient of 0.04 lethal cancer per Sv would give a cancer probability of some 7.2% per year of exposure—of course only apparent after the typical ten-to-twenty-year delay before cancers appear. A five-year experimental period of daily 5 mSv doses, followed by a 20-year observation period, would yield data that would entirely eliminate or validate the threshold hypothesis, even if performed on a population of 50 individuals. This sort of experiment cannot ethically be performed on human subjects, and so surrogate approaches have been used, such as monitoring those children who have had repeated radiation doses for diagnosis of progress of tuberculosis, scoliosis, etc. It is just such treatment that is likely to be the basis for the conclusions of the BEIR-VII report.

At high doses, the induction of cancer cannot be totally linear with dose, if for no other reason than that massive doses kill rapidly and substantially change the intracellular and intercellular environment; our ten lethal pills cause only one death if all are consumed by one person. We don't claim at this point to know what the coefficient is (e.g., 0.04 lethal cancer per Sv, or 0.02 or 0.08 per Sv), but even the lowest doses of ionizing radiation, added to the natural background, must contribute the same amount: for the increase in the probability of induced cancer, we will show that there is a single coefficient for any dose in the sufficiently low-dose range, and that for two reasons.

First, in the developed countries 20% of the population die of cancer, and the cancers do not differ qualitatively from those caused by radiation at relatively high doses. This encourages the view that over the decades there are cells that become cancerous, by the accumulation of spontaneous lesions; when enough damage is done to suitable sites in some random cell (including in most cases disabling the $p53$ gene that produces the cancer-suppressing p53 protein), the cell escapes from normal control, does not commit suicide, and begins a clone of identical cells that constitutes a cancer. For the first time, reliable transformation of human cells into cancer tumor cells has been achieved by the incorporation of three specific genes, including one that prevents the shortening of the "telomeres" at the ends of the DNA chains that normally takes place in every reproduction of a human cell.[5] In other work, a newfound ability to disable the $p53$ gene has been used to permit larger doses of radiation in the treatment of cancer, doses that would otherwise kill surrounding cells they are not severely damaged but would have been condemned to death by the vigilance of p53.

The most important contributor to radiation-induced cancer at low doses, and hence to a linear response, appears to be for an irreparable DNA lesion to occur, which is then reproduced by cell division. This cell or its progeny then contains the first in a series of modifications that can lead to cancer; later modifications would not be caused by radiation, but by the spontaneous events that result in the 20% cancer death toll. Alternatively, in the population of cells in the body that either naturally or under the influence of harmful chemicals accumulate lesions, radiation could be the last event among the five to eight mutations we have indicated are thought necessary to make the cell cancerous.

Any additional lesion due to radiation in a population of normal cells can be the step that leads to a cancer, and can thus contribute a linear factor for the induction of cancer. Quantitatively, we are describing any of the paths to spontaneous cancer, in which the probability p_i of the i'th step is augmented by an amount ε. If the other four to seven steps (of the five to eight total postulated by Vogelstein and Kinzler) are not affected in this particular cell (and it is unlikely that radiation that causes cancer in one cell in a billion will augment the damage in two steps in a particular cell), then the probability of this cell's starting a cancerous clone is $P \times (1 + \varepsilon/p_i)$, where P is the overall (very tiny) probability of spontaneous cancer from an individual cell. Similar terms apply to the other sequential steps, and the composite probability of cancer from a given small dose of radiation is thus a Taylor series; the lowest term dependent on the radiation dose is linear in ε. As we have indicated, it is not known whether this cancer risk that is necessarily linear at the lowest doses gives a cancer incidence in the observable range of 0.04 death per Sv, but there is necessarily this linear relationship. It comes about because radiation produces at least some damage that is similar to the spontaneous damage that accumulates to cause the deaths from cancer in our population, which constitutes 20% of the total death toll. This model of multistep cancer production is compatible with the years of delay commonly observed between the radiation dose and the occurrence of cancer. A similar analysis was already published in 1976.[6] In this way, radiation, no matter when it is received, has a cancer-producing probability that is proportional to the spontaneous rate of occurrence of the disease. This would be in agreement with the so-called "multiplicative model" of radiation-induced cancer.

A second powerful reason for a linear relation between cancer and a dose of artificial radiation that is small compared with the background arises from the very assumption that the added dose is small compared with the background. Then whatever the relationship between total dose and cancer, in the very limited range of dose between the background and the background plus a small added dose, simple mathematics shows that the effect will be linear in the added dose; no matter what the dependence on dose—proportionality to the square, to the square root, or whatever—in this small range beginning with the background dose the effect will be linear.

The linear contribution to cancer, although sure to exist, might be much less than the values adopted, for example, by the International Committee for Radiation Protection. We read in equally authoritative documents that the dose reduction factor, based on epidemiological studies, varies between 2 and 10. The ICRP chooses, for its evaluations, the factor 2; this leads, as mentioned, to 0.04 lethal cancer per Sv. We can consider this factor as an estimate of the real value, not yet measured accurately. Using the ICRP coefficient, we will estimate in Chapter 7 that the Chernobyl radiation caused 29,000 fatal cancers—far fewer than the 475,000 asserted by one analyst, but far in excess of the number put forward by those who maintain that the effect is close to zero.

The effectiveness of low doses of radiation in causing cancer continues to be a matter of controversy and research. One analysis of the relationship between residential radon exposure and lung cancer finds no evidence of an increase of cancer with radon exposure.[7] Because the damage to an individual cell by the alpha particle radioactivity of radon and its daughters in a cell of the lung is so severe compared with that caused by X-rays or gamma rays, this result may be inapplicable to the external and internal electron and gamma ray exposures that we discuss here. Lung cells may die without progeny rather than pass on their radiation-induced DNA lesions.

The difficulty for the man or woman in the street in choosing among the contradictory judgments of experts loaded with titles is exemplified by two extreme positions we will now present.

The Unconditional "Pros"

One view is exemplified by a widely quoted lecture entitled "The Benefits of Low Level Radiation," given by John Graham, an expert on fast-neutron reactors, at the twenty-first annual symposium (1996) of the Uranium Institute. The institute has seventy-eight corporate members and describes itself on its Web page as "a world wide network of those involved in all stages of the production of nuclear generated electricity."[8] The provocative title is not supported by the substance of the lecture. The author writes: "Double strand breaks [of DNA] occur at about 5% of the rate of single strand breaks, so that with a background of about 240,000 breaks per cell per day, about 10,000 to 12,000 are double strand breaks."

Graham begins with a simple but major error. Although 4% of the breaks due to radiation are double, it is only one in fifteen thousand in the case of spontaneous damage.[9] This is not disputed. Graham should have said: "Double strand breaks occur at about 0.007% [not 5%] of the rate of single strand breaks, so with a background of 240,000 breaks per cell per day, about 16 are double breaks." That makes quite a difference—16 instead of "10,000 to 12,000." These double breaks seem to be a necessary step for cancer, and if the figure for those produced spontaneously were as great as Graham thinks, it is clear that the effect of low radiation doses would be negligible when compared to the tidal wave of spontaneous damage. Having overestimated by a factor of 1000 the spontaneous damage with which radiation is to be compared, Graham cites as confirmation an epidemiological study that is instructive (but not definitive) in helping to estimate the hazard of low-level radiation; we consider it here at some length, but find that it does not support Graham's assertion that radiation is good for you. In China, two population groups have been studied, both living under very similar conditions, but for whom the natural background radiation from the soil differs by a factor of 3 to 4.[10] The survey was done by the High Background Radiation Research Group headed by Wei Luxin of the Laboratory for Industrial Hygiene of the Ministry of Public Health.

The groups included 74,000 people in the high-radiation-intensity region and 77,000 in the low-intensity (so-called "control") region. They were studied from 1970 to 1986.[11] They represent a homogeneous population group, descended from more than five generations living in the region. According to Graham, no adverse effect of the higher radiation level is observed; he says that "all types of cancer" are less frequent except that of the cervix/uterus. In fact there were 13 deaths from cancer of the cervix or uterus in the radiation area, vs. 5 in the control area. And in the radiation area 22 cases of Down's syndrome (arising from a chromosomal abnormality) were diagnosed, vs. 4 in the control area. But what can be learned from the survey in China in regard to the truth or falsity of the linear relationship between additional cancer deaths and additional background radiation?

Wei and his colleagues show that the exposed group experiences 1.83 mSv per year of external radiation dose due to the geological environment, while the control group experiences 0.5 mSv/yr; the dose over 40 years would thus be 53 mSv greater for an individual in the radiation area than for an individual in the control area. In each of the two groups of seventy-odd thousand, about 13,000 were between 40 and 70 years old at the beginning of the 16-year study, and these are the group for which Wei et al. provide detailed data. In the control area, 377 of these older people died of cancer; in the radiation area, 299 died of cancer during the 16 years of the study. But what is the excess cancer incidence expected from the high radiation background, according to the ICRP coefficient that Graham claims is refuted by the study?

As we have seen, about 20% of deaths worldwide are due to cancer, with radiation of any type — natural or artificial — apparently contributing little to this toll. Thus of the control area population initially over 40 years old, 20% × 13,000 = 2600 are expected eventually to die of cancer; the observation period of sixteen years was apparently only long enough to observe a fraction 377/2600 = 0.145 of the expected cancer deaths. At the ICRP coefficient of 0.04 cancer death per Sv of exposure, the 692 person-Sv excess (= 1.33 mSv/yr × 40 yr × 13,000 persons) of external radiation exposure among the 13,000 people 40 to 70 years old in the high-background-radiation area should eventually contribute $0.04 \times 692 = 28$ additional cancer deaths, in comparison with the 2600 persons in the control area who are ultimately expected to die from cancer of all causes. But the observation period was only long enough to catch 0.145 of the natural cancer deaths; hence only $0.145 \times 28 = 4$ of the radiation-induced cancer deaths should be expected during this same 16-year observation period.

The reader may recall the familiar "margin of error" in polling a small selection of a large population: pure chance would lead to a likely difference of 25 excess cancer cases in one population or the other during the 16-year observation period. If a given response is obtained from N people, the repetition of a large number of similar experiments will have a spread (a "standard deviation," to be precise) of the square root of N. In this case, the number of cancer deaths in a group, $N = 338$ more or less (299 in one case and 377 in the other), the square root is about 18, so the true number is only slightly more likely to lie in the range $(338-18)$ to $(338+18)$, i.e., 320 to 356, than outside this range. Since both populations are assumed to have the same cancer rates except for any difference caused by the difference in background radiation, pure chance—the fluctuation in a number that on the average is 338—will lead to an apparent difference in cancer deaths of something like 25 in a single such survey of truly identical populations (the difference between two numbers each with a standard deviation of 18 has a standard deviation of 25—the square root of the sum of the squares). The difference in cancer deaths expected from the ICRP coefficient of 0.04 death/Sv is 4, which is certainly masked by a statistical fluctuation with a standard deviation of 25. Thus the study is "statistically insignificant" in choosing between two predicted outcomes: the ICRP prediction of cancer, or no effect at all from enhanced background radiation.

The "statistically significant" lessening of cancer rate in the high-background area (40 fewer cancers, when corrected for different numbers of people in the two groups, compared with the standard deviation of 25 for the difference in number of cancers expected by pure chance between the two areas) cannot be taken to indicate a benefit to health due to these rather low doses of radiation. For instance, the major difference in soil leading to the difference in radiation background might have a significant effect on nutrition.

A report by Wei Luxin from the High Background Radiation Group of the Laboratory of Industrial Hygiene of China's Ministry of Health on an updated study by Yuan (1997), summarizes "newly classified dose groups" observed from 1979 to 1990.[12] Its simple presentation allows us to demonstrate again that this epidemiological study (even if it had been done with perfectly matched populations) just does not have the statistical power to observe the deaths expected from the increased external radiation in the high-background areas. We shall see whether this newer study has the power to preclude even much larger estimates of radiation-induced cancer. Those exposed to "high background radiation" (64,070) received an average of 2.12 mSv/yr, while those in the "control area" (24,876) received an average of 0.68 mSv/yr. The report states that the ratio of "all cancer" risk in the high-background-radiation area to that in the control area was 0.9959, with a "90% confidence interval" extending from 0.864 to 1.148 (this is the span within which 90% of the results would fall if the experiment could be repeated many times, taking into account the fluctuation due to randomness). What would the ICRP coefficient of 0.04 cancer deaths per person-Sv predict? Forty years of exposure would correspond to an additional accumulated radiation dose per person in the high-radiation area of about $40 \times (2.12 - 0.68) = 58$ mSv, which would lead to an individual cancer risk of 0.23%. Compared with normal cancer incidence of 20%, this would be an increase in cancer risk by 0.23%/20% = 1.2%. The epidemiological result of 0.864 to 1.148 evidently cannot be taken to refute the ICRP prediction of 1.012 relative risk of cancer. It makes no sense at all to conclude that the ICRP predictions are in error, as Wei Luxin misleadingly implies on the basis of the 1997 Yuan study:

> Some organizations and authors, based on the "no-threshold, linear" hypothesis estimated the risk for low-dose exposure and stated "no matter how small the dose, there will be increment of cancer induction." But the results obtained in the High-Background Radiation Area of China have not demonstrated any increment of cancer mortality.

The Unconditional "Cons"

We have just demonstrated that it is incorrect to take the Yuan result as a refutation of the conventional estimate of cancer fatalities from low radiation doses. It is easy to see that a real test would require a study of one hundred times the number of people observed in this high-background-radiation-area to begin to distinguish between the ICRP prediction and no induction of cancer at all.

John W. Gofman, a critic of the nuclear industry, is a professor emeritus of medical physics at the University of California at Berkeley.[13] Gofman is also former associate director of the Livermore Laboratory and is coauthor with the Nobel laureate Glenn Seaborg of a patent on the fissionability of uranium-233 and another patent on the separation of plutonium from irradiated fuel. His risk estimates are ten times higher than those of the ICRP. His approach may be in error, but, in any case, he ought not to be ignored or attacked by the nuclear industry without having his arguments addressed in detail.

Gofman's estimate of cancer probability per unit radiation dose is higher than that of the ICRP in part because he interprets the evidence from the Hiroshima and Nagasaki victims, and from medical treatments with X-rays, as strongly supporting a linear model, without any "dose reduction factor." In addition, he assumes that a given radiation dose produces additional cancers in numbers proportional to the natural rate of occurrence of those cancers; this is the "multiplicative model." For instance, in estimating the number of breast cancers caused by radiation in the U.S. population, Gofman uses the increased numbers observed in Japan and then multiplies by the much larger incidence of natural breast cancer in the United States.

It is interesting to apply Gofman's risk coefficients to the study of the effects of the high natural radiation doses in China. If we consider the coefficient preferred by Gofman—0.4 lethal cancers per Sv, instead of the 0.04/Sv of the ICRP—the number of additional deaths to be expected during the 16-year observation period, from the background radiation, is simply 0.4/0.04 = 10 times larger than the increase of 1.2% we have calculated for comparison with the Yuan (1997) update; Gofman would thus estimate an increase of 12% in the high-background-radiation area, while the High Background Radiation Group itself states that its epidemiological result is a 90% confidence interval of 0.864 to 1.148; this epidemiological result does not even seriously contradict the factor 1.12 predicted by Gofman.

If the two populations were truly identical and the diagnosis and reporting done perfectly, and if the study were carried out for a much longer time, its statistical "power" would grow, but it would then be necessary to reckon with the "confounders" of chemical induction of cancer and the adequacy of the survey technique.

THE DARI—A CONVENIENT MEASURE FOR HUMAN EXPOSURE TO RADIATION

As we noted earlier in this chapter, we have created a convenient unit for measuring low doses of radiation exposure—the DARI, standing for the French rendering of "annual dose from internal natural radiation." A DARI is the radiation dose received in a year from the naturally occurring carbon-14 and potassium-40 with which we have become familiar. For a person of 70 kg, this amounts to 0.17 millisievert (0.17 mSv or 17 millirem). This may be compared with the total average background dose from internal radiation, cosmic rays, medical and dental X-rays, and the rocks in the environment, contributing some 2.4 mSv per year for the average American. The DARI, multiplied by roughly 40 years exposure, would contribute 0.17 mSv × 40 × 0.04 lethal cancer per Sv = 0.027%. Therefore 1 DARI (i.e., one annual dose) would be responsible for $1.7 \times 10^{-4} \times 0.04$, or about 7 chances in a million of contracting fatal cancer. If it is assumed that a fatal cancer on the average shortens life by 16 years, 1 DARI thus corresponds to about one hour of life shortening.

The DARI is inescapable—natural radiation from our own bodies. Six hours of flight at 15 mSv per year of flight contributes about 0.06 DARI; a flight every three weeks for a year would provide 1 DARI—adding about as much exposure as a passenger receives from his or her own natural internal radiation. Considering natural radiation and the average medical radiation for Americans, the population is subjected to 2.4 mSv/year (about 14 DARI), or 96 mSv over 40 years, which gives a probability of (96 mSv × 0.04 cancer/Sv) = 0.38% to contract a lethal cancer induced by exposure to this radiation; this 0.38% is included in the overall mortality rate of about 20% by cancer. This figure applies to the populations of all the industrialized countries, although specific types of cancer vary substantially from one society to another. If there is success in significantly decreasing the rate of cardiovascular disease, which is presently the major cause of death, the proportion of cancer deaths will increase, because we all die of something.

RADIATION EXPOSURE FROM MEDICINE, NUCLEAR POWER, AND NUCLEAR WEAPON EXPLOSIONS

Not much can be done to reduce the exposure from natural radiation, with the exception of radon, but it is certainly possible to diminish the radiation dose from diagnostic medical procedures and from nuclear power. In Chapter 7, we will see that the exposure from the nuclear weapons tests that formerly took place in the atmosphere amounts to about 2.3 years of natural background radiation for the average person in the world, and that nuclear power, including the 1986 reactor accident at Chernobyl, in Ukraine, has contributed far less. We shall consider these matters in detail, in comparison with other risks of industrial activity, in order to help plot a course of safety and economy.

In this chapter, we have reviewed some of the health effects of ionizing radiation and have explained in detail why we adopt the ICRP estimate of 0.04 lethal cancer per person-Sv of ionizing radiation. The large-scale epidemiological investigations in China often cited as refuting this linear, no-threshold hypothesis are clearly too small to carry this burden. The detailed calculations of this chapter provide the necessary analysis.

If one values one's life at $1 million, then what would it be worth to avoid one year's exposure to the average medical and dental X-rays that contribute 0.54 mSv? About $0.25. And it would be worth about one-third of that ($0.08) to avoid the DARI—the internal annual dose from potassium-40 in the body. As we shall see, the population exposure from the normal operation of nuclear plants is far lower than these natural exposures. The magnitude of abnormal exposures and their significance is the subject of Chapter 7.

CHAPTER 5

The Civilian Use of Nuclear Energy

IN THIS CHAPTER we shall describe the nature and the scope of the large nuclear power industries in the United States and in France—the reactors, their fuel cycle, and particularly the prospects for appropriate management and disposal of waste from the nuclear power sector.

In France, about 80% of the electricity produced is of nuclear origin; in the United States the corresponding figure is 17%, and worldwide it is 18%. In 1996, 3.33 quads* of primary energy were consumed as electrical energy in France; 3.31 quads were generated from nuclear plants, 0.55 quad from hydropower, and 0.19 quad from conventional fossil-fueled plants. But because of the export of nuclear-electric power, some 20% of the electrical energy actually consumed came from hydropower or fossil-fueled plants. If electrical energy produced by nuclear fission is to play a major role in satisfying the needs of society, nuclear power must be competitive with other means of producing electricity. It must also guarantee an acceptable environmental impact, which includes assuring that catastrophic accidents remain almost impossible. In fact, perfect reassurance does not exist in this world. The adoption of civilian nuclear power requires objective analysis showing that the likelihood of catastrophe is sufficiently small as to be practically negligible.

Civilian nuclear energy has long depended on knowledge gained in the manufacture of nuclear weapons as well as on the production of materials essential to these weapons. In some countries the two are still closely linked. It is therefore important to exercise great care to control nuclear energy in order to avoid the acquisition of nuclear weapons by previously non-nuclear-weapon

---

* A quad is a quadrillion BTU, equal to 1.055 exajoule. See p. xv in frontmatter.

nations or terrorist organizations. For the sale of nuclear reactors to politically unstable countries, certain options are preferable because these reactors are less vulnerable to diversion for military use. The setting of detailed rules and the verification of compliance of nations with those rules is a major responsibility of the International Atomic Energy Agency in Vienna, Austria, whose other charge has been the promotion of nuclear energy.

A layperson considering the panoply of nuclear reactors can't help but be confused by their diversity. In countries whose objective was to provide themselves with nuclear weapons, the approaches to nuclear energy were often determined by military requirements rather than scientific or economic reasons. For example, the means of propulsion of submarines or the type of material needed for weapons influenced the design of civilian reactors. The substantial investment involved required coordination of civilian and military efforts.

Several reactor types are now available for the large-scale production of energy under economically competitive conditions, with less hazard to human life and health and a greater respect for the environment than is the case for coal or oil.

As of the end of 2000, 433 nuclear reactors worldwide—103 in the United States—were converting 30% of their energy into electricity; the remaining 70% were going into warming the atmosphere or the oceans. But nuclear power does not emit carbon dioxide into the atmosphere as does the burning of coal, gas, or oil; in addition to contributing directly to heating the atmosphere and the oceans by the heat ejected by combustion, these fuels represent a long-term menace, because of the increase in the so-called greenhouse effect.

The United States led the development and commercialization of nuclear power, and it has had a major influence on the industry worldwide. Three striking characteristics of the U.S. nuclear energy generation industry are (1) its size—twice that of any other nation's, with France second; (2) its fragmentation, in contrast to France, where all the reactors are operated by a single power generation company; and (3) the fact that no new reactors have entered service since 1978, while France and other countries have since put many plants into operation. Nevertheless, the nuclear power generation programs in the United States and France are similar in most aspects, except in their approach to waste disposal.

BEGINNINGS OF THE U.S. NUCLEAR POWER PROGRAM

On August 6, 1945, and again on August 9, nuclear energy burst on the world with the detonation of nuclear weapons over Hiroshima and Nagasaki. The

Smyth Report on the nuclear weapons program, edited by the Princeton physicist Henry de Wolfe Smyth, was published that month by the U.S. government, revealing something of the nature of the uranium and plutonium bombs, and the existence of a secret industry in the United States that had produced the materials—enriched uranium and reactor-grown plutonium—of which they were built.

Scientists around the world who, before the Second World War, had been at the forefront of nuclear physics then became engaged in the effort to master the nonweapon uses of nuclear energy—in particular, the generation of electrical power. In this, the United States had a commanding lead, because of its experience in building the plutonium production reactors at Hanford, Washington, and its enormous capacity to enrich natural uranium.

The Manhattan Project was directed at the manufacture of nuclear explosives; since it was feared that Nazi Germany might succeed in getting there first, the idea of producing electricity from nuclear energy was of no interest to the United States during the war. After August 1945, there was no immediate need for a large number of new atomic bombs, so the plethora of enriched uranium and the know-how accumulated in the Manhattan Project were immediately directed toward applications of nuclear energy. Some, such as the use of nuclear radiation in industrial measurement and control processes, and in nuclear medicine for diagnosis and for the controlled destruction of malignant cells, continue in use and are highly beneficial. Another of the major postwar applications was for propulsion of military surface vessels and submarines. Having access to enriched uranium, the United States could undertake the development and construction of propulsion systems based on steam turbines powered by compact light-water reactors. These systems were first installed on attack submarines and aircraft carriers, and soon thereafter, following the conversion of these submarines to Polaris missile launchers by cutting the hull and inserting a missile-launch compartment, on the ballistic-missile undersea fleet. At the end of 2000, the U.S. Navy had seventy-three submarines and nine surface combat ships powered by nuclear reactors.

It was natural that the companies involved in these naval nuclear propulsion programs (in particular Westinghouse and General Electric) should turn to the nonmilitary production of electricity using similar reactors. The first of these was put into service in 1957 at Shippingport, Ohio, with a power of 90 MWe. So that they can operate for a long time without reloading, compact naval propulsion reactors use highly enriched uranium, while large civilian power reactors use only slightly enriched uranium.

HOW A NUCLEAR REACTOR WORKS

The design of a nuclear power plant involves expertise in different fields:

• Neutronics—the behavior of neutrons and fission events in the reactor core;
• Thermohydraulics—the transfer of heat and the flow of fluid through the reactor and the generating system;
• Turbogeneration—the mechanical-electrical process that accepts hot steam, delivering cooler steam and electrical power.

Only the first of these involves nuclear physics. The rest are common to most power generation systems. We have discussed in Chapter 2 the relevant aspects of nuclear physics and nuclear fission, and the complexity of the process that enables fission in natural uranium to lead to a steady chain reaction on an enormous scale. It is essential that the neutrons emitted in fission be slowed down (the technical term is "moderated") in very pure graphite or heavy water before returning to the small lumps or rods of uranium fuel to cause fission in the uranium-235 isotope that constitutes 0.7% of natural uranium. In low-enriched uranium fuel, for each atom of uranium-235 destroyed by fission or by capture of a neutron, there is produced, with a probability of about 60%, another fissile atom, owing to the transmutation of uranium-238 to plutonium-239. Most of this plutonium is itself fissioned over the four years during which the fuel rods remain in the reactor.

In providing electrical power, a nuclear reactor is used as a source of heat; the temperature of the reactor must be allowed to rise in order to power a heat engine.

The Carnot limit to efficiency (named after the French physicist Sadi Carnot, 1796–1832) is fundamental. Ignorance of the Carnot limit accounts for the enthusiasm of many novice inventors. Of the heat generated and delivered to a heat engine, no more than a fraction can emerge as mechanical or electrical power, even with perfect machinery—the Carnot limit. With temperature T expressed in degrees above absolute zero ($-273.18°C$), with T_h the temperature at which heat is delivered to the engine, and T_l the temperature at which heat is removed from the engine, the Carnot limit on efficiency is $(T_h - T_l)/T_h$. Measured in Celsius degrees, the melting point of ice is 273.18 K ("K" for Kelvin, named for William Thomson, Lord Kelvin, British physicist, 1824–1907). For a heat source operating at the boiling point of water (100°C) and rejecting its heat near room temperature (20°C), the maximum thermal efficiency would be $((100 - 20)/(100 + 273.18)) = 21\%$, where 273.18 is the temperature of 0°C expressed in degrees Kelvin—i.e., the temperature of the ice point above absolute zero. As we shall see, most nuclear reactors operate near 300°C, which by the Carnot limit would restrict their thermal efficiency to at most 49%, for heat rejection at 20°C. In practice, on the average only about 30% of the heat produced in a reactor emerges as electrical power; 70% of the fission heat is rejected to the environment at the power plant.

What is important in obtaining useful energy is the difference between the reactor temperature and some temperature at which the waste heat can be rejected from the system. The heat is typically transferred to a river, ocean, lake, or atmospheric cooling tower.

The design of a high-performance commercial reactor is a complex and difficult process, involving computation of neutronics on scales ranging from millimeters to meters, and of thermohydraulics over the same range. A combination of theory and experiment is needed to discover the temperature of the hottest spot in the fuel rods and its sensitivity to their swelling. One needs to know and to control the distribution of fission power in the core, so that no part should be excessively hot. The fact that the reactor pressure vessel can become brittle because of exposure to radiation—especially fast neutrons—must also be taken into account. And finally, once the design is drawn up and approved, it is no simple matter to ensure that real human beings in a real industrial setting build the reactor exactly to plan and operate it according to standards.

LIGHT-WATER REACTORS

In Chapter 2, we introduced the nuclear chain reaction in graphite-moderated reactors with natural uranium, heavy-water reactors with natural uranium, and light-water moderated power reactors using slightly enriched uranium. The present chapter concentrates on these last as they are operated in the civilian nuclear power industry for the generation of electricity.

Firms experienced with reactors for naval propulsion during the Second World War turned to civilian power. Eventually Westinghouse was to specialize in pressurized water reactors, while General Electric pioneered the development and operation of boiling-water reactors. Of the 103 civilian reactors capable of commercial operation in the United States, one-third are boiling-water and two-thirds are pressurized-water types. All are provided with a leak-tight containment structure, as part of the reactor building. Today, most of the power reactors in the world are like the first American power reactors—light-water moderated, either boiling-water reactors or pressurized-water reactors.

We have already discussed in principle the behavior of nuclear fission in a system rendered critical by having a large enough mass of sufficiently enriched material in an appropriate moderator. In light-water reactors, the uranium-235 is present as low-enriched uranium—about 4% uranium-235. The moderator is ordinary water, at high temperature and pressure and density lower than normal (because of thermal expansion). The chain reaction is controlled by the hundred or so control rod assemblies containing a potent neutron absorber—boron carbide—that is resistant to the high temperatures; the reactor is kept at the appropriate power level by slow manipulation of the control rod assemblies, although it is largely self-adjusting.

As noted in Chapter 2, it is the delayed neutrons from fission—0.65% of them in the fission of uranium-235—with a typical delay of ten seconds or so, that make it possible to control a reactor in this way.

Under normal circumstances, the power may be varied to maintain the operating temperature as the electrical load (and hence the demand for steam) varies, but shutdown and start-up are gradual in order to avoid sudden changes of temperature in the thick wall of the reactor pressure vessel, which could reduce its life below the design requirement of 40 years. The heat generated by the fissions must flow from the interior of the ceramic fuel pellets to the cylindrical surface of the pellet, then cross a tiny gas-filled gap to enter the thin metal tube in which the pellets are stacked, then cross from the outside of the metal tube to the flowing water within the reactor, and so to the steam generator, the turbine, the waste-heat exchanger, and into the environment. The pellets are typically about 8 mm in diameter, and the tubes in which they are stacked have an outside diameter of 9.5 mm, with a gap between pellet and tube (the "clad") of about 0.05 mm.

In boiling-water reactors, the nuclear fuel is submerged in water, which boils above the core of the reactor at high temperature and pressure. The steam produced is directed to a large steam turbine, which drives an electric generator. The low-temperature steam emerging from the turbine is then condensed in a heat exchanger that transfers the waste heat to a river, lake, or ocean, or to the atmosphere; the condensed water is then delivered by a high-pressure pump back into the reactor vessel, to be reheated and once more transformed into steam.

In a pressurized-water reactor, on the other hand, the water doesn't boil in the core, because, although very hot, the water is maintained at high pressure, greater than the boiling pressure at the operating temperature of the reactor. (Water boiling at sea level has a boiling point of 100°C; that is, at 100°C, the water changes into steam if the surrounding air pressure is that of a standard atmosphere, 1 atm—that is, 14.7 pounds per square inch, or its equivalent 1 kg weight per square centimeter. Every cook in Denver, Darjeeling, Lima, or Bogotá knows that water boils at a significantly lower temperature at high altitudes, where atmospheric pressure is lower. At 322°C, the temperature of the

pressurized-water reactor, the boiling pressure is 69 atm; the reactor is pressurized to 153 atm, so that the water cannot boil in the reactor.)

The nuclear reactor, shown in Figs. 5.1 and 5.2, is simply a heat source, seen
in context in Fig. 5.3; it substitutes for the boiler in a fossil-fuel plant, and in the
pressurized-water reactor the fission heat is transferred to the water "coolant"
or heat-transfer fluid. This water in contact with the fuel rods never leaves its
metallic enclosure. Flowing from the upper portion of the reactor core
through large, heavy pipes, the heat carried by the high-pressure water is transferred through the metal wall of a heat exchanger and steam generator, which
converts the water to steam at high pressure and temperature.

Water impelled by large electrically driven pumps enters at the lower end
of the fuel element and leaves at the upper end. The water enters at 289°C and
leaves modestly heated to 322°C. The water in the reactor vessel is maintained
at a very high pressure—some 153 atmospheres—so that it does not boil; the
boiling point of water at 153 atm is about 350°C.

Compared with Fermi's initial graphite reactor in Chicago, operating at
about 2 watts (2 W), a modern nuclear power plant generates some 3.9 billion
watts of nuclear heat—i.e., 3.9 GWt. Fermi's reactor first operated on December 2, 1942. It was built of bricks of very pure graphite (carbon) roughly in the
shape of a squashed sphere of equatorial diameter 7.76 m and vertical diameter

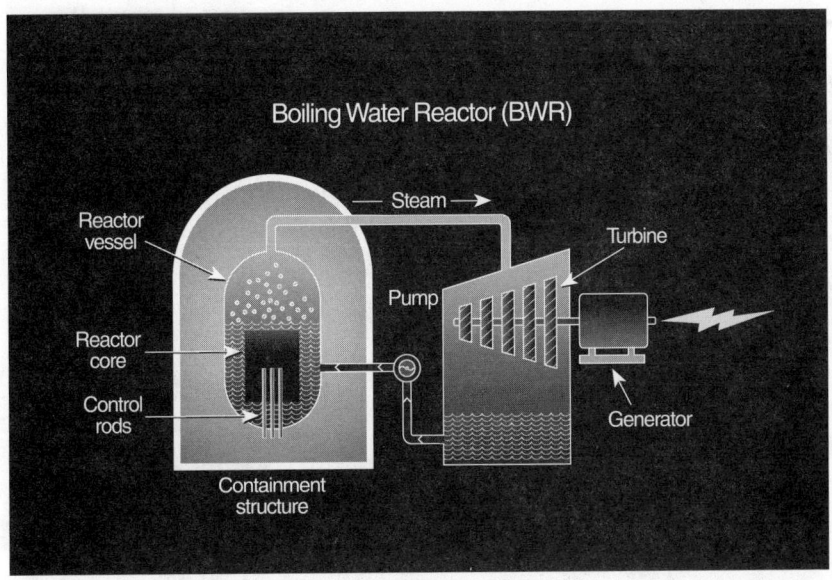

Fig. 5.1. A boiling-water reactor.

The pressurized-water reactor, like other light-water reactors, operates with slightly enriched (4% to 5%) uranium fuel in the form of uranium oxide ceramic pellets that are stacked in zirconium alloy tubes some 5 m long and 9 mm in diameter—about the diameter of a pencil. Exactly 264 of these tubes ("fuel rods") are mounted (with spacers) in a 17 × 17 square fuel assembly (or "fuel element") that can be inserted into the reactor pressure vessel when the lid is removed for refueling. Twenty-five of the 289 tube positions in the element are devoted to neutron absorbers or control elements. Typically 25% of the 50,000 fuel rods that represent one hundred tons of fuel in a reactor are replaced each year—that is, about 40 fuel elements, each containing 264 fuel rods. This is about 25 tons per year.

6.18 m. It contained 385 tons of graphite, 6 tons of uranium metal, and 40.5 tons of uranium oxide. The uranium and UO_2 were in squat cylinders in a cubic "lattice" 21 cm on a side; the uranium metal lumps each had a mass of some 2.7 kg, while the UO_2 quasi-spheres each had a mass of 2.14 kg. This first reactor had no provision for removal of the fission heat. In contrast, heat in the typical power reactor is transferred by the water flow in the primary circuit to the steam generators, producing dry steam in the secondary circuit, where the

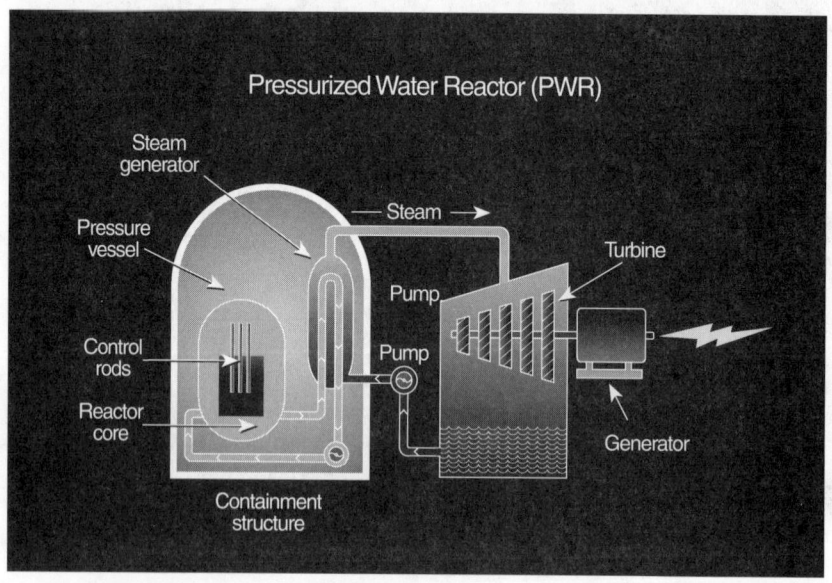

Fig. 5.2. A pressurized-water reactor.

pressure is maintained at some 69 atmospheres. ("Dry steam" means steam near 322°C, which would need to be cooled to 286°C before it would condense to water at that high pressure of 69 atm.) The steam then flows to a multistage turbine driving an electrical generator, which produces power for the grid (and, incidentally, power to run the reactor coolant pumps and the other machinery in the plant). The steam is returned to the steam generators by three feedwater pumps, each of 17,500 horsepower (13 megawatts), which force the water, condensed in the secondary loop by the action of the waste-heat exchanger, from its subatmospheric pressure back to the steam-generator pressure of 69 atm.

The difference from the boiling-water reactor is that in the pressurized-water reactor the water and steam of the secondary loop do not come in contact with radioactive materials, which simplifies maintenance and the containment problem. However, these advantages come at the price of a complex and expensive system of heat exchangers. Boiling-water reactors are somewhat cheaper, because they don't need heat exchangers. But the maintenance of the turbines is more costly, because they are exposed to radioactivity.

Some idea of the scale can be obtained from looking at Fig. 5.3. The reactor pressure vessel is some 14.6 m high and 7 m in diameter, with a forged steel wall 23 cm thick. It contains about one hundred tons of fuel in its 50,000 fuel rods.

In summary, in a light-water reactor, the fission heat is removed by flowing water coolant, which in a boiling-water reactor converts to steam at the top of the reactor pressure vessel or in a pressurized-water reactor is transferred at high pressure in the primary loop to the steam generator, where water in the secondary loop is converted to steam for operating the turbine.

THE CANDU HEAVY-WATER REACTOR

During World War II, Canadian scientists played an important role in the Manhattan Project, but no enrichment plants were built in their country. After the war, it was therefore natural that Canada developed a reactor that uses natural, nonenriched uranium (0.7% uranium-235 content); but this type of reactor needs a moderator of heavy water (deuterium oxide rather than hydrogen oxide). (As was explained in Chapter 2, heavy water absorbs fewer neutrons than ordinary light water, and this permits a chain reaction with a lower concentration of uranium-235 than in light-water reactors, which require uranium fuel enriched to a uranium-235 concentration of 3 to 5%.) This type of reactor is called CANDU (CANadian Deuterium Uranium).

Twenty-two CANDU reactors are currently operating worldwide, at powers of 500 to 900 MWe. They use a large reservoir of heavy water at room tempera-

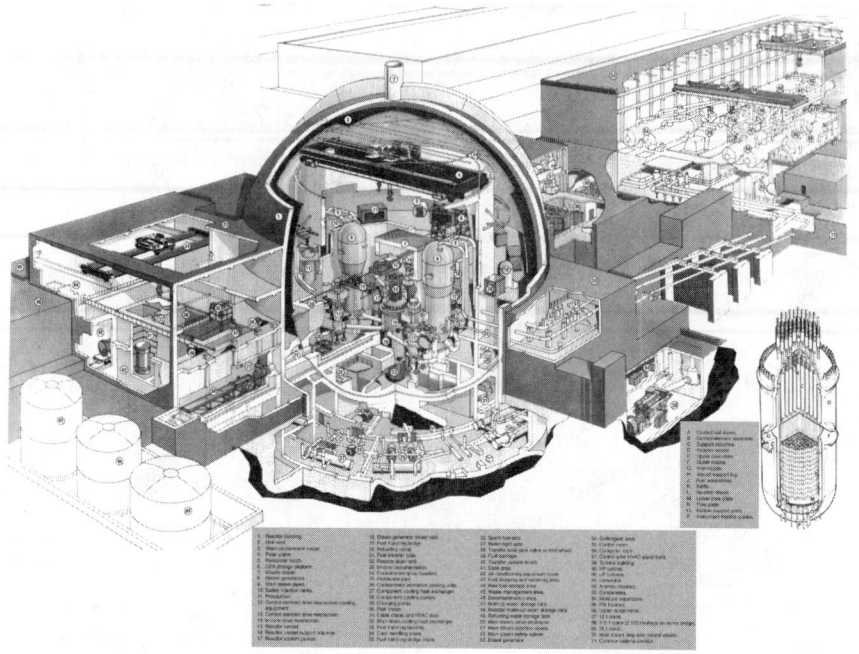

Fig. 5.3. A pressurized-water reactor as part of an electricity generating plant.

ture and atmospheric pressure, which is traversed by a substantial number of horizontal, double-walled tubes containing the uranium fuel, arranged as in a radiator grill. The hot nuclear fuel is separated from the heavy-water moderator by the walls of two concentric tubes each sufficiently strong to withstand the pressure of the heat transfer fluid, also heavy water, which flows along the fuel rods in the inner tube. The space between the hot inner and the cool outer tubes is filled with gas, which thermally insulates the cool heavy-water reservoir from the hot fuel elements. In contrast, the fuel rods in a light-water power reactor are immersed in the ordinary water that is used both for heat transfer and for moderating the neutrons and that is therefore contained by a single massive steel pressure vessel. In the CANDU reactor, because the heavy-water moderator is at room temperature and pressure, there is no need for a heavy container.

Over the life of the fuel elements, CANDU reactors produce only 20% as much heat per kilogram of uranium as light-water reactors, since there is five times less uranium-235. To avoid production losses due to frequent shutdowns, CANDUs are reloaded while operating at full power, the fuel elements being in individual pressure tubes rather than in a common pressurized vessel. An automatic loading machine connected to one end of a fuel conduit pushes in a short bundle of fuel rods and ejects used bundles at the other end, while maintaining the pressure and flow of heavy water to prevent the fuel from melting and destroying the reactor.

For a CANDU reactor, there is no enrichment cost involved in the price of a kilowatt-hour of electricity. In addition, the CANDU sidesteps the risk of nuclear proliferation associated with enrichment plants, which might be used for the clandestine production of highly enriched uranium. On the other hand, continuous loading facilitates the diversion of plutonium from the reactor because the spent fuel is extracted continually rather than at predetermined, specially monitored intervals. There would be little economic penalty for discharging fuel after only brief irradiation, when it would contain ideal weapon-quality plutonium, and this increases the proliferation risk. (As noted in Chapter 3, extended use of fuel in the reactor produces a heavier isotope of plutonium, plutonium-240, which spontaneously gives off neutrons. In the simplest implosion weapons, these neutrons can cause a premature initiation of the explosive chain reaction, thus resulting in a yield of 1000 to 2000 tons of high explosive rather than, say, the intended yield of 20,000 tons.)

Because CANDU reactors are bigger than light-water reactors, they are not well adapted to military needs such as naval propulsion, which were the priority during the Cold War. Bomb manufacture had required the construction of gigantic uranium-235 enrichment plants, so the need for enriched fuel was then only a slight handicap in U.S. commercial reactors, considering that the light-water reactors could also be used for submarine or aircraft carrier propulsion; and enriched uranium makes possible a more compact core and more design flexibility in power reactors—at lower investment cost. That is why commercial power reactor types based on natural uranium were not considered seriously by the United States in 1945 at the beginning of the nuclear era, although Fermi's first reactor and the large Hanford plutonium-production reactors did use natural uranium with a graphite moderator.

THE NUCLEAR FUEL CYCLE

The uranium fuel for all light-water reactors is produced in a similar fashion the world over. Since low-enriched uranium fuel is a world market commodity, it is essential to have standards that are internationally harmonized.

The sequence of steps in the fuel cycle, as shown in Fig. 5.4, is:

- Mining and milling of uranium ore.
- Beneficiation (purification), resulting in impure uranium oxide U_3O_8—"yellow cake."

The Open Fuel Cycle

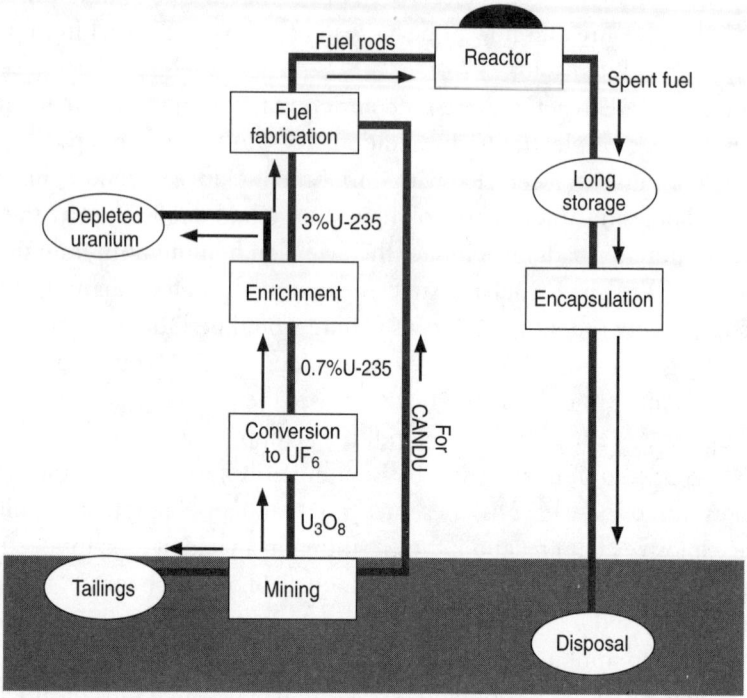

Fig. 5.4. The "open" nuclear fuel cycle (with direct disposal of spent fuel).

• Conversion to uranium hexafluoride, UF_6. This is a solid at room temperature but a gas at slightly elevated temperature.

• Isotope enrichment. This is done in the United States and in France by gaseous diffusion, and elsewhere in Europe and in Russia by gas centrifuges.

• Conversion to oxide. The UF_6, which is chemically reactive with air or water, is converted to UO_2—a black powder of grain size appropriate for pressing into ceramic pellets. These pellets are then baked at very high temperatures (so that the fine particles "sinter," or form strong bonds through surface diffusion) and ground to size, and are then ready to be fabricated into fuel rods.

• Fuel fabrication. Fuel fabrication is completed by loading the accurately ground pellets into thin-walled zirconium alloy tubes, to constitute the fuel rods, some 5 meters long and about the diameter of a pencil. Several hundred such rods are carefully mounted with spacers in a fuel assembly, for insertion in the reactor. In turn, several hundred of these fuel assemblies (also called fuel elements) side by side constitute the fuel load of a light-water reactor.

• Burning in the reactor. A typical fuel assembly is inserted into the reactor and operated at essentially full power for four years. Typically, every twelve months the reactor is shut down for a month so that one-fourth of the core can be replaced with fresh fuel. Other assemblies may be "shuffled" in order to equalize the neutron exposure and the fissions obtained from each gram of fuel. The fuel is typically burned to produce an energy of 40,000 megawatt-days per metric ton of heavy metal (MWD/MTHM), at a steady heat output of 30 megawatts per ton. Some reactors are operated for eighteen months, with one-third of the fuel replaced during the shutdown.

Of the typical 100 tons of fuel in a reactor, the 25 tons removed each year, after four years of fission heat production, can be stored safely without overheating by a transfer, under water, to the storage "pool" at the reactor. There the fuel assemblies are immersed, together with other fuel elements, for a period ranging from 6 months to 5 years. The fission-product decay heat is such that the fuel assembly would glow red-hot if it were removed from the water, in which it is cooled by natural circulation. In addition to fission products such as cesium-137, strontium-90, iodine-131, and scores of others—which are the inevitable ashes remaining from the fission itself—most of the initial uranium-238 remains. The spent fuel also contains plutonium-239 from the capture of neutrons by uranium, together with substantial amounts of heavier plutonium isotopes from the capture of successive neutrons by plutonium-239, such as plutonium-240, plutonium-241, and so on. After this temporary storage period, the fuel can be "reprocessed" to recover plutonium for fueling light-water reactors; the fission products and the transuranic elements are then cast with glass into a large "log" into a stainless steel container which is then welded closed and stored for ultimate disposal in a mined geologic repository. The closed fuel cycle, involving recycling of plutonium into the reactor, is shown in Fig. 5.5. Short storage times before reprocessing mean that the reprocessor gets the business and the profit earlier. Also, less storage capacity is required at the reactor. Long storage times mean less cooling is required in transport, and reprocessing is simpler with less problem from radiation destruction of solvents.

Each of these steps contributes to the cost of the fuel cycle, and is regulated in order to maintain an acceptable level of protection against hazard to the public and the workers. Some residual radiation exposure, to the workers and to the public, is associated with each of these steps, as described in Chapter 7.

The Closed Fuel Cycle

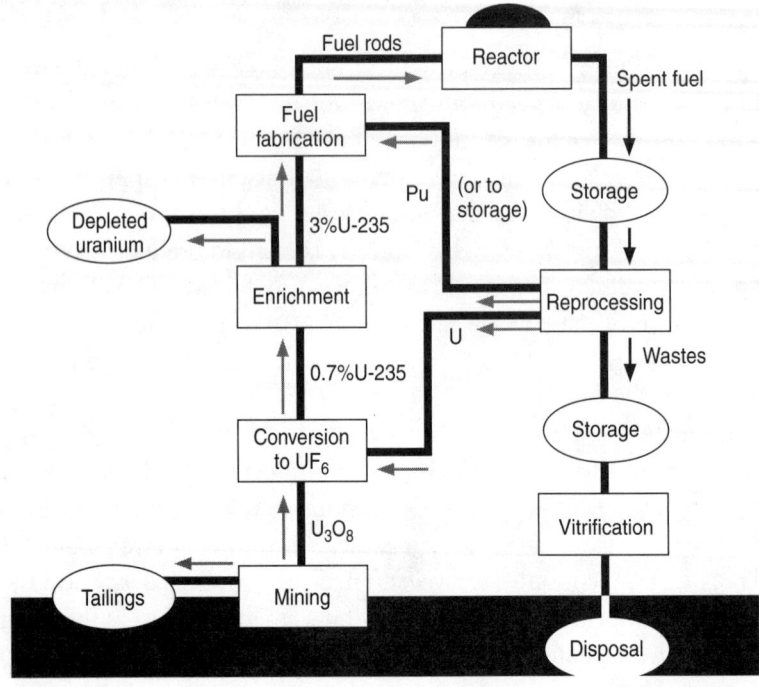

Fig. 5.5. Closing the fuel cycle with reprocessing and recycling of plutonium and uranium.

THE TASK OF MANAGING SPENT FUEL

The spent fuel from a year's operation of a nominal 1-Gwe power reactor constitutes a radiotoxicity of about 10^{10} Sv. If one part in a billion of this radioactivity were actually ingested, it would pose a 40% risk of an additional cancer. All of the spent fuel from the one hundred reactors in the United States is destined for direct disposal by placing a number of fuel elements in large casks to be buried in a mined geological repository in Yucca Mountain, Nevada. In contrast, all of the fuel from the 58 reactors in France is reprocessed, with the plutonium destined to fuel some light-water reactors; the fission products and other radioactive materials are vitrified (fused with glass) and will be similarly entombed in a mined geologic repository. Either of these approaches can be satisfactory. Nevertheless, reprocessing and recycling are often attacked for their potential contribution to the proliferation of nuclear weapons, and as an additional complication and cost; similarly, advocates of reprocessing criticize the direct disposal route as wasting a valuable resource and putting long-lived radioactive elements into the repository, when they might otherwise be consumed in the reactor.

In the Chernobyl catastrophe, a fraction of the radioactive fission products escaped from just one reactor and was widely dispersed. At the end of 2000, 433 power reactors were in operation in the world, generating an enormous quantity of radioactive products.[1] Even in the absence of any catastrophe, the problem of managing these radioactive materials remains; it requires a special sense of responsibility, not only to this generation, but to future generations as well. Today's solutions must stand up for thousands, tens of thousands, even hundreds of thousands of years; either that or our descendants must be ready to revise the plans made for them in the light of experience, even to reburying or reprocessing the spent fuel. Both economics and common knowledge show that delaying a costly expenditure is often good business, so it may be preferable to handle the spent fuel again in the future, provided that it will not be much more costly to do so, and does not produce irreversible effects.

Coal-fired power plants use a "fuel cycle" in which coal is mined, cleaned, transported, burned, and the combustion products (ash, carbon dioxide, sulfur oxide, minor constituents) either discharged to the atmosphere or disposed of in other ways. Nuclear plants produce highly radioactive spent fuel containing many "actinides," which are elements with chemistry similar to that of actinium—the element with atomic number 89—including uranium, plutonium, and elements present in lesser amounts in spent fuel, the "minor actinides" such as americium (element 95) and curium (element 96). If spent fuel is reprocessed to recover the uranium and plutonium—as is done in France for all the fuel from commercial reactors—the resulting "high-level" nuclear waste takes the form of glass "logs" encased in welded stainless-steel canisters; these contain most of the heat-producing radioactive materials.

A one-gigawatt electric, water-moderated reactor discharges, each year, 21 tons of radioactive fuel with the following inventory:

- 20 tons of uranium containing 0.9% (180 kg) uranium-235
- 200 kg of plutonium (of which only 63% is fissionable)
- 21 kg of minor actinides (i.e., not uranium or plutonium): 10 kg of neptunium, 10 kg of americium, 1 kg of curium
- 760 kg of fission products in which the long-lived elements are: 18 kg of technetium-99, 16 kg of zirconium-93, 9 kg of cesium-135, 5 kg of palladium-107, and 3 kg of iodine-129.

The accompanying table lists the principal long-lived radioelements present in the spent fuel from a power reactor, as well as their half-lives—the time necessary for half of their atoms to disintegrate.

One sees that for many of these radioisotopes, the half-life is greater than a million years, and, of course, the longer the half-life, the less the radioactivity per gram. (For instance, the 9 kg of long-lived cesium–135 per GWe-yr can be compared with the 20 kg of cesium-137, of 30-year half-life, in the same 21 tons of spent fuel; the Cs-135 is 100,000 times less radioactive per gram.) Still the radioactivity is so great that it would be irresponsible to store these materials in accessible sites or in those exposed to flowing groundwater that can return to the earth's surface. In addition to the scientific and technical difficulties of a safe disposal of nuclear waste, the challenge in dealing with this issue is to reach a consensus on the solutions, considering the pervasive fear of radiation.

In France, nuclear waste totals about 1.2 kg per person per year, in comparison with 100 kg of industrial toxic waste (and 15 kg per year of hospital waste, 3000 kg of industrial waste, and 7300 kg of agricultural waste). The options to manage this small quantity of nuclear waste will be discussed in more detail, to see whether it may be safely and economically disposed of in a way that keeps it away from the biosphere, or whether it might be possible and desirable to destroy it in whole or in part.

The radioactivity of spent fuel evolves over time as the various elements decay. The same is true of its radiotoxicity: that of fission products decreases very rapidly in a few hundred years and then persists at a low level for millions of years, because of the presence of long-lived fission products as listed in Table 5.1. The radioactivity of plutonium (mostly of half-lives 24,000, 6500, 14, and 87 years) represents less than 10% of the total toxicity of the spent fuel when it comes out of the reactor. With the passage of time and the disappearance of the short-lived products, this proportion increases. After several thousand years, plutonium dominates and represents nearly 90% of the radiotoxicity.

If, as is standard French procedure, the plutonium is removed from the fuel to be recycled one or more times in reactors in the form of MOX fuel (the ceramic mixed oxide of plutonium and uranium), the radioactivity of all the other long-lived substances—fission products, minor actinides, and residual uranium—after 60,000 years or so comes down to that of the natural uranium from which the fuel was made; this time increases to 600,000 years for untreated spent fuel. To this, however, must be added the radiotoxicity of the spent MOX fuel itself.

Fig. 5.6 shows the evolution, over thousands of years, of the composition of nuclear waste buried without reprocessing. Only the minor actinides and the

TABLE 5.1. HALF-LIVES AND DOSES PER UNIT OF THE PRINCIPAL
LONG-LIVED RADIOELEMENTS IN POWER-REACTOR SPENT FUEL,
AND OF NATURAL POTASSIUM

	ISOTOPE	HALF-LIFE (YEARS)	DOSE PER UNIT INGESTED (Sv/Bq)	DOSE PER UNIT INGESTED (Sv/g)
Minor actinides	Neptunium-237	2.15×10^6	1.1×10^{-7}	2.9
		432	2.0×10^{-7}	2.6×10^4
	Americium-241*	7380	2.0×10^{-7}	1500
	Americium-243*	18	1.2×10^{-7}	3.6×10^5
	Curium-244	8532	2.1×10^{-7}	1300
	Curium-245			
Fission products	Technetium-99	2.13×10^6	6.4×10^{-10}	0.04
	Zirconium-93	1.5×10^6	1.1×10^{-9}	0.10
	Cesium-135	2.3×10^6	2.0×10^{-9}	0.09
	Palladium-107	6.5×10^6	3.7×10^{-11}	0.0007
	Iodine-129	1.5×10^7	1.1×10^{-7}	0.91
Natural radioelement	Potassium-40 (1 part in 10,000 in normal potassium)	1.28×10^9	6.2×10^{-9}	0.0016

Note that Am-243 decays to Pu-239 (via Np-239), and Am-241 to Np-237.

one most active fission product, technetium-99, are included. The complicated shape of the decay curves is due to the fact that the disintegrating products, like plutonium, give birth to other radioactive atoms like uranium-235 that have a very long lifetime. For example, if typical spent fuel, containing 1% of uranium-235 and 1% of plutonium, disintegrates for 250,000 years, it ends up with a concentration of 2% of uranium-235. Our descendants will have the pleasant surprise of finding mines enriched to 2% in uranium-235, lowering the price of enrichment unless, of course, the fuel had been reprocessed and the plutonium had been used for the production of energy.

Long-lived fission products and transuranic elements (those beyond uranium, such as neptunium, plutonium, curium, etc.) contribute to the difficulty of evaluating the solutions proposed for nuclear waste management over many millennia. In as little as a few tens of thousands of years (distant in comparison with the 70-year life span of the human in modern developed societies, but not a long time in comparison with the 24,000-year half-life of plutonium or the fifteen-million-year half-life of iodine-129), the conditions for the return of some of the material to the biosphere may be radically different from what they are now, owing to geological evolution that is very difficult to foresee. For instance, it is only 13,000 years since glaciers a kilometer thick covered much

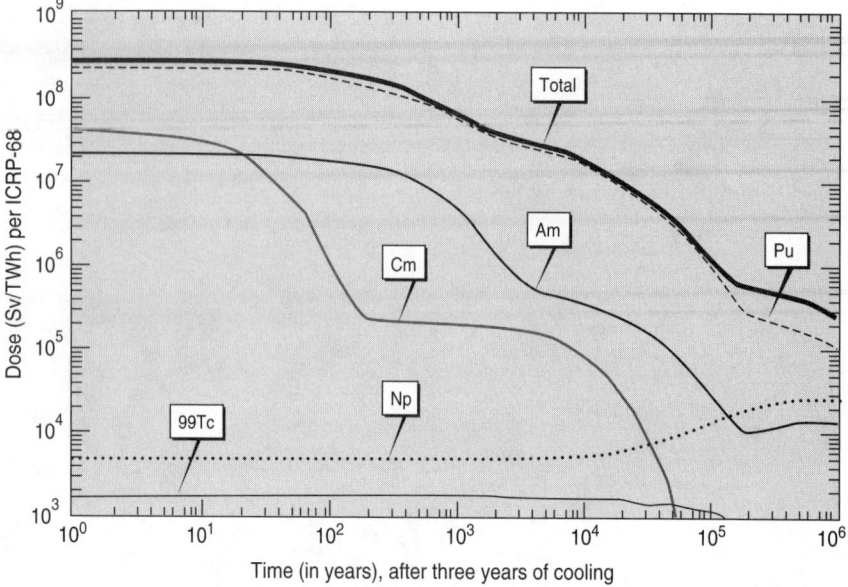

Source: Dissertation of S. Sala, Université de Provence, 26 June 1995.

Fig. 5.6. Fuel irradiated at 33,000 MWd/t. Contribution of the initial inventory to radiotoxicity (by ingestion). One reactor-year of operation is about 8 terawatt hours (TWh).

of the northern hemisphere; if, on the other hand, the earth's existing glaciers melt and raise the sea level, seawater will inundate many low-lying locations, and such eventualities must be taken into account in the siting of repositories of spent fuel or of vitrified fission products.

THE BACK END OF THE FUEL CYCLE:
DIRECT DISPOSAL OF SPENT FUEL

The United States plans to store spent fuel elements as they are—the so-called direct disposal option. In this scheme, after being discharged from the reactor, the fuel elements are suspended in a pool on the reactor site, for a first cooling-off phase. They are then placed in heavy steel casks, about 24 fuel elements to a cask, and transferred to an above-ground storage facility for 10 to 50 years before being disposed of as nuclear waste in a mined geologic repository. The period of above-ground storage allows much of the fission-product radioactivity to decay before the waste is emplaced underground. This is important because the amount of fission products that can be put in a repository is determined primarily by the gradual heating of the rock, so cost is reduced by allowing much of the decay heat (about 1 kW per fuel element at the time the elements are loaded into the cask) to dissipate before emplacement.

For its mined geologic repository, the U.S. has selected Yucca Mountain, Nevada, adjacent to the Nevada Test Site, where more than nine hundred nuclear explosives have been tested. The emplacement tunnels will be about 300 m below the surface of the mountain and some 300 m above the water table. The groundwater is trapped within a closed basin and does not flow into any rivers that reach the sea.[2]

The casks containing spent fuel are expected to have a 10-cm-thick outer wall made of carbon steel over a 2-cm-thick layer of corrosion-resistant high-nickel steel; they will measure about 6 m high by 2 m diameter, and weigh about 70 tons each. Between the containers and the tunnel wall there will be still-unspecified materials such as gravel, perhaps supporting a kind of sloping rock-slab roof to prevent groundwater from dripping on the casks.

Unlikely scenarios must be taken into account in considering a strategy of burying considerable quantities of fissionable materials or fission product waste for tens of thousands of years.[3] Burial sites for spent fuel or for vitrified waste must be largely free from infiltrating water, because fission products are soluble. Ceramic or granite slabs or (the authors' preference) overlapping tiles may be suitable for diverting underground water flow, as the roof on a house diverts rain. (Titanium drip shells have recently been proposed, but we believe that tiles will prove preferable.) But, of course, earthquakes can challenge these precautions.

We shall not discuss in detail the criteria for choosing geological sites for deep burial that offer guarantees of stability for millions of years. Some experts think it is difficult to provide an absolute guarantee beyond a thousand or ten thousand years, but that should not prevent us from doing our best to understand scientifically the risks associated with the various strategies well beyond these time scales.

In a Viability Assessment, the Nuclear Regulatory Commission comments, ". . . not knowing what the nature of the biosphere at Yucca Mountain will be 10,000 years from now (e.g., glaciation, rain forest, a large city surrounding Yucca Mountain) it is difficult to license in accordance with a dose criterion that is to govern for this length of time."[4] The Viability Assessment concedes this difficulty and simply presents the example of a settlement 20 km from the site, obtaining water from wells and with "lifestyle similar to the average person living today . . . about 30 km from the site." According to the Viability Assessment, for 10,000 years after closure of the repository, the mean added individual radiation dose per year to this population would be about 0.001 mSv, compared with the current annual average dose per person in the United States of 3.6 mSv. During the first million years, however, the mean annual dose to an individual in this community is forecast as 2 mSv, because of leakage from the repository; it is estimated to take about 100,000 years for the leakage to begin. Similar effects are predicted for other repositories studied in Sweden, Finland, Canada, and Japan.[5] In a society at the technological level of today's, it would be no major problem for the small number of people exposed to such radioactive contamination of their water to be provided with alternative sources of water. But that possibility, and not absolute ignorance, should guide decisions in this field.

The U.S. disposal program has encountered political and technical difficulties, and the Yucca Mountain repository will not be ready to store any waste before 2010 at the earliest. As a consequence, reactor operators have had to modify their storage pools to hold more fuel elements, and they complain that they are running out of storage capacity.

The opening of the repository in 2010 depends on a presidential decision to formally recommend the site to the Congress, on a license application to the Nuclear Regulatory Commission in 2002, and on favorable action on the license. Waste emplacement would end in 2033, but the repository would be kept open and observed for one hundred years after the start of emplacement; closing and sealing would begin in 2110 and take six years. Yucca Mountain is planned to hold 63,000 tons of commercial spent fuel, 2333 tons of defense spent fuel (irradiated at Hanford or Savannah River, but not reprocessed to obtain plutonium for nuclear weapons), and 4667 equivalent tons of high-level waste from reprocessing operations. Some $5.9 billion has been spent during the period 1983–98 on characterization of nine sites and on activities at Yucca Mountain, including a 5-mile exploratory tunnel. Future costs through 2016 are estimated at $36.6 billion and would all be covered by the fees paid by utilities (0.1 cent per kWh) into the Nuclear Waste Fund. The 100-year total is now estimated at $58 billion.

The required lifetime of the container materials is still being debated in other countries: should one use steel (durability of five hundred to one thousand years) or copper, or even noble metals (i.e., chemically nonreactive metals such as gold or platinum) that can last for a million years? Precious metals add another vulnerability, since it is expected that they will be valuable in any society and may actually provoke intrusion into the repository, just as ancient tombs have been plundered.

The difficulty in finding and characterizing a disposal site within each country leads us to the firm judgment that it would be beneficial to the world and to the nuclear industry to create competitive, commercial mined geologic repositories in the most suitable locations, to which the nuclear industry could send approved forms of vitrified fission products or properly packaged spent fuel. A proposal to create just such an enterprise in Western Australia is dubbed Pangea and would make use of deep burial in a mined repository in an enormous region in which groundwater does not migrate. The Pangea International Repository would be integrated with a purpose-built Australian port, a dedicated railroad line, and a fleet of special ships for safe, secure transport of casks of spent fuel or vitrified fission product waste.[6]

A MATURE PROGRAM: THE CASE OF FRANCE

As we have seen in Chapter 2, before the Second World War, France was a leader in research in radioactivity and fission. At the end of the war in Europe, France, which had been defeated and occupied by the Nazis, had the task of rebuilding its society, its economy, and its spirit. After the Atomic Energy Commission of France (Commissariat à l'Energie Atomique—CEA) was formed in 1945, France's support for nuclear energy was driven not only by its desire to produce plutonium for nuclear weapons and by the technological challenge, but also by its lack of indigenous energy resources and a recognition of the uncertainty of alliances and even of territorial possessions that might provide fuel.

The first French nuclear reactor, or "pile," began its operation on December 15, 1948. Its name, Zoé, derives from its characteristics, *zéro/oxyde/eau*, meaning zero (power), (natural uranium) oxide, and (heavy) water. We have seen that there is the choice between heavy water and purified graphite for such an experimental pile using natural uranium, and the French had experience with heavy water before the war. Zoé was the first of a line of heavy water (*eau lourde*) reactors and thus was given a second name, EL1. A second heavy-water moderated, natural-uranium reactor, EL2, began operation on October 21, 1952; this reactor was maintained at room temperature by the circulation of high-pressure carbon dioxide gas. On July 4, 1957, a third reactor, EL3, began to operate.

At the same time, the CEA began to build reactors named G1, G2, and G3 ("G" for *gaz* or gas-cooled) for the express purpose of producing plutonium for nuclear weapons. G1 began to operate on January 7, 1956, G2 in June 1958, and G3 in June 1959, producing both plutonium and electrical power.

The French Pressurized Water Reactor (PWR) program started with a company formed in 1960 owned equally by the public utility for electricity in France, Electricité de France (EDF), and a consortium of Belgian electric utilities. A 305-MWe power plant was built in France, based on the Yankee Rowe (Massachusetts) plant; this plant had been constructed by Westinghouse and had begun its operation in 1961. This first PWR in France was put into service in 1967. Additional PWRs were built in France, under license from Westinghouse, by Framatome (Société Franco-Américaine de Construction Atomique). By 1981, Framatome and Westinghouse replaced the license arrangement with an agreement for cooperation that allows Framatome to build reactors independently. In 1982, EDF launched the construction on a new (N4) series of PWRs, of 1450-MWe capacity; the first, at Chooz on the border between France and Belgium, began operation in August 1996.

With its commitment to nuclear power arising from the oil shocks of the 1970s, France has created a highly centralized industry for the supply of nuclear fuel. Electricité de France owns and operates the reactors; COGEMA (General Company for Nuclear Materials—COmpagnie GÉnérale des MAtières nucléaires) is responsible for all aspects of the nuclear fuel cycle; and ANDRA (National Agency for the Management of Radioactive Waste—Agence Nationale pour la gestion des Déchets RAdioactifs) has managed radioactive wastes since 1991. EDF, COGEMA, and CEA generate 95% of the radioactive waste. CEA oversees the development and manufacture of nuclear weapons, as well as much research and design of commercial nuclear reactors; Framatome collaborates with other technical companies of the European community; and France builds power reactors in other countries. France employs a highly standardized design in all 58 reactors currently operating there and producing electricity for EDF.

We shall describe now another type of reactor—the "breeder" reactor, which produces more nuclear fuel than it burns. The fuel for this reactor is a composite of uranium-238 with about 10% plutonium. There is no moderator such as graphite or water in the core, so that fast neutrons that do not cause fission in plutonium are largely captured in U-238 to yield more plutonium-239. The French breeder program was designed at the time of the oil shocks in the 1970s, when a shortage of uranium fuel at low prices was envisaged. The United States had built experimental fast-neutron reactors, and a single such commercial power reactor (Fermi I) that operated in Michigan for a few years beginning in 1969.

DESIGN PRINCIPLE OF FAST-NEUTRON REACTORS

Innovations were necessary in order to make use of fission produced by fast neutrons, but the benefit could be enormous. The reward is nothing less than the ability to generate fission energy by using the abundant isotope of uranium—uranium-238—in a two-step process that involves a "cycle" rather than the simple burning of uranium-235. We have noted in Chapter 2 that neutron capture in uranium-238 leads inevitably to an atom of plutonium-239, which is fissile with thermal neutrons, and that it is easy to burn plutonium-239 in a water-moderated reactor.

To maintain a chain reaction, one neutron per fission must go on to cause another fission. But to *regenerate* the burned fuel, a *second* neutron from each fission must be captured in uranium-238. A significant fraction of the plutonium-239 or uranium-235 fission at thermal energy is accompanied by the parasitic capture of a neutron by the plutonium-239 or uranium-235 fuel to form the nonfissile isotopes plutonium-240 or uranium-236. To avoid this loss so as to regenerate fully the burned fuel (and even a bit more—that is, to breed more fuel than is necessary to sustain the reaction indefinitely, hence the name "breeder" for a reactor of this type) it is necessary to avoid excessive slowing down of the 2-MeV fission neutrons.

The cross sections for neutron-induced fission and capture in uranium-235, uranium-238, natural uranium, and 4.1% U-235 are shown graphically in Chapter 2. Fig. 2.3 shows that for uranium-238, neutron capture rather than fission dominates at energies below 1 MeV. Of the neutrons produced by fission, some 12% manage to fission directly uranium-238; the others arrive at an energy where they are captured by the uranium-238 or go on to fission plutonium-239. Uranium-238 nuclei are not like hard billiard balls, but like billiard balls covered with moss: collisions with fission neutrons, called "inelastic," abruptly slow the neutrons. Most of the energy lost in these inelastic collisions reemerges as gamma rays emitted by the uranium-238.

In the breeder reactor, the fuel core is surrounded by a uranium blanket so that the escaping neutrons will be captured and form plutonium. Uranium depleted in U-235 is normally used, obtained when much of the U-235 is removed from natural uranium in an enrichment plant. Because, by design, there is no hydrogen or carbon in a breeder, the neutrons are prevented from arriving at an energy sufficiently low for maximum fission probability of uranium-235 or plutonium-239, and fuel enriched to 4% in uranium-235 or plutonium-239 will not permit the reactor to become critical, i.e., will not sustain a chain reaction. Therefore, a fast-neutron reactor requires an enrichment at the level of 10% to 12% to achieve criticality.

Fig. 5.7 shows the neutron absorption probability as a function of the atomic number of the nuclei for several elements, at a propitious energy for absorption by uranium-238. The extraordinary variability of the nuclear appetite for neutrons can be seen. It reflects the fact that in the nucleus, the nuclear matter is organized like the electrons in atoms, with protons and neutrons in "shells" located at determined energy levels, so that the addition of a neutron or a proton to a shell can substantially change the properties of the nucleus. The choice of a material that scatters the fast neutrons without

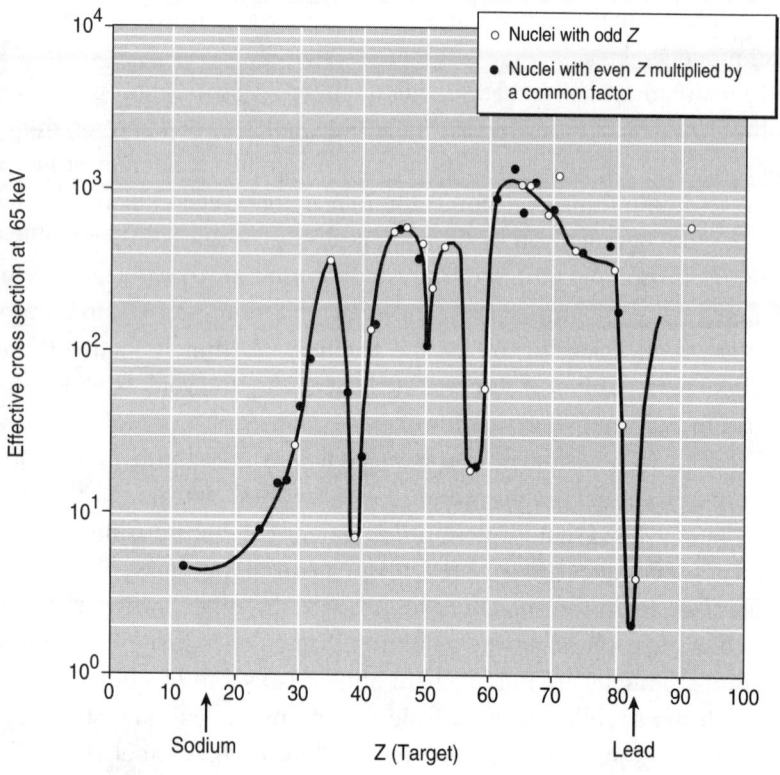

Fig. 5.7. Absorption probability for 65,000 eV neutrons in various nuclei as a function of atomic number, Z. Nuclei with odd Z are plotted directly on the graph, while those with even Z have had their cross-sections multiplied by a common factor before being plotted.

absorbing them is limited to a few elements that are in the valleys of the curve. As we shall see, only sodium and lead have, in addition, the thermal properties required to serve as a coolant. Bismuth, also suitable, is far more expensive. Sodium and lead (or lead-bismuth alloy) are suitable coolants.

Like thermal-neutron reactors, fast-neutron reactors depend for their control, in normal circumstances, on the small proportion of neutrons emitted from fission products with a delay of 10 to 20 seconds. This allows plenty of time to reposition the control rods. As we have seen, in a pressurized-water reactor that uses uranium-235, the fraction of delayed neutrons is 0.65%. But in fast-neutron reactors, fission occurs mainly in plutonium-239, and the proportion of delayed neutrons is only 0.21%, which makes both the design of the reactor and the manipulation of the control rods more delicate. In addition, the life of a prompt neutron is of the order of 25 microseconds in a pressurized-water reactor, but about a thousand times shorter in a fast-neutron reactor. This apparently vast difference (25 microseconds vs. 25 nanoseconds), however, has no importance for the normal operation of a reactor, since the delayed neutrons in each case allow the controls to operate with a delay of seconds or minutes rather than microseconds or milliseconds.

The Rise and Fall of Superphénix

A fast-neutron reactor, Superphénix, operated at Creys-Malville, France (on the east bank of the Rhône River, some 50 km east of Lyon) from 1985 to 1997, delivering up to 1200 MW of power to the grid (but a small fraction of the time, because of sodium leaks and other problems).[7]

The first French fast-neutron reactor, Rapsodie (derived from *rapides-sodium*), began to operate in January 1967. On the basis of the experience from Rapsodie, a fast-neutron reactor power plant named Phénix was built and operated from August 1973 to 1990, and was soon delivering 250 MWe; its name is taken from the mythical bird that renews itself by fire every 500 years, with a new, young phoenix springing from its ashes. In the case of Phénix, the fresh plutonium-based reactor fuel arises from the spent-fuel "ash" of the reactor.

The initial objective of Superphénix was to produce in a full-sized nuclear power plant more fissionable material than it consumed. In such a design, for one hundred fissions in the plutonium, thanks to the capture of neutrons by the fertile material, uranium-238, an excess of 13 atoms of plutonium are produced.

Sodium was chosen as the heat exchange fluid for Superphénix, as for all of the fast-neutron reactors in the world, because in spite of its unpleasant chemical property of catching fire when exposed to air or water, its thermal and nuclear properties are very suitable. It is liquid at 98°C. The engineers initially thought that the considerable mass of 3000 tons of liquid sodium in the reactor would be easy to manipulate. The sodium warms up as it is circulated in the core by pumps (Fig. 5.8, next page); it arrives at 545°C from the reactor core and cools in the exchangers in which it passes its heat to the sodium of the intermediate circuit.

The general characteristics of a fast-neutron reactor's cooling system are similar to those of a pressurized-water reactor, but in Superphénix an intermediate sodium loop and an additional heat exchanger were used to avoid the possibility of contact between the water of the steam generator and the radioactive sodium that circulates in the core of the reactor. There were also considerable technological differences due to two essential factors. On the one hand, the pumping and circulation of the cooling fluid, sodium, posed problems that had never before been encountered on this scale in any industry, and that

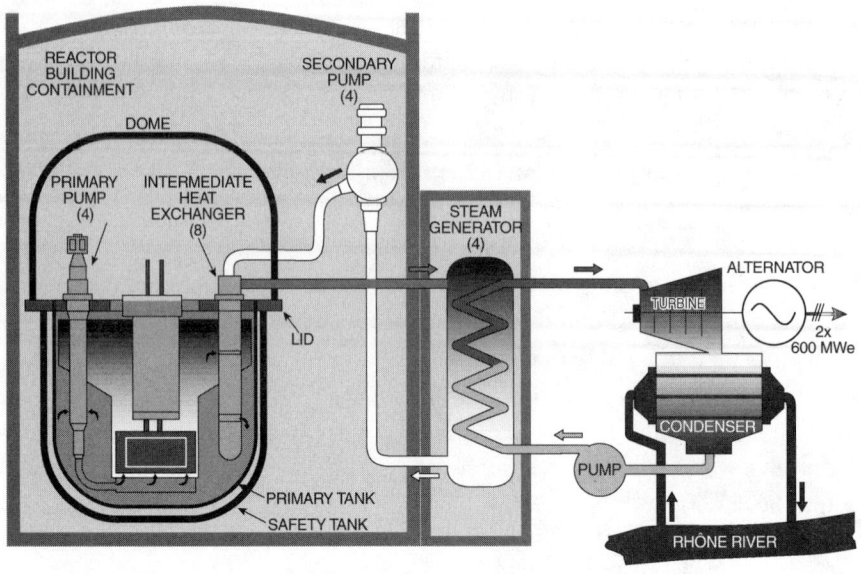

Fig. 5.8. A sodium-cooled fast reactor—Superphénix. Shown is the pool of liquid sodium in the reactor vessel itself, with the core submerged under the sodium. Four primary pumps circulate the sodium through the heat-producing core. Eight sodium-to-sodium heat exchangers immersed in the pool transfer the heat to the secondary circuit. The circulation in this secondary circuit is maintained by four pumps. Four steam generators transfer heat from the liquid sodium of the secondary circuit to water, which as steam drives the turbine that in turn drives two 600-MW alternators. The steam is condensed and the heat rejected to the Rhône River. The condensate is forced back into the steam generator against the steam pressure by a high-pressure feedwater pump.

caused substantial delays. On the other hand, the accidental loss of sodium coolant would have led to enhanced criticality and potentially to an uncontrolled chain reaction. The design of the reactor aimed, therefore, to make this eventuality impossible.

The ensemble, including the core, the primary pumps, and the intermediate exchangers, were enclosed in a primary stainless-steel vessel welded to a cover slab from which it was suspended. This primary vessel was itself enclosed in a second stainless-steel safety vessel, in order to avoid the possibility of loss of coolant. The primary sodium, which was in contact with the core, never left the main vessel. These measures were taken to satisfy concerns that had been raised about fast reactors much earlier in their history, exemplified by a comment from Edward Teller, a great supporter of nuclear power:

I have listened to hundreds of analyses of what course a nuclear accident can take. Although I believe it is possible to analyze the immediate consequences of an accident, I do not believe it is possible to analyze and foresee the secondary consequences. In an accident involving a plutonium reactor, a couple of tons of plutonium can melt. I don't think anybody can foresee where one or two or five percent of this plutonium will find itself and how it will get mixed with some other material. A small fraction of the original charge can become a great hazard.[8]

The experience with Superphénix has been mixed. The fact that sodium is opaque has made in-reactor inspections much more difficult than is the case in a water-moderated reactor in which a simple TV camera can be lowered to inspect the structures. According to Bernard Magnon, director of the power plant, as quoted in an interview in 1996, "I think that running Superphénix is a good lesson in modesty." Many supporters argue that unwarranted obstacles were placed in the path of the fast-neutron reactors and whether or not we accept this view, it is important to hear it. It is instructive to read these lines from an impassioned plea in defense of Superphénix, R. Barjon, "Nécessité et sûreté de la filière des réacteurs à neutrons rapides en général et de Superphénix en particulier" (Need for and safety of fast-neutron reactors in general and of Superphénix in particular), *Bulletin de la Société Française de physique,* July 1996, pp. 15–17:

During the years 1970–80, under the pressure of ecology movements, supported behind the scenes by the coal and oil lobbies (shameless exploitation of the Three Mile Island accident, which had no victims and dispersed no radioactivity), a certain number of countries (USA, Germany, Sweden, and Great Britain) halted their nuclear development, although a general consensus remained in favor of the exploitation of the existing plants and the completion of plants under construction.

The present situation can be characterized by noting that if the principal industrialized countries are divided into two categories, the coal-producing countries and the non-coal-producing countries, the coal-producing countries (USA, Germany, Great Britain . . .) are anti-nuclear and the non-coal-producing countries (France and Japan) are pro-nuclear. This is an obvious fact which, unfortunately, is not well enough understood.

The safety of fast reactors has been studied and organized according to the same concepts, designed by the same teams, and approved and verified by the same safety authorities as has that of thermal reactors.

The main differences in favor of the fast reactors are: the absence of pressure in the core, the boiling temperature of sodium being much higher than the nominal operating temperature; the containment of the primary cooling circuit; a particularly large Doppler effect, which is fundamental for the kinetics; the stainless-steel cladding of the fuel elements, which improves safety and leads to a yield of 100,000 megawatt-days per ton [of fuel] instead of 50,000 for thermal fuel clad in zircalloy; a much larger thermal capacity in the primary circuit, allowing longer safety system response time, which is decidedly favorable to enhanced security.

In the eyes of many people, French nuclear specialists are not credible when they proclaim the safety of Superphénix because—and this is often true—they are acting as both judge and judged.[9]

Reflecting responsible official opinion of the time, Bertrand Barre, Director of Nuclear Reactors, CEA, stated in an interview in late 1996, before the decision to close Superphénix, "I don't want people to judge fast reactors based only on Superphénix results. . . . Some factors are due to the complexity of engineering Superphénix and probably due to moving too fast when extrapolating the size (something to which everybody agrees on today), a fact explained by the psychological context of the oil crisis." However, the technical leadership felt that, given the immense capital invested, there was much to be learned from the operation of Superphénix. Toward the end of its short life, it was advocated as a source of neutrons that could be used for learning whether one could ease the problem of disposal of long-lived elements of reactor waste by transmutation of some of the plutonium and heavier elements in the fast reactor.

Reducing the uranium-238 blanket or replacing the uranium by steel would make the core a consumer of plutonium, since most of the plutonium is born in the blanket and not in the core of the reactor. This could be useful for burning up some excess plutonium from weapons—as would light-water reactors.

The thermal-neutron reactors that today make up the world's entire nuclear power installation were, when they were planned, considered largely as a stopgap to accumulate a sufficient stock of plutonium for the future fast-neutron breeders. In reality, it would have been more effective to bypass this first step by fueling each breeder initially with enriched uranium, which would support a much more rapid deployment of breeder reactors than an approach that depended on the prior accumulation of plutonium for providing the initial fuel loads for the breeders.[10] The reason is simple: a light-water reactor each year converts a ton of U-235 to 250 kg of plutonium, whereas a fast reactor of the same power would produce almost a ton of plutonium from a ton of U-235 each year.

The decision to build an expensive industrial prototype breeder in France was premature; it was due in part to technological optimism on the part of the participants, coupled with a lack of appreciation for alternatives. In 1997 the new government of Prime Minister Lionel Jospin announced that it would shut down Superphénix; it has been closed and its fuel is being removed. Through 2000, the plant cost some $8 billion, and continues to cost $100 million per year; its ultimate dismantling is expected to cost a further $2 billion. Despite its limited lease on life, Superphénix was the only fast-neutron reactor in the world capable of producing electricity at a power level comparable with that of the largest reactors in the worldwide nuclear power system, and therefore it is an important benchmark. The 600-MWe sodium-cooled breeder (the Beloyarsk nuclear power plant near Sverdlovsk, in Russia) is the next-largest in the world after Superphénix, and has been operating since 1981—with a core, however, of enriched uranium rather than plutonium.

The technology of the fast-neutron breeder reactor is one of the few approaches to future energy in which there can be confidence. We build upon it in our consideration of the future of nuclear power and world energy supply.

REPROCESSING: THE FRENCH SOLUTION

In France, much of the fuel cycle, in which the fuel removed from light-water reactors is allowed to cool in at-reactor pools and then transported to intermediate storage, is identical with that used in U.S. reactors. But in France the fuel irradiated in water-moderated reactors used for the production of electricity is reprocessed to recover plutonium for recycling in the reactor, rather than held for burial in a mined geologic repository. This decision was intended as a first step toward a breeder-reactor vision of the future.

France, Japan, and some other countries with few domestic sources of energy reprocess spent nuclear fuel to extract energy-producing components and to separate the wastes. Each component of the waste (such as the small pieces of metal tubes of the fuel rods, and the intensely radioactive fission product wastes) presents a different processing problem. The fission products can be rendered somewhat more compact than the spent fuel itself, and can be immobilized by incorporation in glass and cast into welded stainless-steel canisters.

Reprocessing was initiated to obtain plutonium, which can substitute for some of the uranium-235 in a nuclear power plant, and also to recover uranium from the spent fuel, in which, even after four years in the reactor, the uranium-

235 concentration is higher than in natural uranium—some 0.9% as compared to 0.71%. But spent-fuel uranium is not worth as much as raw uranium because of its content of uranium-236 and uranium-234. The uranium-234 increases somewhat the radiation dose to the workers who fabricate reactor fuel, while the uranium-236 acts as a neutron absorber and requires that the uranium-235 content of the reprocessed fuel be larger than that of normal low-enriched uranium if it is to be substituted for it in the reactor. The 1994 report of the Nuclear Energy Agency of the OECD (Organization for Economic Cooperation and Development) assumes a worth of "70% of the cost of new uranium at the same enrichment."

Spent nuclear fuel from a typical French power plant, after about a year of cooling in underwater storage at the reactor (in the swimming pool, or *piscine* in French), is shipped in radiation-shielding casks to a reprocessing plant. There, after two more years of cooling in pool storage, the fuel elements are disassembled by machine, the rods are chopped automatically (all under water), and the "hulls" (the sheaths and ends of the rods and the other fuel element hardware) are cleaned and collected for disposal. The uranium-oxide ceramic fuel pellets—now containing about 1% plutonium and 5% fission products by weight—are then dissolved in strong chemical reagents, with a view to purifying to an extreme degree the uranium and the plutonium from the fission products; less than one part in 10 million of the other fission products remain in the separated uranium or plutonium, and 99.9% of the plutonium is removed from the fission products. France does not process this "recycle uranium" but ships it to Russia in the form of uranium oxide, where it is enriched by powerful centrifuges. Other clients of the reprocessing plant, to whom the reprocessed uranium is returned, have the enrichment done in the Netherlands. France thus avoids contaminating its own gaseous diffusion uranium enrichment plants with traces of plutonium. The enriched uranium is returned to France for fabrication into fuel for the reactors.

The reprocessing plant at La Hague, France, operated by COGEMA, has the capacity to process 1600 tons of spent fuel per year in two large modules of 800 tons. Annually, the French reactors provide about 21 tons each, or about 1200 tons of spent fuel altogether. The spent fuel from low-enriched uranium contains about 1% plutonium—about 200 kg to 250 kg from the annual download of 20 to 25 tons from each reactor. La Hague thus separates about 16 tons of so-called reactor-grade plutonium per year. This is produced as plutonium oxide, a dry powder, which is then welded into small steel cylindrical containers each holding about 2 kg of plutonium. Compared with weapon-grade plutonium, which contains typically 94% Pu-239 and no more than 7% Pu-240, reactor-grade Pu is composed of some 60% Pu-239, 25% Pu-240, 9% Pu-241, 5%

Pu-242, and 1% Pu-238. Only the odd isotopes 239 and 241 (i.e., those with an odd number of nucleons) of Pu are fissile — that is, are subject to fission by slow neutrons.

Five of these small containers are sealed into an outer steel cylinder for protection and storage. They are stored in this form and eventually transported to the fuel fabrication plant, where the plutonium oxide is mixed with uranium oxide and fabricated into "mixed oxide" (MOX) ceramic-fuel pellets for use in light-water reactors. The fabrication process for MOX fuel is considerably more costly and potentially more hazardous than that for uranium fuel. We have seen that the half-life of Pu-239 is 24,000 years, in comparison with 4.5 billion years for uranium-238 and 0.7 billion years for uranium-235. In a gram of Pu-239 there are thus $4.5 \times 10^9/2.4 \times 10^4 = 200,000$ times as many disintegrations per second as in a gram of uranium. The Environmental Protection Agency regulations mandate that there should be less than one gram of reactor uranium (with a concentration of 4.4% uranium-235) in 38 million cubic meters of air (about 50,000 tons of air). The corresponding limit for reactor plutonium is one gram in 2.9×10^{13} cubic meters of air — a million times more stringent. The resultant protective measures in a MOX plant lead to a fabrication cost of some \$2000 per kilogram of fuel — \$2000/kgHM, where "kgHM" is read "kilograms of heavy metal."

To provide, for the MOX fuel, a fissile content comparable with that in normal uranium, about 5 kg of spent uranium fuel must be reprocessed to yield enough plutonium for 1 kg of MOX. At an estimated \$1000/kg of spent fuel for reprocessing, this results in a cost of \$5000/kgHM for reprocessing; therefore, the total MOX fuel cost, including fabrication, is about \$7,000/kgHM, to be compared with a current purchase price of about \$1400/kgHM for uranium fuel ready to load into a reactor. Since each kilogram of uranium fuel requires about 8 kg of natural uranium, this cost excess of \$5600 per kg of fuel would correspond to a cost of \$5600/8 = \$700/kg above the current cost of natural uranium. In other words, even if natural uranium were \$700/kg more expensive than it is now, using it would still be as cheap as or cheaper than reprocessing.

France sells the MOX manufacturing know-how and the service to other countries that implement recycling, such as Japan, Germany, Belgium, and Switzerland: the spent fuel is reprocessed in France and the uranium, plutonium, or MOX, as well as the vitrified waste, is returned to the country of origin, as required by French law.

The reprocessing and recycling of the fuel allow the recovery, in the form of uranium and plutonium, of about 30% of the energy initially in the fuel element — but at a high price. In contrast, fast-neutron reactors, in principle, permit the consumption of nearly all of the stock of natural uranium by fission, with the disadvantage that the highly radioactive nuclear fuel has to be reprocessed and recycled many times.

France is the pioneer in reprocessing and recycling spent fuel from its power reactors. Of its 58 pressurized light-water reactors, about 20 are already burning recycled plutonium as MOX in one-third of their fuel elements, and 28 are eventually to be licensed to do so. In Britain, at Sellafield, in West Cumbria (near the Lake District), a reprocessing plant with an annual capacity of 700 tons of fuel from light-water reactors and gas-cooled reactors began to operate in 1997. The Thermal Oxide Reprocessing Plant (THORP) receives fuel from light-water reactors 6 months or more after discharge, and reprocesses the fuel after 5 years of storage at THORP. Britain has only one light-water reactor. Its 27 other nuclear power plants have graphite moderators; for these reactors the heat is transferred by a gas coolant from the fuel rods to the steam generator. The country has no reactor licensed to be loaded with fuel fabricated with plutonium. Japan has under construction at Rokkasho a reprocessing plant similar to the one in France.

When MOX fuel is loaded into a reactor and burned for four years, only part of the plutonium is consumed. France has not decided whether the remainder is to be disposed of as unreprocessed spent fuel or reprocessed and mixed with plutonium from uranium spent fuel for further recycle. If so (and this seems the likely approach), this multirecycle plutonium becomes a burden on the light-water reactor, because spent MOX fuel has a larger fraction of nonfissile plutonium isotopes and yields less energy per kg of fuel reprocessed. Only fast-neutron reactors have the intrinsic capacity to totally consume the reprocessed plutonium and to burn the minor actinides. That property gave rise to the ambitious French program CAPRA (Concept to Amplify Plutonium Reduction in Advanced fast reactors), which was conceived to use the now-closed Superphénix fueled not only by plutonium but by americium and other minor actinides that would have been separated by an augmented reprocessing activity at La Hague. In a February 1998 report of the French Prime Minister's Office, it is stated that experiments along this line will now be done at the research reactor Phénix, since the closure of Superphénix.[11] Phénix, in turn, needs about $100 million to be restarted, following a sodium leak in 1998 that passed the second containment barrier.

NUCLEAR WASTE IN FRANCE

It is estimated that in 2020 the French stock of nuclear waste will contain 150 tons of plutonium and 70 tons of so-called minor actinides (compared with 526 tons of plutonium and 55 tons of minor actinides if the fuel had not been reprocessed and recycled), and 2000 tons of fission products, assuming that degraded plutonium from spent MOX fuel is not recycled into light-water reactors. On the other hand, if plutonium is routinely extracted from MOX fuel and recycled together with plutonium from uranium fuel, the amount of plutonium in waste will ultimately be only about 0.5 ton, assuming that the loss of plutonium to waste at La Hague remains at about 0.1%. But if spent MOX fuel is not reprocessed, about 50 tons of plutonium will remain as waste. For comparison, if the United States continues to have about one hundred operating power reactors, its wastes by the year 2020 will contain about 4000 tons of fission products and 1000 tons of plutonium for its direct-disposal approach.

In France, as mentioned, the quantity of radioactive waste is 1.2 kg per year per inhabitant against 100 kg per year per inhabitant for toxic industrial waste, a portion of which has infinite life—that is, consists of cadmium or lead or other heavy metals that do not decay with the passage of time.

Only 1% of this 1.2 kg of radioactive waste generated per person per year, that is, the weight of two quarter coins, represents the real hazard. It consists of both very radioactive short-lived and less radioactive long-lived radioelements. Mixed with melted glass by COGEMA, they make up the 12 g of products per person-year that must be managed for 100,000 years or more if society is to make use of nuclear energy with a clear conscience.

Plutonium and uranium are recovered from the brew that is kept in liquid form at La Hague for five years, and the remaining material containing essentially all of the solid fission products is dried, vitrified with ground glass, and poured into stainless-steel containers, which are then welded closed. Each glass "log," of about 0.15 m³, weighing some 400 kg, contains the fission products from three fuel assemblies, of which there are about 250 in a water-moderated reactor. They are to be kept in air-cooled pits at La Hague for thirty or forty years, covered with a concrete shielding plug; those not originating from reactors in France are then to be sent to their country of origin for storage and ultimate disposal—probably in an underground repository.

In order to provide shielding of the public from the intense nuclear radiation, the glass logs in their welded stainless-steel containers will be transported and probably emplaced in large casks that may be very similar to those used for spent fuel. Indeed, Germany has received some of its vitrified waste in the form of large CASTOR containers (CAsk for Storage and Transport Of Radioactive materials). Each cask is 5.5 m long by 1.5 m diameter and weighs 65 tons; it holds ten fuel elements from a pressurized-water reactor and it is used both for shipping spent fuel to La Hague and for receiving vitrified waste in return. COGEMA is forbidden by French law from retaining any of the vitrified fission products from the fuel that it reprocesses for reactors of foreign countries.

In France, the future of vitrified waste has not yet been decided. For the time being, it is being stored at La Hague, where only material of French origin is to be kept. In fact, however, much material has not yet been returned to its country of origin; for instance, in December 2000, COGEMA had about 27 tons of Pu from reprocessing of Japanese fuel.

Underground Storage

A law passed in the French National Assembly on December 30, 1991, defined a research program on the possibilities of reversible or irreversible storage in geological formations. This statute also calls for an exchange of views among the government, the major nuclear actors, and the populations concerned in order to avoid the misunderstandings that have arisen in the past. In particular, it defines the legislative role in the choice of the ultimate solution.

The law foresees the construction of underground laboratories for studying sites proposed for geological storage. Three have been identified, of which two have a clay environment at 500 to 800 m depth, while the third has granite; all three are below the water table. The law envisages, also, the possible transmutation of radioactive substances to reduce their radiotoxicity. Various scenarios are under study. Some would use the slow neutrons of the present pressurized-water reactors; others would use fast neutrons from reactors such as the now shut-down Superphénix or hybrid reactors based on an accelerator with an intense beam, such as the design that will be described in Chapter 6. An evaluation report on the status of the research is submitted, annually, to the government. The law provides that on receipt of a final report in 2006, a decision will be taken, to be translated into legislation when voted upon. The complexity of this procedure is commensurate with the difficulty of the problem, which is both political and technical in nature.

The French government and the nuclear industry make every effort to speak of an "underground laboratory" rather than a "mined geologic repository" or a radioactive "waste" repository. In a February 1998 report from Prime Minister Jospin's office, it is emphasized that the law of December 30, 1991, required a decision by the year 2006 on the disposition of high-level nuclear waste, that it was by no means clear that transmutation was an industrially feasible option, that the underground laboratories would certainly not receive any waste until the decision was made, and that the option of surface or near-surface storage for high-level waste, including both vitrified fission products and unreprocessed spent fuel, was not receiving enough attention to be a contender in the race.[12] Accordingly, the surface/near-surface engineered storage research would receive a considerable increase of funds.

A fast reactor might be an option for the management of nuclear waste, in competition with other approaches such as the use of powerful accelerators to produce neutrons for this purpose, or the option of direct disposal of the waste. In all these applications, cost is an important consideration; thus far, fast-neutron reactors have been considerably more expensive than the usual light-water type.

Other Wastes

In addition to the vitrified fission product waste from reprocessing of spent fuel, other wastes are generated. For instance, the metallic ends and chopped sheaths of the fuel rods are imbedded in concrete for disposal, while the weakly radioactive and short-lived wastes are sealed in metal barrels and preserved at an aboveground site under surveillance. The management of even this short-lived or slightly radioactive waste poses problems that are far from trivial in the political and practical sense. A recent example is found in the report of a commission charged by the French government to evaluate the waste storage situation at La Manche storage center, located next to the COGEMA plant at La Hague.[13] It is the first French center for the surface storage of moderately and slightly radioactive waste, and between 1969 and 1996, nearly 500,000 cubic meters of residue left by the civilian and military nuclear industries were piled up there. These wastes contain long-lived radioelements such as plutonium, but also toxic heavy metals such as lead, whose toxicity does not decrease with time.

According to the original plan, surveillance of the site was scheduled to terminate in 2294. Assuming that most of the radioactivity of the waste would have disappeared in three hundred years, the shallow burial of radioactive packages under a waterproof covering capable of protecting them for three centuries was undertaken in 1991. After 300 years, the site would no longer be regulated. But the commission decided otherwise, concluding that the site would retain its regulated character even after this lapse of time, a decision the government immediately accepted. At the request of environmentalists, a court order had already been obtained in 1995 to discontinue burial.

What happened then? The commission considered that the center did not present a significant risk of polluting the environment with radioelements. Moreover, chronic pollution observed due to tritium and radon is less by a factor of 10 than that allowed by the most severe standards, and in the short term the present covering system appeared acceptable. But the commission also noted a danger associated with possible intrusion of people on the portions of the site with high concentration of radioactivity; the original estimates were based on a uniform distribution. In the present state of affairs, the commission proposes, during a period of five to fifty years, to prepare a more permanent covering, which would cost $50 million. After that, reduced surveillance would be allowed.

CURRENT POLLUTION IN NUCLEAR CENTERS

The problems at La Manche are very small compared to those at the U.S. site of Hanford, Washington. Hanford is a nuclear complex of 1500 km^2 created in 1943 for the manufacture of plutonium for bombs. Its design and management suffered from the normal lack of experience in a newborn technology, as well as the haste to get it built to provide nuclear weapons for use against the Nazis. Portions of the site are highly contaminated with radioactivity. The fission products that accompanied the plutonium created at Hanford are in a mudlike sludge stored in steel tanks (either single- or double-walled; some already leak), and a small fraction is in shallow trenches. However, Hanford is in an arid region, so that the radioactivity that penetrates into the soil descends slowly, instead of being carried to neighboring regions to reemerge with the surface water. It is estimated that it will take fifty years to clean up the site, at a cost of some $50 billion. The situation is far worse in Russia, in the secret centers where Hanford's Soviet counterparts forged their weapons.

By virtue of its extensive experience in reprocessing and vitrification, COGEMA will be able to profit from Hanford's misfortune to the tune of $300 million. Between now and 2028, Hanford's liquid waste will have to be vitrified. The United States has now implemented vitrification on a massive scale at the former plutonium production plant, at Savannah River, South Carolina, and plans to do so also at Hanford. British Nuclear Fuels Limited — BNFL — was also hired by the Energy Department sites — e.g., the gaseous diffusion plant at Oak Ridge that has been used since 1944 for enriching uranium. But there are problems. In late 1999, BNFL admitted falsifying data regarding MOX fuel that had been shipped to nuclear reactors in Japan, and that called into question BNFL's contracts with the Department of Energy. In spring 2000, BNFL's chief executive resigned, and there were discussions in the press of possible bankruptcy. Some Department of Energy contracts with BNFL have now been canceled.

The United States is not without experience in vitrification. For instance, a site at West Valley, New York, filled its two hundredth radioactive glass canister in May 1998. The West Valley Demonstration Project employed about 850 people to fill with vitrified waste some 260 to 280 such canisters that weigh some 2.5 tons each — much larger than those used in France. The waste was generated by an early attempt to reprocess spent fuel from commercial power reactors.

Altogether, eliminating the pollution at American defense nuclear sites is expected to cost $230 billion. The French nuclear site at Marcoule is also polluted, with the cleanup cost estimated at $6 billion. Vitrification of defense wastes started in France, whereas it is only now about to begin at Hanford. In part the lower cost at Marcoule arises from the fact that far less plutonium was produced there than at Hanford, and more recently—so that tanks have had less time to corrode.

The nuclear industry is emerging from a childhood that was all the more difficult because of its military relatives, and its future development must be completely open, which has not been the case in any of the world's arsenals. Openness will be a necessity for the continued existence and growth of nuclear power. Fortunately, present knowledge far exceeds the modest amount that was known at the birth of the industry.

The industry is regulating itself more strictly. In contrast to La Manche, a second, more recent French storage site, at Soulaines, in view of its higher standards does not seem to have elicited criticism. This points up both the need for and the difficulty of establishing a new environmental culture, capable of taking responsibility for the effects of our present actions into the distant future, a future farther removed—maybe ten or a hundred times more so—than is the current century from the construction of the Great Pyramid at Giza, 4600 years ago. Who would be willing today to pay local taxes intended for the maintenance of a dangerous waste site left by the ancient Egyptians? Our descendants will need to be concerned about our nuclear waste sites—at least to the extent of not entering the ones they have inherited from us.

DIRECT DISPOSAL VS. REPROCESSING

For an engineer, a ton of plutonium represents one gigawatt of electricity for a year; for an economist, a ton of plutonium has a negative value of $25 million (this is the additional expenditure required to enable a ton of plutonium in 20 tons of MOX fuel to be sold at the same price as 20 tons of enriched uranium fuel of the same energy value in a light-water reactor); for Saddam Hussein, it can make 200 nuclear bombs. For the industrialist responsible for the management of spent fuel, the cost of reprocessing and final burial depends on complex industrial processes, but also on the approach selected, which may vary considerably from one country to another. These disparate values of energy, cost, and potential nuclear-weapon use of the material underlie the choices and the arguments over reprocessing and recycling spent fuel from commercial power reactors—together with the real or perceived health impact of normal reprocessing activities or of accidents. The choice of direct burial without reprocessing is motivated by an effort to minimize cost and to avoid having separated plutonium, unprotected by radioactive (or "radiotoxic") fission products, stored in countries where it might be diverted or stolen to make nuclear weapons.

Energy and Cost

In support of their decision to reprocess spent fuel, the French point out that uranium and plutonium are a source of energy and can be recycled; in a ton of irradiated fuel, before reprocessing, there are nearly 10 kg of uranium-235, as much plutonium, 32 kg of fission products, and 7 kg of minor actinides. As we have discussed, about 21 tons of spent fuel are removed, on the average, from each reactor every year; the annual total in France is about 1200 tons. The French reproach the Americans for their legendary wastefulness. They say disparagingly that Americans are ready to junk their automobiles when the ashtrays are full. They claim it is more reasonable to recover the fissionable materials and separate the dangerous products to reduce the volume to be stored.

French industrialists emphasize that reprocessing-recycling exists and is operational. Its costs are known, they say, whereas direct burial is still at the planning stage and its cost is only estimated and, therefore, not guaranteed; it is well known that the costs of large projects grow inevitably between the first estimates and the final realization.

The U.S. approach to waste disposal appears cheaper, in view of the fact that reprocessing is an additional step before disposal and that the amount of space needed in the repository is determined primarily by the heat produced by the fission products and the need not to overheat the rock. The advocates of direct disposal note that the French have yet to commit even a single ton of their thousands of tons of packaged vitrified waste to any permanent disposal site, so that the cost of reprocessing-recycling is no better known than that of direct disposal.

The Weapons Connection

Following the lead of his predecessor, Gerald Ford, President Jimmy Carter in 1977 issued a directive forbidding reprocessing of civilian power reactor fuel in the United States and attempting to lead other nations to the same goal, primarily to avoid the contribution that separated plutonium could make to proliferation of nuclear weapons. At the time, it was clear that direct disposal of spent fuel was less costly than reprocessing fuel and recycling the plutonium and uranium. Nevertheless, the costs of the two methods are still being disputed by their partisans. A 1994 study of the Nuclear Energy Agency of the Organization for Economic Cooperation and Development claims no significant difference between the two options, but the study uses for direct disposal the very high cost per unit of spent fuel estimated for the small Swedish program, rather than those for the massive U.S. activity. Nevertheless, it estimates the reprocessing and recycling approach to be about 0.06 cent per kWh more costly than the direct disposal of spent fuel. This is a small fraction of the average price paid for electrical energy in the United States—some 6 cents per kWh—but that same report indicates that 0.01 cent per kWh is about $1 million per year per reactor. For the hundred or so reactors in the United States, 0.06 cent per kWh additional cost would total some $600 million per year. Other estimates that involve the construction of new reprocessing facilities, such as the one being built at Rokkasho-Mura, Japan, result in much higher costs for reprocessing—some 0.5 cent per kWh.

United States power reactor operators have been paying 0.1 cent per kWh into a special fund for disposal of spent fuel, amounting to some $8 billion by the end of 1998. However, in 1998 the Department of Energy, which has the disposal responsibility for spent fuel under this contract, claimed that events beyond its control had delayed the availability of the spent fuel repository at Yucca Mountain. The American reactor operators maintain that it is the government's responsibility to take the spent fuel in any case, even if it needs to be held in interim aboveground storage. At-reactor spent-fuel cooling pools are reaching their limit of fuel holdings. In June 1999, the Department of Energy announced that it would take title to the spent fuel at the reactors, thus easing tensions between the department and the reactor operators. The opposition to the Yucca Mountain repository would probably not have been reduced had there been fission products in glass logs to be disposed of there rather than intact spent fuel.

If the American companies producing electricity today had to convert to reprocessing and bear the capital cost of construction of the necessary industrial complexes, they would be spending about 0.65 cent more per kWh than the 0.1 cent per kWh they pay today, to get rid of the spent fuel, which disposal is to be managed, not by themselves, but by the Department of Energy.[14] With an average cost of electricity of some 5.9 cents per kWh (e.g., in 1997, 8.5 cents/kWh for residential customers, 7.6 for commercial, and 4.6 for industrial), this would increase the cost of electricity by about 9%, and it would not ease significantly the problem of disposal of nuclear waste.

In France, the cost of producing a nuclear kWh is about 5 cents. Electricity is sold to the public for approximately 11 cents/kWh, plus 30% tax. The margin of price over cost is much more than in the United States. Nuclear power generation in France is a monopoly of Electricité de France. This extra price, paid in part for the improved handling of nuclear waste, is accepted by the majority of the population, which, however, has had no choice in the matter.

In creating the breeder program that led to the industrial scale prototype, Superphénix, the French were betting on a major expansion of nuclear power worldwide, on the likelihood that uranium was much less abundant than it has turned out to be, and hence on the early skyrocketing of the cost of uranium fuel. The unexpected drop in the price of oil and uranium has made reprocessing for plutonium recycling an expensive route for the short term and has put off by at least fifty years its possible economic profitability.

A report on the economic outlook for nuclear power, requested by Lionel Jospin, prime minister of France, was published in September 2000 by Jean-Michel Charpin, Benjamin Dessus, and René Pellat.[15] For the first time, the world public and the nuclear industry have an open and unbiased comparison of France's choice of reprocessing and recycle with the choice of direct disposal of spent nuclear fuel. The three authors of the report are, respectively, Commissioner for Planning, Director of the Ecodev (energy and development) program at the CNRS, and High Commissioner for Atomic Energy — all of the French administration. The report assumes extension of the life of existing French reactors to forty years, so that major replacement sources of electrical energy need not come online until 2025. It considers future options of high demand for electricity, low demand, elimination of nuclear power, reprocessing of all fuel for recycle in nuclear plants, and the once-through (direct disposal) approach. This report determines for each program choice the timing of activities and expenditures required to deliver the required electrical energy (kWh).

In comparison with reprocessing and recycle, direct disposal would have saved France some 164 billion French francs (more than $20 billion). Of course it was not guaranteed, at the start of the French nuclear program, that the price of uranium would fall to the current $25 per kg, or that the price of crude oil would fall to some $10/bbl before its rise in 2000 to over $30/bbl, so the extra cost could be considered as the premium on an insurance policy. The question then arises, "Insurance against what?," and the report provides some tentative answers. For instance, if the $20 billion extra cost is divided by the amount of plutonium that need not be entombed (because it was recycled and consumed in the reactors), the cost per ton of plutonium disposal avoided is about 1.2 billion French francs, about $150 million per ton, or $150,000 per kg of Pu, in comparison with the cost of U-235 of some $24,000 per kg.

Direct disposal would have required some 38,000 tons more of natural uranium than would the option of recycle; allocating all of the $20 billion extra cost of recycle to the saving of this uranium would value natural uranium at $526 per kg—some twenty times its current price. (This is to be compared to the estimate in this book, p. 137, of $700 spent per kg of natural uranium saved.) French reactor operators, industry, and consumers—as well as those in Japan, which has not yet begun to recycle plutonium on a large scale—will have much to consider in this report.

Should reprocessing be abandoned because the breeder program has been closed? Major reprocessing plants in France and Britain were largely built with contracts from Japan for the reprocessing of 7120 tons of spent fuel and from Germany for the reprocessing of 5615 tons. Whether the major plants were economically justifiable at the time they were planned is a different question from whether expanded reprocessing facilities can pay their way. If the French were to adopt the direct disposal strategy, they would have to put the containers in coffins 500 meters underground, with the standard measures (e.g., neutron absorbers such as boron in steel) to avoid creating a critical assembly.

Although there is a minor advantage in terms of volume of material stored—vitrified fission products from reprocessing occupy about half the volume of the unvitrified waste—advocates of reprocessing argue that recovery of the uranium-235 and plutonium fuels from the waste has long-term benefits for the economics of nuclear energy, even if the immediate effect on the price is slightly adverse. The same goes for automobiles with better gas mileage, they say, beneficial not only because of pollution problems but also for the long-term management of oil resources. One day the price of oil will go sky-high and the advantage will become obvious. One day the price of uranium will also climb; when it is sufficiently high, then it will be economical, they argue, to use breeder reactors, to reprocess light-water reactor fuel, and to recycle the plutonium and the uranium.

But would it be cheaper then to extract uranium from low-grade ores or even from seawater? While it is clearly of no interest to industry right now to learn how to extract uranium from seawater or from low-grade ore, it is exceedingly important for society to establish now that this technology is available for the future. We have already sketched an estimate of the "breakeven price" for raw uranium ($700 per kg, compared with $30–40 per kg current price) that will make recycling competitive; in Chapter 8 we shall return to what is known of the cost of uranium from seawater.

LONG-TERM SAFETY

As was mentioned in the previous chapter, choosing a direct disposal strategy implies taking measures to ensure that the nuclear waste will not reenter the biosphere because of earthquakes, volcanism, meteorite impact, or human intrusion. This argument is made in France, in support of the strategy of reprocessing waste to eliminate plutonium.

In a sense, it is necessary to compare the danger of something that is possible and cannot be controlled, but is unlikely to occur even over thousands of years, with the danger due to possible accidents or sabotage over the decades, centuries, or millennia, while reprocessing is taking place, and where radioactive materials once they have been separated will be much closer to the biosphere. These two risks can be analyzed and appropriate action taken in the face of uncertainty. What is needed is probabilistic risk assessment, as has been instituted for possible accidents to operating reactors, but not yet applied to the back end of the fuel cycle. But the risks of entombment of spent fuel and those of entombment of reprocessing waste are not so different.

Radiotoxicity

Evaluating the hazard posed by the radioactivity of the various components of nuclear waste is a complex matter. The reduction of radiotoxicity of the waste, accomplished by reprocessing, is often used to justify the French reprocessing program. However, it is important to consider the specific routes by which radioactive materials can reach individual humans rather than considering simply the effective whole-body exposure in sieverts per gram of nuclear waste if actually ingested (which is the implication of "radiotoxicity"). Studies have been conducted and more are now under way to estimate the extent of possible migration of the radioactive elements far from their initial storage places, carried along by underground water, for example. Unfortunately, the mechanisms depend strongly on the nature of the underground environment—whether it is acidic or basic, oxidizing or reducing. Further, the long-term danger of the waste varies considerably according to its chemical composition—e.g., whether it is in the form of a ceramic or else vitrified as glass—as well as the nature of the environment.

For about one thousand years, the radioactivity of fission products dominates both the radiotoxicity of nuclear waste and the potential for transfer to the biosphere. The vitrification of waste after removal of 99.9% of the plutonium, which is the procedure used in France, is claimed to reduce by a factor of 20 the radiotoxicity transferred to the biosphere by water intrusion into the buried waste—again, not including the disposition of the spent MOX fuel itself.

The "Relative Dose Index" is a more comprehensive measure of hazard to people from radioactive waste than is radiotoxicity. It differs from simple radiotoxicity by including the important factor of the rate at which the material dissolves in the ground water for transport to the biosphere. In this regard, recent studies show that americium (element 95, mass number 241, half-life 432 years, created by beta decay of Pu-241 before reprocessing) dominates the radioactive hazard of buried waste for 4000 years because it is not removed from the waste before vitrification.[16] For direct disposal of spent fuel, or vitrified fission products, it is important to note that the hazard posed by plutonium is far less than that of the soluble fission product technetium-99, of 212,000-year half-life, because of the very low solubility of plutonium in groundwater. Thus there is little benefit to safety in removing plutonium from the material before it is buried. As we have seen in the Oklo reactor, plutonium stays where it is emplaced—at least in Gabon.

Reprocessing and Long-Lived Wastes

Reprocessing can allow the volume of long-lived radioactive wastes to be reduced by a factor of 4, while adding additional volumes of intermediate and low-level waste. The ultimate goal of some who favor reprocessing is to separate the long-lived high-level wastes and then transmute them with fast neutrons. The purpose is to produce electricity while totally consuming the plutonium and all other long-lived radioelements. Although Superphénix is dead (at least temporarily—remember the origin of its name!), "separation and transmutation" have some firm supporters in the technical community the world over.

Fig. 5.9 shows four curves for the radiotoxicity of about one-eighth of a reactor-year's output of spent fuel. This is the dose that would be given to a population ingesting the entire twenty to twenty-five tons of highly radioactive spent fuel. Obviously, this will not happen, and the curves indicate the relative risk if a small fraction of the material is ingested—the same fraction for each of the scenarios. Reprocessing as practiced now by COGEMA and BNFL is shown as R1, with radiotoxicity varying between one-half and one-tenth that of the unreprocessed fuel. Of course, the material removed from the fuel to be buried is present in the fuel cycle until it has been destroyed or, in turn, buried. If the reprocessing is augmented by removal of americium, neptunium, and curium to the same 99.9% level as is done now with plutonium, the curve R3 results—approximately one hundred times lower than state-of-the-art reprocessing, R1. This is about the same result that would be expected from separation and transmutation, and Fig. 5.9 shows that this strategy would lead, after a thousand years, to a degree of radiotoxicity equal to that of the initial natural uranium.

To advocates of burial without processing and those who propose maximum processing, the curves seem to argue in favor of reprocessing. After a

Evolution of radiotoxicity

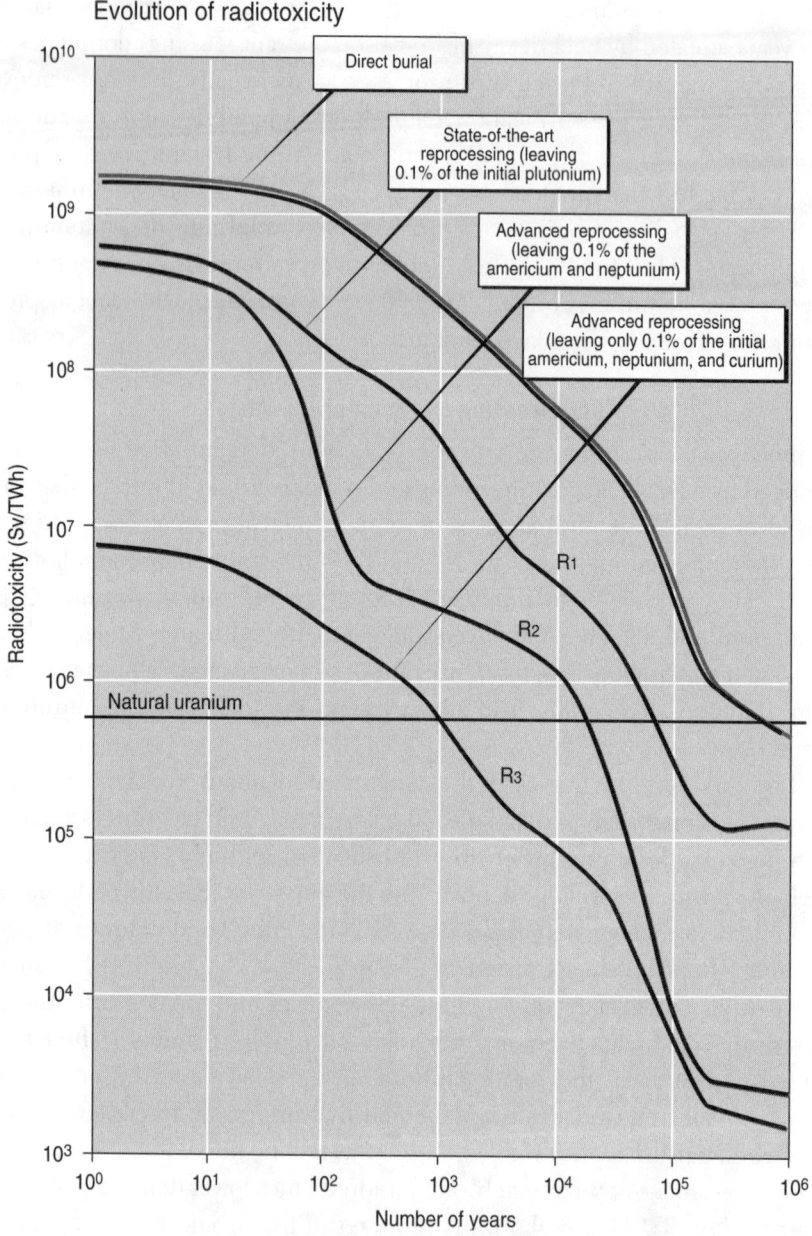

Fig. 5.9. Evolution of radiotoxicity with time for nuclear wastes after several types of treatment.

thousand years, the radiotoxicity of the directly buried waste would be nearly a thousand times greater than that of the reprocessed waste, and it would take nearly a million years to arrive at the level of natural uranium. However, the most recent comprehensive analysis of separation and transmutation published by the National Research Council—the research arm of the National Academy of Sciences complex—concludes that separation and transmutation does not have sufficient merit to abandon the direct disposal of spent fuel.[17] The report also emphasizes the loss to waste of some of the radioactive material in the many reprocessing operations that are required for effective transmutation—a problem not considered by some of the advocates of separation and transmutation.

It takes between fifty and a hundred years to process and dispose of the waste produced by a reactor. At the end of its service life of thirty to forty years, it is planned that each reactor in operation will be shut down, will be defueled on a schedule for which there is technical and economic justification, and will be decontaminated and decommissioned, the site being then available for other uses. During the operating life of American reactors, funds are set aside for this purpose.

An Advanced Reprocessing Scenario

The advanced reprocessing scenario (R3 in Fig. 5.9), which would massively reduce the radiotoxicity, would require an extensive installation of fast-neutron reactors as well as specialized installations for the reprocessing and manufacture of fuel—all at substantial cost, in order to begin to reduce the long-lived minor actinide radiotoxicity. It is far from certain that the transformation of the breeder into an instrument destined to consume plutonium and fission products would be a sufficiently convincing element to allow a number of them to be constructed in the present European climate; all the more so, since it may be possible to adopt a nuclear chain reaction based on thorium rather than uranium, which would avoid the production of most of the long-lived heavy metals under operating conditions that promise to exclude any possibility of an uncontrolled chain reaction.

Nevertheless, spent fuel must be managed, and there is wide agreement that the safest place for disposal of spent fuel is a suitable mined geologic repository, whether the fuel has been reprocessed or not. The Pangea proposal would accept vitrified wastes from reprocessing or unreprocessed spent fuel—equally for disposal in a stable mined geologic repository. This is an imaginative commercial activity that we favor, both from the point of view of the nuclear reactor operators and for the protection of the biosphere. We are hopeful that the U.S. Nuclear Energy Research Initiative, begun in early 1999, and European activities in fission will provide choices for nuclear power that are at once cheaper, safer, and more reliable than existing plants; but in the meantime plants of existing designs, properly operated, have an important role in satisfying human needs.

CHAPTER 6

A Glimpse of the Future of Nuclear Power

THE METHODS CHOSEN for the production of energy for the next fifty years will not be determined by the needs of the inhabitants of our planet a thousand years hence, let alone ten millennia in the future. A crucial period will open up in the century's second decade (twenty years in France), when many reactors will reach the end of their lives and will have to be replaced. It is unlikely that nuclear energy will be abandoned worldwide, but it is possible that new strategies for nuclear power will be adopted in the longer term. We first consider in some detail an approach based on the coupling of a very-high-intensity proton accelerator with a nuclear reactor, to gain a perspective on variants of the breeder reactor. We close with a discussion of two approaches to nuclear power using tiny self-contained fuel pellets, with helium coolant driving a gas turbine.

THE "ENERGY AMPLIFIER"

Physicist Carlo Rubbia, former director of the European particle-physics research establishment in Geneva, Switzerland (CERN), received a Nobel Prize for his discoveries in particle physics and for his role in introducing a particularly audacious type of particle accelerator. Recently Rubbia has been leading a project at CERN to develop an accelerator-driven reactor, with collaborators in many European institutions, and supported since May 1998 by the European Community.

In the heaviest nuclei in nature, such as lead, gold, and uranium, protons and neutrons are less strongly bound together than they are in nuclei of average size, such as iron; for this reason, the mass of the heavy nucleus is greater than the sum of the masses of its middleweight counterparts. Breaking up heavy nuclei will accordingly yield more energy ($E = Mc^2$) than is required to provoke the breakup if an effective means of disruption can be found. Thanks

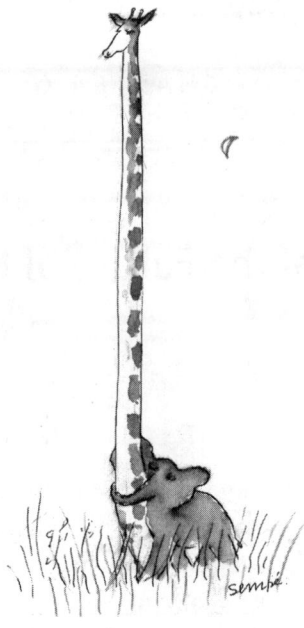

Fig. 6.1. Linking an accelerator with a nuclear reactor: an unnatural coupling or the future of nuclear energy?

to neutron-induced fission, this can sometimes be done at modest cost; a neutron, even at rest, can slip into certain heavy nuclei and make them break apart, allowing not only the harvest of the energy difference evident in the deficit in mass of the products of the nuclear reaction with respect to the initial mass, but also the exploitation of the multiplicity of fission neutrons to provide a self-sustaining chain reaction. But only a few types of nuclei can be fissioned with slow neutrons, and aside from uranium-235, they are rare or nonexistent in nature; like plutonium-239 or uranium-233, they must first be produced artificially.

The Rubbia system is to burn thorium, whose terrestrial reserves are three times more abundant than those of uranium. It presents considerable advantages concerning radioactivity of the waste. But thorium is not fissionable with slow neutrons, nor to a significant extent with fast neutrons. It is, however, *fertile*, in that neutron capture converts the thorium-232 (which constitutes 100% of natural thorium) into thorium-233, which (like uranium-239 undergoing two successive beta decays to produce plutonium-239) produces uranium-233, an excellent fissile isotope. Uranium-233 has characteristics similar to uranium-235 and fissions readily with thermal neutrons, yielding about the same energy release. However, it is more difficult to make a fast-neutron breeder with the thorium-232/uranium-233 cycle than with the uranium-238/plutonium-239 approach. There are barely enough neutrons liberated per fission in uranium-233 to use one to cause further fission and another to be absorbed in thorium-232, while losing an occasional neutron to capture in the steel structure of the reactor, or to nonfission absorption in uranium-233.

The approach, then, is to have a reactor core that is frankly subcritical (with a reproduction factor k of 0.95–0.98) so that it has an amplification of externally injected neutrons by a factor $1/(1-k)$, which is a factor of 20 or 50 in the two cases. Thus, 1/20 or 1/50 of the neutrons causing fission in the energy amplifier at full power are to be supplied from an external source—a particle accelerator. If the Th-232/U-233 breeder can effectively be achieved, only 70 tons of ore per year would be required (for a common ore containing 10% thorium). This compares with 100,000 tons of uranium ore per year (for ore containing a typical 0.2% uranium) to feed the light-water uranium reactor that uses uranium-235, which represents only a fraction of 0.71% of natural uranium. Of course, a U-238/Pu-239 breeder would need only about 500 tons of 0.2% uranium ore per year—not so different from the requirement for thorium ore.

Nuclear physicists know how to break up all heavy nuclei. All that is needed is to strike them with sufficiently accelerated particles so that the binding forces of the protons and neutrons will be negligible with respect to the energy of the projectiles: this phenomenon is called nuclear "spallation." It is a way of obtaining neutrons from electrical energy (that used for acceleration), but the neutrons are costly; one must pay not only the cost of the energy, but also the capital investment for the accelerator, irradiation facility, and cooling system to remove the energy deposited as heat by the particle beam.

In 1949, in response to the concern that there was inadequate uranium for a burgeoning American nuclear weapons program, a substantial development project (code-named the Materials Testing Accelerator, MTA) was created to use accelerated protons or deuterons to produce plutonium—from uranium that might already have been depleted of its uranium-235 by having served in a reactor for producing plutonium for nuclear weapons. The success of geologists and entrepreneurs in discovering domestic uranium resources put an end to the MTA program. For a long time there was no serious prospect of using this approach to produce energy. Because of accelerator inefficiency, the energy needed to accelerate the particles was so much greater than that recoverable from the favorable balance of masses that it seemed to present no practical interest. Spallation has also been considered for the transformation of the most noxious radioactive species into species of shorter half-life, but this approach was judged uneconomical and unnecessary in the 1996 "Separations and Transmutation" report of the National Research Council, to which we have already referred.[1]

Accelerator technology has progressed since 1949, when plutonium production via the accelerator route was first considered. Technical advances now allow the conversion of 50% of the energy invested into that of the accelerated particles. This results from the physicists' need of more and more intense particle beams for their experiments, which, therefore, consume more and more electrical energy. Passing from the 1% efficiency with which they were previously satisfied to 50% leads to an enormous reduction in the operating costs of their machine and in the required investment. More important for the present discussion, coupling the accelerator to a fission reactor that is not quite critical would allow the creation of a device that produces energy in sufficient quantity to supply the power necessary for the operation of the accelerator, using only 10% of the energy produced. These advances are analogous to those that were necessary to make practical aviation jet engines, in which the efficiency of compressors and turbines was initially so low that the engine was not able to keep itself running, let alone allow energy to be extracted for propulsion.

The goal of the Rubbia team is to produce energy at a competitive price, to keep the fissionable materials at a subcritical level (that is to say, at a level at which there can in no case be a runaway chain reaction as at Chernobyl), and to come out, at the end of the cycle, with far smaller quantities of long-lived radioactive waste than in a classical reactor. As with any fast-neutron reactor, Rubbia's approach allows the consumption of plutonium and the actinides produced in the light-water reactors that require disposal in one way or another. The configuration of the fuel and the cooling system remain similar to those of a conventional reactor. The coupled system has been dubbed an "energy amplifier"—a term we retain not because it is apt, but because Rubbia continues to use it. This nuclear system is largely passive, coupled to an accelerator that is not potentially dangerous. That is enough to draw the attention of the nuclear power industry to the virtues of this approach.

The energy amplifier is a mutual fertilization of independent technologies: particle accelerators like those used at CERN in Geneva, or Fermilab in Illinois, fast-neutron reactors like the now-closed Superphénix, and the reprocessing of spent fuel analogous to that done by COGEMA in France.

DESIGN AND OPERATION OF A PROTON-BEAM-DRIVEN SUBCRITICAL REACTOR

In the first stage, molten lead is bombarded with protons at 1000 MeV, which is five hundred times the average energy of a fission neutron. The lead nucleus splits into several nuclei and very fast neutrons, which, in turn, provoke nuclear reactions in the lead. At the end of the series of reactions provoked by a single proton of 1000 MeV energy, about thirty 3-MeV neutrons remain. These neutrons then bombard the surrounding core made of fissionable material, producing fissions that release energy. It will not be necessary to have a critical assembly, in which each neutron released can generate another neutron available to continue the reaction, because the spallation supplies the additional neutrons necessary in the case of a subcritical assembly—typically, 2% of the neutrons needed.

A large assembly of uranium-235, plutonium-239, or uranium-233 would be supercritical and need no accelerator; it would be an ordinary reactor, with fissions caused by fast neutrons or slow—respectively without or with a moderator. When thorium is added as fertile material, so that the fissile uranium-233 constitutes only about 10% of the core and thorium 90%, an injected set of neutrons causes a long chain of fissions (100 neutrons are replaced on average by 98, 98 by 96, 96 by 94 . . .) before it dies away. It is this 2% deficit for a fast-neutron U-233/Th-232 core that is supplied by neutrons from lead spallation induced by the deuteron beam.

Under these circumstances, thorium resources are sufficient for thousands of years, since all of the thorium can be used—in contrast to the consumption in light-water reactors of only a fraction of the 0.71% of natural uranium that is uranium-235. Specifically, because of the large absorption of thermal neutrons in the water coolant of an LWR, and in the fission products, only about 80% of the U-235 present in its core load can be consumed. But this is partially compensated by the fission of the Pu-239 produced by capture on U-238; for a ton of U-235 loaded into a reactor core (as 4% U-235 and 96% U-238), about one ton of fission products are created.

The Rubbia reactor design specifies a core made of 27 tons of mixed thorium-232 and uranium-233, but stocks of uranium-233 are insufficient to provide the 2 to 3 tons that the design specifies for the core. The reactor would accordingly start by using thorium mixed with more conventional fissionable materials such as uranium-235 or plutonium-239, which become excess and surplus as the number of U.S. and Russian nuclear weapons is reduced. The energy amplifier's spent fuel might require a cooling-down period of only a few months before reprocessing to remove the fission products and adding thorium to make new fuel elements. The American breeder programs at the Argonne National Laboratory in Illinois and at the Idaho National Engineering and Environmental Laboratory have studied a suitable separation approach, dubbed "pyroprocessing," in which electrolysis in molten salt is used rather than chemical reactions in solution to extract the fission products from the heavy metal that is eventually to be returned to the reactor. Such approaches have much merit and would be considered in conjunction with the energy amplifier. Pyroprocessing can handle fuel much more radioactive than can be used with the organic solvents employed in processing light-water reactor fuels. The reprocessing does not need to reach the decontamination factor of one in 10 million or 100 million (the fraction of the fission-product radioactivity that remains with the recovered uranium or plutonium in the product of La Hague or Sellafield), because it is envisioned to refabricate the fuel by a totally automated process. In a fast reactor, fission products that accumulate in the core do not absorb a significant fraction of the neutrons in the system, and that would be true also for a normal U-238/Pu-239 fast-neutron breeder. Removing all but 0.1% of the fission products would be adequate for Pu recycle into light-water reactors, but would require automated fabrication and handling of the MOX fuel—which we advocate in any case.

Leaving a significant fraction of the fission products with the reprocessed metal to be fabricated into a new core is not only an economy in reprocessing, it also averts the production of separated weapon-usable fissile material largely free of the intense gamma radiation from fission products that constitutes a barrier to theft or diversion. These virtues of tolerance of fission products, ability to transmute any of the heavy elements, and the barrier to theft constituted by substantial amounts of fission products remaining in the fuel for recycle are common to all fast-neutron reactors.

The fissionable core of the energy amplifier is to be immersed in 10,000 tons of molten lead, contained in a well 30 meters deep and 6 meters in diameter. The lead serves to cool the core. It has the same desirable nuclear properties as the sodium chosen to cool the core of Superphénix: it does not absorb neutrons, and has the further advantage of not reacting rapidly in contact with air or water. It circulates at 560°C and rises from the core toward the top of the vessel because of its thermal expansion (i.e., lower density for hotter lead), without the need for pumps, and supplies heat that will be extracted to power the turbines and drive the electric generators. When it cools, the lead returns to the bottom of the vessel, under the core. With all of these attractions, why was lead not used as coolant in Superphénix? In part because it must be heated to a considerably higher temperature to melt, in comparison with sodium, but also because lead is more corrosive toward steel structures and other materials in the core. Substantial research is needed to determine whether this problem can be overcome.

The core structure of the energy amplifier is designed so that its criticality with prompt neutrons is 0.98. An ordinary fast-neutron reactor operates with criticality of 0.998—very close to 1—and it is with the delayed neutrons, which make up only 0.2% of the total of the neutrons emitted in plutonium, that the designer must play to approach the fateful value 1. The situation of the energy amplifier is, in a sense, ten times easier to live with (2% instead of 0.2%).

It is a simple matter to turn off the accelerator at short notice. Moreover, in case of excessive heating of the core, the expansion of the lead can be arranged to cause it to flow into the proton injection tube, thus stopping the protons from penetrating into the reactor. In an equally passive manner, overheating causes the sudden rise into the core of boron carbide rods, which absorb neutrons and stop any runaway that might occur.

A major feature of the project is illustrated in Fig. 6.2, which shows the radiotoxicity of the waste of such a reactor as a function of time. It is seen that after five hundred years the radioactivity of the waste is less than that of coal ash (coal always contains a bit of uranium and thorium). This would represent, therefore, considerable progress with respect to the radioactive waste from light-water-moderated reactors, also shown in the figure.

A French Atomic Energy Commission document compares the radioactive residues of thorium/uranium-233 with a standard uranium/plutonium fast-neutron reactor and with various types of light-water-moderated reactor.[2] Although after a thousand years the thorium nuclear waste toxicity is 300 times less than that of a normal fast-neutron reactor, after 100,000 years the two types of waste have equivalent toxicity—comparable with that of natural uranium. Nevertheless, in an economy stably based on nuclear power, the combined toxicity of all materials is dominated by those that are in the reactors, undergoing reprocessing, or waiting to be processed. Furthermore, these materials still in the fuel cycle can penetrate the biosphere more easily than those that are buried, so the benefit of reduced radiotoxicity of waste may be more apparent than real.

With 27 tons of fuel for the net production of 600 MWe of electric power, the energy amplifier can operate for five years. Then it might be allowed to cool down for some days before moving the old core to a corner of the lead-filled pit, where it could cool for a year while the fresh core is exposed to the proton beam. In five years of operation, only 12% of the initial thorium will have been transformed to U-233, while an equal amount of U-233 will have been fissioned. At the end of five years, the core will therefore still contain about three tons of uranium-233 produced by neutron capture in the thorium-232 followed by successive radioactive disintegrations.

The concept is attractive because the design relies on familiar technologies, although from two disparate fields. Because of the somewhat reduced power level envisaged in this design (600 MWe vs. 1200 Mwe from a typical light-water reactor), a hundred reactors of this type would be needed to supply electricity equivalent to that produced by the fifty-six reactors operating in France.

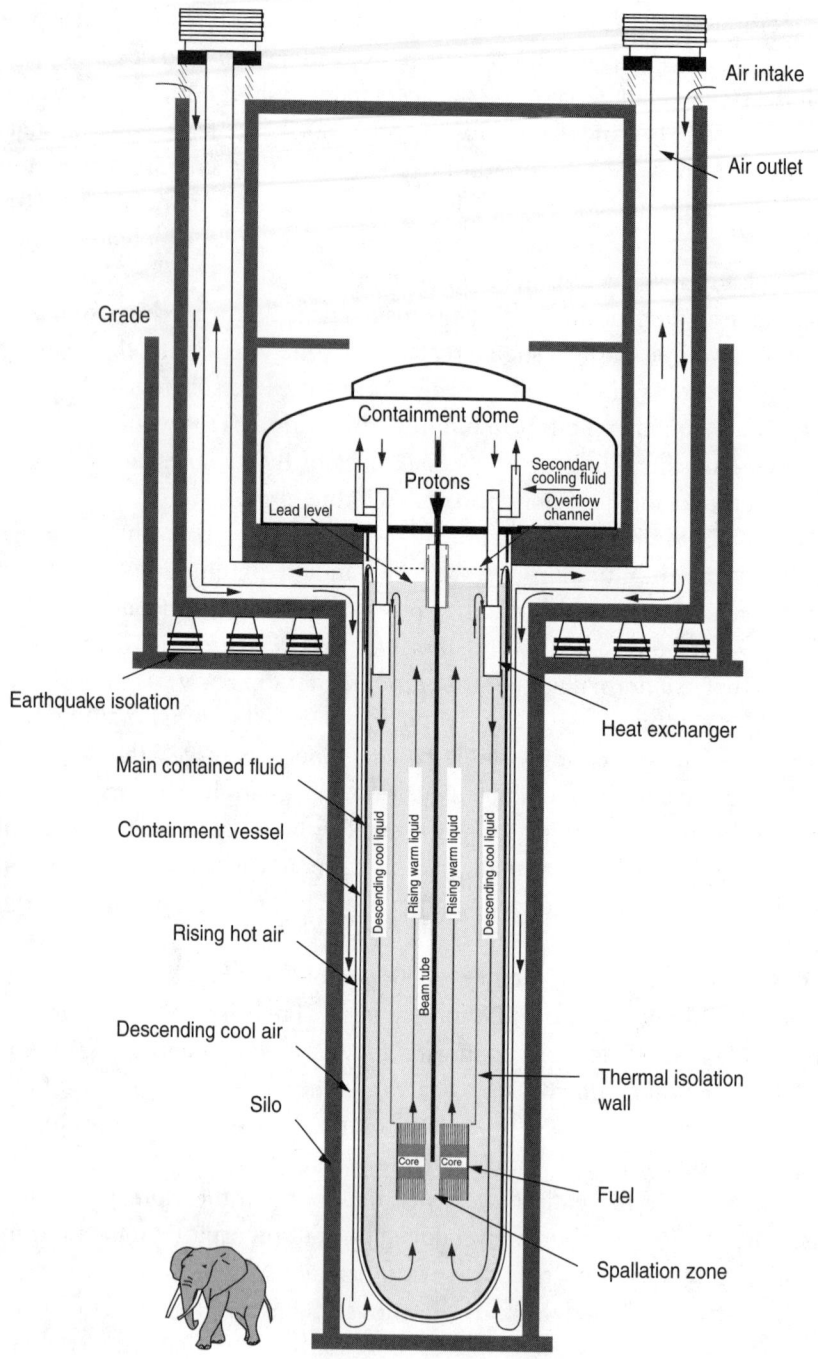

Fig. 6.2. An energy amplifier.

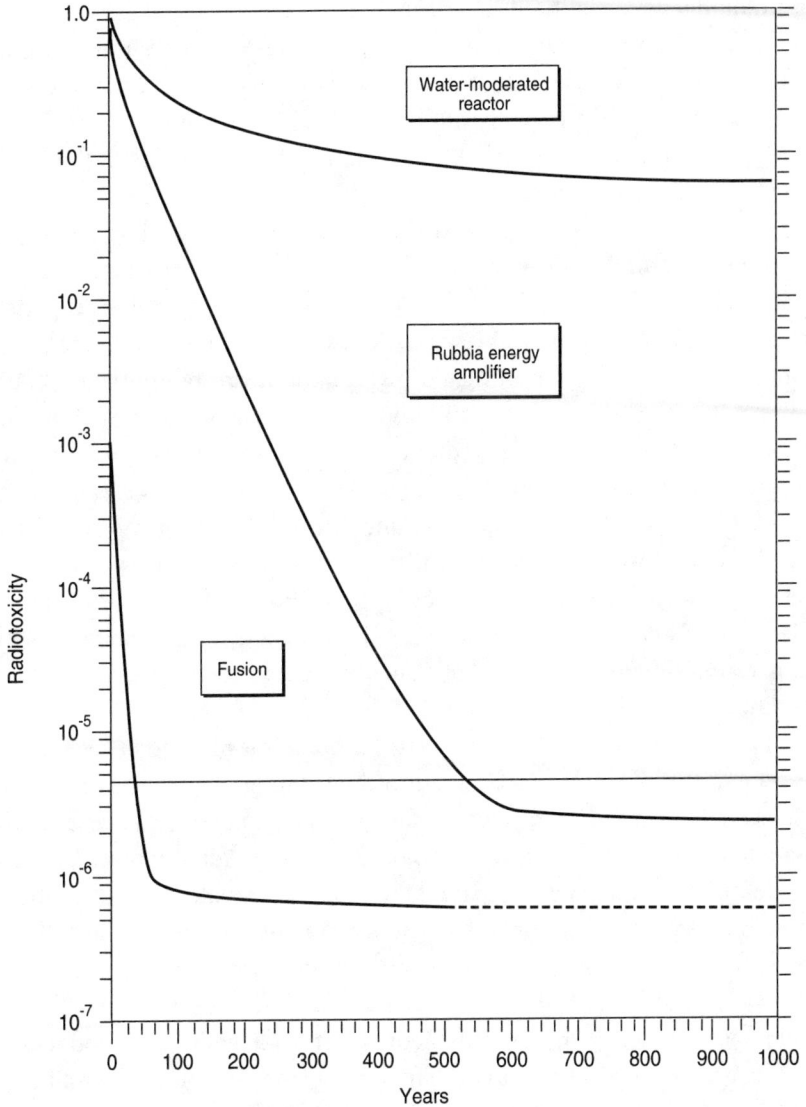

Fig. 6.3. Relative toxicity of a water-moderated reactor and the energy amplifier.

The cost of the accelerator is estimated by the Rubbia group as some 20% of the plant, with 5–10% of the plant power being used to drive the accelerator.[3] Its proponents argue that it will produce power at lower cost than the existing reactors.

The capital and operating cost of the reactor need to be evaluated. The energy-amplifier team estimates the cost of energy produced by their system at a third of that for a coal-fired system and half of that of natural gas. That would lead to a cost for electricity of 2 cents per kilowatt-hour, or about half that of other sources, despite the cost of the accelerator and the power to run it. These estimates recall early enthusiasm for nuclear electricity as "too cheap to meter." However, we believe that the *possibility* of reduced cost is the primary reason to pursue this approach, and much of the work should be focused on determining the cost of the system if it can be made to work reliably, rather than on the technology irrespective of cost.

It is necessary to keep in mind the cost of reprocessing. Past experience urges a certain humility in this respect. For example, plans for the reprocessing plant at Rokkasho, Japan, estimated investment at 570 billion yen (about $5 billion). It is now reckoned at $26 billion, and this new figure includes some simplification and substantial reduction in the size of the system, based on experience with the center at La Hague, whose performance is exemplary. This large capital expense leads to a much higher cost for reprocessed fuel than if the fuel were directly drawn from ore at present prices. But we have seen that the choice between reprocessing and burial of spent fuel not only is determined by costs but is also part of an overall nuclear waste strategy that was defined in a different era and that may or may not be appropriate to current conditions of uranium supply.

The cost of reprocessing for an energy amplifier is likely to be less than that for a light-water reactor per kilogram of fuel; the energy yield per ton being twice that of a light-water reactor, it follows that for each kilowatt-year, it is necessary to reprocess only half the mass of spent fuel as for a light-water reactor. Besides, there is no need to separate the fissionable material (that is, the uranium-233) from the thorium; the heavy elements can be kept together. As opposed to the quasi-perfect extraction of plutonium from spent fuel by the process used at La Hague (the Purex process), the residual material for the refabrication of the energy amplifier core is very radioactive and produces penetrating gamma rays. As a result, the fuel manufacturing process must be automated, without human intervention, and cannot be done with the simple "glove boxes" and manual processing in laboratories where merely toxic products (and plutonium) are handled. The highly radioactive fuel reduces, however, the potential for diversion or theft of the separated fraction containing the U-233.

One goal of this project is to greatly simplify the problems of safety, but that is not its only attraction. Some of the innovations could be useful even with more conventional reactors. But much remains to be done—for example, to evaluate the ability of the steel used for the reactor vessel to stand up to possible erosion caused by the molten lead, stimulated by the intense radiation. Other potential metallurgical problems must be investigated.

Some reactor technologists disagree with the choice of molten lead as the coolant for the fast reactor, primarily because of the tendency of hot lead to corrode steel and other structural elements. They also emphasize the difficulty of designing a reliable "window" to close the lower end of the long tube through which the proton beam in vacuum enters the lead at the center of the reactor. This window must be made of thin, strong metal in order to withstand the pressure of the thirty-meter-deep vat of molten lead—some 34 atm or 34 kg per square centimeter (500 pounds per square inch). If the window is made thicker to ease the pressure stress on its material, the temperature rise within the window, because of the heat deposited by the 50-megawatt proton beam, increases as the square of the thickness. But we are optimistic about the window problems: for instance, the window can be thinner if the beam spot is smaller in diameter, although the temperature rise is the same; so the solution may be found in using the virtuosity of accelerator designers to divide the beam among an array of N small windows in a thick egg-crate-like frame, thereby reducing the temperature rise by a factor N. There is also the alternative of a windowless design.

The accelerator-driven reactor shows that the resources of scientific invention have not been exhausted in the nuclear energy area. The necessary research will require considerable investment before achieving an optimal reactor design, which will in turn take decades before it can be widely deployed.

A FUTURE FOR FAST-NEUTRON REACTORS?

The United States has considerable experience with fast-neutron reactors, among them the Fast-Flux Test Facility at Hanford (not operating at present, but with its sodium maintained molten by electrical heating), the ill-fated Fermi-I commercial reactor near Detroit, and the Experimental Breeder Reactors EBR-I and EBR-II in Idaho. The Fermi-I commercial reactor (which had nothing to do with Enrico Fermi) began part-power operation in 1963 and was damaged in 1966 by overheating caused by a loose piece of metal blocking a portion of the sodium flow. France's experience with operation of Phénix and Superphénix—both sodium-cooled reactors—is extensive. The fast-neutron reactor has the advantage of allowing effective fission of all the minor actinides, those nuclei that are produced by neutron capture in plutonium. They would be kept with the plutonium for recycling after separation and possible transmutation.

One can imagine a variant on fast-neutron reactors in which the liquid sodium coolant, whose technology has posed many problems and inspired some mistrust, is replaced by liquid lead. The Russians have for some time been actively studying this substitution for fast-neutron reactors. But as compared with the sketch of the energy amplifier in Fig. 6.2, we go one step farther (or backward); the accelerator, which bombards the lead and produces neutrons by spallation, is eliminated, and thus a richer combination of fuel and fertile material must be used. This might be uranium-233 plus thorium; or plutonium-239 plus uranium-238 could maintain criticality without using the accelerator. One could consider mixed-oxide fuel of plutonium and depleted uranium (uranium that has had its uranium-235 removed by isotope enrichment for use in light-water reactors—having a content of 0.2% of uranium-235 instead of the 0.71% in natural uranium) with, for example, 20% of plutonium. In such a system, the plutonium is burned—but regenerated from uranium-238, with some excess. Expensive primary pumps are not needed; the lead coolant flows by natural convection to heat exchangers immersed in the upper portion of the lead pool. The natural convection of the lead is then used to extract the heat from the core. In reality, this is a lead-cooled, pool-type fast-breeder reactor with uranium-238 breeding plutonium-239.

The essential difference between a reactor of this type and the energy amplifier concept would be the removal of the accelerator. The subcritical nature of the core, which prevents any possibility of runaway in case of loss of the coolant, the molten lead, is then abandoned. But it is difficult to lose 10,000 tons of lead enclosed in a pit, and voids created by boiling would be prevented by the natural circulation of the coolant. One can benefit from the considerable progress made in safely controlling present-day classical reactors. If the pumps were to fail and the reactor were to shut down, the radioactivity of the accumulated fission products would continue to heat the reactor but, even assuming a total failure of the lead cooling system, the specific heat and the thermal conductivity of lead are sufficient for it to accept simmering gently in its pot.

At a slightly more innovative level, the visionary view of an accelerator-driven reactor operating on a core that at any time contains about 24 tons of thorium and 3 tons of uranium-233 (and is claimed to be able to produce electrical energy for half the cost of a common reactor) leads us to consider a more ordinary fast reactor with natural-circulation lead cooling of a thorium/uranium-233 system in which the neutron deficit is compensated not by an accelerator but by the addition of some 20–40 kg per year of excess weapon plutonium or waste civil plutonium. The benefit would be a considerable reduction in long-lived radioactivity, compared with light-water reactors, for instance—so long as excess fissile material was available, and the reactors would do a great service in consuming dangerous and hazardous material.

The energy amplifier has great flexibility. If the purpose were to consume plutonium, it could be loaded with as much as five tons of the element, and with no initial U-233, so most of the energy produced would come from the fission of plutonium. The illustration with 20–40 kg per year of excess weapon plutonium (perhaps 150 kg of plutonium at the beginning of use of a 5-year core) shows how long one might manage without an accelerator, if one were willing to use a U-233/Th-232 fast-neutron near-breeder, with its deficit in production of U-233 made up by the addition of plutonium. Fifteen hundred tons of plutonium already accumulated in the spent fuel of light-water reactors would thus suffice for some 45,000 years of operation of a single energy amplifier, or 45 years of one thousand energy amplifiers.

These concepts open a vast domain for research, and it would be presumptuous to believe that in the years 2050 to 2060 (when the French suggest that they will review the viability of the breeder) reactor designs will be limited to solutions thought up a quarter of a century ago when fast-neutron breeders were first designed. Let us assume that all of the problems associated with nuclear energy have been solved and that, faced with the threat of the greenhouse effect, humanity decides to exploit to the limit this source of energy to generate electricity. It will be necessary to expand the number of reactors from the present value of about 440 to a figure of the order of 2000 to supply all current electrical needs. With present technology, 2000 1-GWe plants would consume 2000 tons of uranium-235 every year, which would require 400,000 tons of natural uranium. (Since the enrichment process cannot, in an economical fashion, extract all of the uranium-235 from the natural uranium, it leaves a concentration of 0.2% in the reject uranium—meaning that only 0.5% of the natural uranium is available as uranium-235 in the enriched product.) The known ore stocks available at near-current prices would allow these needs to be satisfied for only a decade. Given that the cost of nuclear energy derives largely from the capital invested in the power plants, and that they have a lifetime of the order of thirty to forty years, or now even sixty years, it would be absurd to build them if they could be supplied with fuel for only ten years. The solution commonly considered would be "breeding," using as fuel either uranium-238 or natural thorium, thorium-232.

But it is important to realize that, in general, resources far exceed the quoted "reserves." An analysis of the uranium resource indicates that instead of the 4.5 million tons of uranium often cited as the "reserve," the exploitable resource is likely to be more in the range of 100 to 250 million tons at prices below $350 per kg.[4] This tremendous difference comes from the fact that it is not in the economic interest of companies that produce uranium to explore now and to assess definitively the availability of this material at costs ten times more than are profitable today.

The same is true with thorium. As we have by now explained many times, a 1-GWe nuclear power plant requires about 1 ton of U-235 per year, from about 200 tons of natural uranium. This assumes enrichment to 4.4% U-235, with "tails" containing 0.24% uranium-235, compared with 0.71% U-235 in natural uranium. Higher uranium prices would encourage lowering the tails assay to 0.1%, so that only 160 tons of uranium would be needed per ton of uranium-235; at a sufficiently high price for raw uranium, the increased cost of enrichment would be more than compensated for by the smaller quantity of raw uranium needed. A world with 2000 nuclear reactors instead of the present 434 would need 0.4 million tons of uranium per year; so a resource of 100 million tons would serve such a population of light-water reactors for 250 years.

It is also important to consider uranium from seawater, where there are 4 billion tons; two billion tons of this could be extracted without significant change of cost. We note that Japanese work on the extraction of uranium from seawater resulted in a 1998 estimate of $100 per kg, which would ensure the availability of inexpensive nuclear fuel for thousands of years. The seawater uranium resource would supply a population of 2000 light-water reactors for 5000 years and breeders for 500,000 years. We discuss uranium from seawater more fully in Chapter 8, in the context of avoiding global warming.

MODULAR HIGH-TEMPERATURE GAS TURBINE REACTORS

In addition to advanced reactors such as the evolutionary EPR (European Pressurized Water reactor) of 1500 MWe, the energy amplifier of Carlo Rubbia—a fast-neutron reactor using molten lead coolant and a high-energy proton beam to make up the neutron deficit of the thorium/U-233 cycle—there are two ventures that use no liquid coolant or steam turbine. These both aim at passively safe reactors with enriched uranium fuel in a carbon matrix.

One approach is a cooperative effort among the U.S. firm General Atomics Corporation, Russia, and France for a 400 MWe reactor. The other is a development of the Pebble Bed Modular Reactor under development by the South African national utility ESKOM. Each module is to produce 110 MWe and would consist principally of the reactor and a power turbine, with the heat transferred by high-pressure helium driving the gas turbine in a manner similar to modern gas-fired turbine plants. In both the General Atomics and the ESKOM reactors, the fuel in its containment is fundamentally a tiny sphere of enriched uranium dioxide coated first with porous carbon, enclosed in a thin coating of pyrolytic carbon, in turn covered by a layer of silicon carbide, and a final layer of pyrolytic carbon. The silicon carbide is very strong and forms a pressure vessel to contain the fission product gases that themselves are accommodated in the pores of the porous carbon layer. Pyrolytic carbon prevents the emergence of the radioactive fission products from the fuel particle. The diameter of the entire pellet is about 0.5 mm.

In the General Atomics approach (based on similar reactors — Peachbottom, in Pennsylvania, and Fort Saint Vrain, Colorado), the encapsulated fuel is formed into "compacts" that are inserted into prismatic carbon fuel elements, which make up the reactor. Helium at high pressure and temperature is heated by the reactor and flows to a gas turbine, where it is expanded, delivering power to the alternator that supplies electrical energy to the grid. The expanded helium is then cooled, recompressed, and reheated before it enters the reactor.

Hot helium leaves the reactor at a pressure of 1000 psi (70 atm) and a temperature of 900–950°C. It leaves the power turbine at about 530°C and 26 atm and is recycled to enter the reactor at 70 atm and about 530°C. The control rods for the reactor move vertically within a layer of graphite blocks 30 cm thick that line the reactor pressure vessel. The reactor will need to be shut down for refueling.

The ESKOM approach forms the 0.5-mm fuel particles into balls of 50-mm diameter. Adding a 5-mm layer of carbon and pressing, sintering, and grinding the "pebbles" provides tennis-ball-sized fuel elements of 60-mm diameter, each with 9 grams of enriched uranium. The reactor core would contain about 330,000 of the fuel pebbles, plus 100,000 pure graphite balls. The outside diameter of the reactor pressure vessel that would contain the high-pressure helium is some 3.5 meters, and the height 8.5 meters.

The Pebble Bed Reactor is to be refueled while under pressure and delivering power, with pellets entering the top, while used fuel pellets are to be discharged from the bottom. They would be measured for their reactivity and loaded back into the reactor for an average of ten transits through the core (over three years). About three tons of enriched uranium would be in the reactor at any time, with fresh fuel entering at 8% and being discharged at perhaps 2% enrichment.

The great benefit claimed for the modular high-temperature gas turbine reactors is passive safety, achieved by low power per module and the absence of water. The fuel is claimed to remain intact under all foreseeable circumstances, even if the helium coolant vanished through a leak. Fort Saint Vrain and Peachbottom used heat exchangers to transfer the heat from the helium coolant to water, which then drove a steam turbine, which was limited in temperature by the pressure achieved by water. We have discussed in Chapter 5 the Carnot efficiency, which, together with practical losses, limits a typical light-water reactor to a thermal efficiency of 33%. The MHTGTR is expected to achieve an efficiency of 42% to 50%, thus economizing on fuel, cooling towers, and capital investment.

ESKOM intends to order ten modules (1100 MWe altogether) of the Pebble Bed Reactor after a demonstration module is ready in 2006. The essential criterion is inherent safety with cost totaling 1.5 cents per kWh. Such a small module, or even the General Atomics module of some 400 MWe, would be much more suitable for developing countries than are reactors of 1300–1500 MWe.

THE LONG-TERM FUTURE FOR NUCLEAR POWER

Radioactivity was discovered a hundred years ago, and now 18% of the world's electricity is of nuclear origin. But the Chernobyl accident brought to light dangers that had been thought to be overcome. Burying nuclear waste requires continuous isolation for hundreds or thousands of centuries. Is our society prepared to manage such challenges? Discordant voices rise among the experts with respect to the choice of strategies and are often fueled by intense commercial competition between firms or even nations.

The next hundred years can be thought of as a period of grace, during which further research will allow us to make dispassionate choices. If society avoids the temptation to impose irrevocably for a million years what it now considers wise solutions, it will be easier to arrive at the goal, which is to find energy sources to fill a major portion of the energy needs of mankind, at acceptable hazard, pollution, and cost. Whatever the solution adopted for the ashes of the nuclear furnace—direct burial of spent fuel or reprocessing—the quantity of radioactivity that will need to be buried in a hundred years represents, in a layer of earth five hundred meters thick, a very small fraction of the radioactivity that exists already because of the natural radioelements and their daughters—uranium, thorium, radium, etc. All the same, we are not suggesting scattering the waste about or diluting it in the earth's crust or in the oceans.

During the coming century, scientists will study, with the use of underground laboratories, the best way to store waste and to keep it out of contact with the biosphere. They will examine the various forms of nuclear reactors, some of which seem to be able to considerably reduce the amount of long-lived waste. They will seriously investigate alternative energy sources.

In order for a major expansion of nuclear power to take place with acceptable safety, concerted efforts must be made to dispose of the spent nuclear fuel expeditiously, whether by dry cask storage followed by burial in a mined geologic repository, or by reprocessing, with the burial of the wastes in similar fashion.

Furthermore, considering the problems of terrorism and warfare, it would be desirable for nuclear reactors to be placed underground, where they would be protected against conventional attack and would have adequate containment of any pressurized gases that might escape in an accident. Protection against earthquake could be augmented not only by design considerations but also by sensors that would allow the reactor to be rendered subcritical before seismic waves could reach the plant from afar.

There remains one hazard to which we offer no solution. If a nation has a considerable number of reactors, and assuming that the fuel cycle is managed in such a way that after the fuel is removed from the reactor and placed into the swimming pool for some years before it is transferred to dry-cask storage and then to the mined geologic repository, there is still the problem that arises if there is societal disruption or a major war that disrupts the system of operating and caring for the reactors. At a time in which individuals are forced to leave the country or are subject to genocide, it is not reasonable to expect them to do their professional jobs of caring for nuclear power, at the risk or cost of their lives. Thus it must be imagined that a considerable fraction of the reactors would be abandoned and fall into the hands of those whose interest may be more toward damaging society than preserving an essential element that, besides, can pose a very significant danger. One can assume that the reactors shut down safely, but may be left with their entire inventory of radioactivity after days of shutdown procedure and entering into a mode of passive cooling. Furthermore, over the centuries or millennia, the reactor pressure vessels will corrode, the swimming pools will leak or the water evaporate, and the fission products and plutonium in the spent fuel will become more available to the biosphere than it was planned to be in the normal course of operation.

While it is not easy to specify the date of a revolution or severe societal disintegration, it is less difficult to specify a probability. And since one is dealing with like hazards—radiation soon vs. radiation later—one can have a sounder comparison than might be apparent at first sight between these two hazards. That analysis has yet to be performed.

CHAPTER 7

Safety, Nuclear Accidents, and Industrial Hazards

A FUTURE that may require a massive consumption of energy, and specifically of nuclear energy, forces us to understand the causes of accidents in power plants and the means to avoid their repetition. Society needs to know whether nuclear power can be used to provide an abundant source of energy, one that is neither excessively polluting nor dangerous, and whether it can manage, at the same time, to deal with the prospect that the use of nuclear power will increase the apocalyptic threat of a nuclear war. This second question and the proliferation of nuclear weapons are addressed in Chapter 11.

For a nuclear power plant, it is essential to consider abnormal accidental conditions—even catastrophes—in order to develop designs that prevent the escape of radioactive materials present in an operating reactor, even after the fission process has terminated. The insertion of control rods stops the fissions, but a reactor continues to produce heat because of the decay of the radioactive fission products—initially at a rate of 7% of the reactor in full operation, decreasing after a day to 0.6%, and after two weeks to about 0.2%, which is still some 6 megawatts of decay heat for a typical large power reactor.

Among the possible accidents are local electrical failure, bursting of the high-pressure piping, earthquake, an airplane crashing into the reactor building, a short circuit in the internal electrical power system, the possibility of human error, an enemy bomb, and even sabotage. Since a nuclear reactor is potentially hazardous and constitutes an investment of several billion dollars, it is essential to reduce the likelihood of accident to an extremely low level.

An indication of the quality of plant operation is the number of unplanned automatic "scrams." A scram is an old term from the very first nuclear reactor (Fermi, 1942), when the scientists wanted immediately to shut down the reactor and (themselves) "scram"—that is, rapidly depart the area. When a power reactor is scrammed, the event is noticed, since other sources of electricity suddenly have to take up the load; it must be reported as well. In 1980, U.S. civilian reactors had an average number of scrams of 7 per plant per year, which fell to about one per plant-year in 1999.

During normal operation of the reactor, the operators as well as the general public are protected from the intense radioactivity of the reactor core and the primary circuit by reinforced concrete radiation shields several meters thick.

Several successive safety features are generally required in power reactors. First is a redundancy of control rods—typically three independent clusters of control rods and independent mechanisms that provide for the reliable insertion of the clusters by gravity into the core; any one of the three would terminate the chain reaction in an operating reactor. All U.S. power reactors (and most in the rest of the world except for some in the former Soviet Union) have reinforced concrete containment shells which constitute the exterior of the reactor building, or else a steel pressure containment building surrounded by a concrete shell.

To prevent melting of the reactor core by the afterheat of the fission products, modern reactors using water as a moderator have emergency core cooling systems that inject cold water into the reactor under various circumstances, including hypothesized massive "loss of coolant accidents," in which it is supposed that one of the main coolant pipes feeding the reactor is severed in two places and displaced, so that high-pressure water gushes from the reactor vessel and from the coolant supply system. Many other engineered safeguards are present in modern commercial reactors, including water spray systems within the containment for condensing steam in order to avoid steam pressure that could burst the containment shell, and systems for dealing with a potentially explosive mixture of hydrogen and air that could also burst the containment and liberate the reactor's load of radioactivity.

Among the factors that play an important role in the safety of nuclear power plants are, on one hand, the capacity of the reactor to tolerate, after an accident, the disappearance of the cooling fluid (which provides for the transfer of heat produced by fission in the fuel rods), and, on the other hand, the failure of the safety systems intended to stop the chain reaction. In the case of accident, it is necessary to avoid an evolution toward an uncontrolled chain reaction that would lead to a rise in temperature, melting all the structures and leading to the possible ejection of an enormous quantity of radioactive material into the plant's environment.

A reactor in which the reaction is quenched when some coolant is lost, therefore creating a void, is said to have a "negative void coefficient." In the case in which loss of coolant makes the chain reaction more vigorous, the void coefficient is positive. It was positive for the Chernobyl plant; it was positive for the fast-neutron reactor Superphénix. But a positive void coefficient is not enough by itself to indict a reactor type, if there are self-regulating features—such as, for instance, a slowing of the chain reaction, thanks to the Doppler effect, discussed in Chapter 2, if the fuel temperature increases.

Reactors like those at Chernobyl (the RBMK type, in which the "B" is for *bolshoi*, "large," and the "K" stands for "channel") were built without a containment structure. Pressure-tube reactors, such as those at Chernobyl, are produced without enormous pressure-resistant steel vessels like those used in modern water reactors, such as the VVER-1000 series in the Soviet Union (in which "VV" stands for "water-moderator" and "water cooling"). The graphite in the RBMK operates at the high temperature of the steam produced, but the CANDU's heavy-water moderator must remain at room temperature. The CANDU reactors also use pressure tubes, but there are two concentric tubes, and the outside tube can withstand the pressure even if the inner tube breaks. Thermal insulation is provided by the gas that separates the two concentric pressure tubes.

We shall discuss two catastrophic accidents that have permanently destroyed reactors—at Three Mile Island and near Chernobyl. It may not seem fair to consider on the same footing these two different events, one of which caused only financial loss to the reactor operator—although that was considerable—while the other was the source of serious damage, death, and injury to hundreds of thousands of people. What these two cases have in common is that human error played the primary role in both; and that good luck as much as skill averted the evolution of the 1979 Three Mile Island accident into a catastrophe.

THE THREE MILE ISLAND ACCIDENT (1979)

On March 28, 1979, an accident occurred at four o'clock in the morning at the Three Mile Island, Unit 2, nuclear power plant, located on the outskirts of Harrisburg, the capital of Pennsylvania. It was a pressurized-water reactor generating electrical power of nearly one gigawatt, in full power operation since late 1978 in a plant considered one of the safest. The accident resulted from a combination of equipment failure and the inability of the plant operators to understand the reactor's condition. The description below follows closely the account presented by the Nuclear Energy Institute (NEI), an association of U.S. nuclear reactor operators—an account that our independent evaluation has convinced us is generally accurate and reliable.[1] NEI may or may not agree, of course, with our comments and views concerning the accident.

The accident began when the main feedwater pumps stopped, leading to an increase of pressure in the vessel containing the reactor core, where heat is generated. The rise in pressure caused a relief valve to open and the reactor to shut down automatically; rods containing boron, a neutron absorber, rapidly entered the core, thus terminating the neutron chain reaction. Fission products in the reactor fuel continued to decay, liberating about 200 megawatts of heat right after shutdown. Pressure in the reactor vessel was prevented from rising excessively—as was intended by design—as steam escaped through the open safety relief valve into the reactor containment building. However, this valve did not shut at the designated setpoint, so the pressure continued to drop and much cooling water was lost as steam through the stuck-open valve.

The reactor operators did not understand what was happening. Their instruments showed that the safety valve had been commanded to close, and they assumed that it had done so. In addition, they interpreted other control room indicators to mean that there was already too much water in the reactor coolant system. Consequently, they did not replace the lost coolant. As the pressure continued to drop, more and more of the coolant turned into steam. Eventually, the main coolant pumps began to shake violently, because they were not designed to pump a mixture of water and steam.

Still not recognizing what was happening and afraid of damaging the pumps, the operators shut them down. This made matters worse. Eventually, a large steam bubble formed inside the reactor vessel, further preventing the flow of coolant through the core.

Emergency feedwater pumps that should have provided additional coolant were blocked by valves that should have been open; the reactor was being operated in violation of its fundamental safety rules. The overheated fuel, not covered by water in the upper portion of the reactor, caused the metal tubes containing the fuel pellets to melt and to react chemically with the steam and water, so that the steam bubble was replaced by a hydrogen bubble at high pressure, which for days prevented the water in the reactor pressure vessel from fully covering the core. Eventually, the bubble was dispersed and coolant flow restored, but by then about half of the fuel had melted. The colder cooling water also shattered some of the hot fuel rods. All the fuel was damaged or destroyed.

As a result of the accident, 700,000 gallons of radioactive cooling water spilled onto the floor of the reactor building and auxiliary building, contaminating them. In addition, a small amount of radioactive material went up a stack and through a charcoal filter into the atmosphere.

For 11 hours after the beginning of the accident, loud alarms continued to sound in the control room, simple evidence of poor preparation for an accident. Some hydrogen escaped to the containment dome of the reactor, where it burned or exploded with the air normally contained in the building. Fortunately, the pressure pulse was not strong enough to burst the dome and liberate the enormous amount of radioactivity that had emerged from the partially melted core of the reactor.

The formation of a hydrogen bubble had not been foreseen in any of the accident scenarios imagined by the governmental regulatory authorities. The accident was compounded by the ability of the operator to intervene with actions that made the situation more dangerous; and even in a reactor considered to be the safest, venting to the atmosphere of much of the reactor's burden of radioactive materials was not impossible. This plant was one of the most modern in the United States; the same accident occurring in an older reactor might have produced a major catastrophe. One can imagine that the psychological and economic consequences of an explosion in the United States would have totally ruined the nuclear industry. As it was, despite the absence of harm to the citizenry or to the environment, the Three Mile Island accident significantly sapped public confidence and was a substantial contributor to a halt in reactor orders.

Tens of thousands of people evacuated the area around the reactor for the several days of the acute phase of the accident, and it was not until April 27, almost one month later, that the reactor was put into a natural-circulation, cold-shutdown mode—the fission product decay heat was then some 4 MW, about 0.13% of the heat produced in the reactor in full operation. The safety of the reactor against further damage no longer depended on pumps to cool the reactor vessel.

One of the authors of this book (Garwin) had served in 1975 on a panel of physicists studying the safety of light-water reactors, together with Robert Budnitz, who was at the time of the accident on the staff of the Nuclear Regulatory Commission (NRC). As a result, Garwin had some insight into the ongoing accident via telephone. He provided preliminary analyses of various options for removing the hydrogen bubble. At the time there was great concern about oxygen being generated by the fission-product radiation ionizing and decomposing the water in the reactor. Mixing with the hydrogen at the top of the reactor pressure vessel, oxygen could form an explosive mixture, which, if it detonated with the energy release equivalent to three tons of high explosive, could burst the reactor pressure vessel and perhaps the containment building and thus release much of the fission product content of the reactor, as was later to happen at Chernobyl. One option was to scour the industry for experts on hastening the recombination of hydrogen and oxygen (catalysis), and to urgently obtain a stock of such a catalyst. Another approach would have been to snake a flexible hose through a pressure-tight fitting that could be installed outside the immediate vicinity of the reactor pressure vessel (somewhat like the catheter snaked in from a vein in the leg to a person's heart in an angiography procedure). Later analyses showed that no significant amount of oxygen would have grown into the hydrogen bubble, because the radiation-induced decomposition of the water would have been inhibited by the high pressure of hydrogen already present. The hydrogen itself was gradually removed over a period of days by "degassing" the cooling water after circulation had been restored.

Although 40% to 60% of the major fission product isotopes xenon-133, iodine-131, cesium-137, and cesium-134 were released from the core, less than 1% escaped from the reactor complex to the environment outside the plant. The total radioactivity released into the atmosphere was 4×10^{17} Bq, mostly xenon-133, and 6×10^{11} Bq of iodine-131, leading to a collective dose inflicted on the United States of 20 to 40 person-sieverts. Using the 0.04-per-sievert estimate of the induction of cancer, that would lead to one or two fatal cases. The prevailing winds from the west carried the radioactive cloud over the Atlantic ocean; there was little impact outside the United States, because xenon-133 has a 5-day half-life and it had largely decayed before the radioactive cloud reached Europe.

The owner of the Three Mile Island plant, General Public Utility Nuclear, removed the reactor fuel by 1990, and between 1991 and 1993 evaporated 2.2 million gallons of radioactive water, using filters to retain the radioactivity. In December 1993, the operator placed the unit in monitored storage. It will be decommissioned when its sister reactor, Three Mile Island 1, is taken out of service at the expiration of its operating license, in the year 2014. (TMI-1 might, in principle, apply to be operated beyond 2014 on the basis of a license extension, but no such intent has been indicated). The accident contributed greatly to public apprehension about nuclear power and to the subsequent disinclination of public utilities to acquire nuclear reactors. Concerns about the competence of the firm operating Three Mile Island have led to many reactors being kept out of operation for some years. Today, two decades later, it is clear that the nuclear industry has taken the lessons of Three Mile Island seriously.

U.S. reactors are built and operated under license from the NRC, a federal agency independent of the Department of Energy or of business organizations with an interest in encouraging nuclear power. Nevertheless, the NRC does not require that nuclear plants withstand all conceivable threats; for instance, in 1994, the NRC director testified that nuclear power plants were not protected against trucks loaded with explosives, which as we have seen from the 1995 bombing of the Federal Office Building in Oklahoma City are part of a possible scenario. Since that year, however, truck bombs have been a specified threat. In its assessment of "risk" of accident in the nuclear industry, the NRC considers not only the consequences of a possible event but also its probability (in fact, risk is defined here as the numerical product of probability and consequences). As better estimates of each become available, NRC requirements will change, beyond current requirements—e.g., that plants be able to endure any earthquakes foreseeable for the site.

The accident served as a wakeup call to the industry, which had not readily accepted analyses of the mid-1970s cautioning that a meltdown of a reactor core was to be expected in a population of one hundred reactors operated for thirty or forty years, i.e., once in 3000 to 4000 reactor-years.[2] This contrasted with the results of previous studies concluding that a meltdown was to be expected only once in 20,000 reactor-years.[3] Faced with a modern U.S. reactor core that was largely melted, the NRC chairman and President Carter created two investigatory commissions.

In response, the industry established the Institute of Nuclear Power Operations (INPO), which sets standards within the United States, evaluates the performance of plants, and provides a formal link to the entire industry even for operators of a single reactor. The institute trains reactor operators and management and is involved with emergency exercises. INPO is taken seriously in the industry; for instance, its ratings of a plant can affect the price of the stock of the operating company. In a constructive change, INPO has extended its classification so that a nuclear plant can receive a rating of "unacceptable." In analogy to INPO, there is now a World Association of Nuclear Operators, which operates in a similar fashion with nuclear power plants in the rest of the world.

Can the public be confident that there will be no more accidents like Three Mile Island? More needs to be done. The case of Boraflex is instructive. This plastic material loaded with boron is used in some storage tanks for spent fuel ("swimming pools"). Fuel elements removed after four years in a reactor are supported vertically in pools 8 meters deep, so that decay heat is removed by natural circulation of the water. The liquid is filtered and regulated so that it corrodes the fuel clad as slowly as possible. The fuel elements that undergo a chain reaction in the reactor are prevented from doing so in the pool by the inclusion of boron, a strong neutron absorber. Stainless-steel tubes filled with boron powder are used in some pools, but in others Boraflex has been used. Unfortunately, the plastic degrades rapidly under the gamma radiation from the spent fuel, and the material crumbles and falls to the bottom of the pool, where it is ineffective in absorbing neutrons.

Some reactor operators have instituted frequent surveillance of pools controlled with Boraflex rather than replacing the material with metallic structures containing boron. A storage pool going critical and sustaining a chain reaction does not present the same hazard as would meltdown of an operating reactor, but such an accident would erode confidence in the industry; more rapid progress in eliminating this risk is evidently to be desired.

In general, the NRC is moving toward "risk-informed regulation," based on probabilistic risk assessment. This is welcome, but NRC capabilities have been much impaired by the decline in its research budget from $200 million in 1981 to $43 million in 2000, without inflation being taken into account.

THE CHERNOBYL CATASTROPHE (1986)

The Chernobyl accident, which occurred in what is now Ukraine in the early morning of April 26, 1986, resulted from a safety experiment conducted in violation of the plant's technical specifications. The site housed four reactors operating independently in adjacent halls. Plant managers were testing the ability of equipment to provide electrical power as the reactor shut down, until diesel generators could start up. The reactor was being run at very low power, with the core's reactivity depressed by "xenon poisoning" as described in Chapter 5. Accordingly, the control rods had been largely withdrawn from the reactor core, to prevent quenching the chain reaction. The team in charge of the test had not coordinated the procedure with the personnel responsible for the safety of the reactor, and the experiment was performed without adequate safety precautions. The operators took a number of actions that deviated from established safety procedures, and this led to a dangerous situation.

There were also several significant flaws in the design of the plant, which made the reactor potentially unstable and susceptible to loss of control in case of operator error. The Chernobyl reactor design had a positive void coefficient, as explained above. This means that the nuclear chain reaction and power output increase when cooling water is lost. The large value of the positive void coefficient caused the uncontrollable power surge that led to the destruction of Unit 4. The surge resulted in a sudden increase in heat, which ruptured some of the fuel-containing pressure tubes. The hot fuel particles reacted with water and caused a steam explosion, which lifted the 1000-ton cover off the top of the reactor, rupturing the rest of the 1,660 pressure tubes, bringing about a second explosion and exposing the reactor core to the environment.

Chernobyl did not have the massive containment structure common to most nuclear power plants elsewhere in the world. Without this protection, radioactive material escaped to the environment. However, because the energy released by the explosions was greater than most containment designs could withstand, it is unlikely that such a structure would have prevented the release of radioactive material. The crippled reactor is now enclosed in a hastily constructed concrete shelter, which is growing weaker over time. Ukraine and the Group of Seven industrialized nations have agreed on a plan to stabilize the existing structure.

We now give some details of the impact of the Chernobyl accident and some eyewitness reports of its evolution, and discuss its health impacts on the population in the surrounding area.[4]

The Origins of the Chernobyl Disaster

A year before the accident, the Soviet Minister of Energy, Anatoli Mayorets, had promulgated a decree stipulating, among other things, that "Information about the unfavorable ecological impact of energy-related facilities on operational personnel, the population, and the environment shall not be reported openly in the press or broadcast on radio or television."[5] Tight secrecy was maintained regarding all accidents, with the result that personnel were unable to profit from the experience of others to improve safety, while the power of the responsible chiefs was reinforced at all levels. The Soviet Union had set up a system in which propaganda replaced information—an extreme example of the "spin" we see even in the United States, where it is practiced by government, the press, and some public interest groups.

And yet there did exist honest, energetic, and responsible individuals who did what they could. In *The Truth About Chernobyl*, Grigori Medvedev, the former associate director of construction of nuclear power plants in the USSR, writes:

> On 20 February 1986, while attending a meeting in the Kremlin of nuclear power plant directors and the officials in charge of building the stations, I had observed a most curious pattern: when submitting their reports, the directors and construction heads spoke for 2 minutes each, while Shcherbina [the deputy chairman of the Council of Ministers] interrupted them constantly and spoke for at least 35 to 40 minutes.
>
> The most interesting statement was made by the senior construction officer of the Zaporozhiye nuclear power plant, R. G. Khenokh, who plucked up the courage to state, in a deep bass voice—a bass voice, at such a meeting, was regarded as tactless—that the No. 3 reactor at Zaporozhiye could not, in the best of circumstances, be started up before August 1986 (it was actually started up on 30 December 1986) because equipment had been delivered late and the computing complex, on which assembly had only just begun, was not ready.

Shcherbina was indignant. "Well, that's just great! Here's a man who sets his own deadlines!" And, his voice rising to a shout, he went on, "Who gave you, Comrade Knenokh, the right to set your own deadlines instead of the ones set by the government?"

"Deadlines are dictated by the technology in use," the construction chief stubbornly replied.

Shcherbina interrupted him. "Forget it! Don't evade the issue! The government deadline is May 1986. Kindly start up in May!"

"But they won't finish delivering the special reinforcing bars until late in May," Khenokh rejoined.

"Have it delivered earlier," Shcherbina urged him, and, turning to Mayorets, who was seated next to him, went on, "See, Anatoly Ivanovich, your construction chiefs are blaming everything on the absence of equipment and have been failing to meet deadlines."

"We'll take care of that, Boris Yevdokimovich," Mayorets promised.

"I fail to understand how a nuclear power station can be built and started up without equipment," Khenokh muttered. "After all, I'm not the one who supplies the equipment, it comes from industry through our client," and, thoroughly disgruntled, he sat down.

After the meeting, in the lobby of the Kremlin Palace, he told me, "There you have our whole national tragedy in a nutshell. We ourselves tell lies, and we teach our subordinates to lie. Lies, even for a worthy cause, are still lies. And no good will come of it."

These remarks were made two months before the Chernobyl disaster.[6]

These pages reveal the quality of the directors of the Soviet nuclear industry. They are the ones with whom the political leaders set up a program intended to provide the nation with one-third of the world's nuclear power. Public opinion, drugged by the soothing declarations of these officials, had accepted what they were told until the lightning stroke of Chernobyl.

Fifteen years before the catastrophe, Medvedev was assistant to the chief engineer of the Chernobyl plant. He went to Kiev, where he was asked to make a report to the Ukrainian Minister of Energy. He was asked if the type of reactor had been well chosen, considering especially the fact that Kiev was not far from it. He answered without hesitation that it would have been preferable to choose a pressurized-light-water reactor, of the type extensively used in the West. He could find no explanation for the selection made other than that it was the recommendation of the Academicians rubber-stamped by the Council of Ministers. In 1972, the authorities could still have changed their minds and acted differently. But a bureaucratic chief would have needed a very strong personality to swim against the tide.

Sequence of Events on April 25 and 26, 1986

The plant was located in a lightly populated region: about 110,000 citizens lived within a radius of thirty kilometers. On April 25, 1986, preparations were underway for the shutdown of one of the four reactors for maintenance work. Strict rules existed for such operations: in practice, the protocol adopted by the chief engineer respected none of them.

The workers were badly trained and badly prepared, and, not knowing the dangers, they did not follow orders. They did not realize that the reactor's design was such that a sudden loss of coolant would lead not just to overheating but to an uncontrolled chain reaction. The conspiracy of silence on accidents that had occurred in power plants during the preceding thirty-five years contributed to the unawareness of the operators. For convenience, they turned off the safety devices.

There followed a chain of events in the course of which the specialists and the chief engineer committed a series of glaring blunders. It is interesting to note a comment by Medvedev on an important character in the drama whose promotion he had opposed: "He was slow-witted, quarrelsome, and difficult."[7] The chief engineer is described in these terms: "He was . . . quick to take offense, pushy, vain, vindictive, spiteful, though he was occasionally fair. His voice was a usually pleasant baritone but sometimes shot up an octave when he got excited."[8]

In 1986, electronics should have been used to assure the safety of a reactor, even if the operators had falsetto voices, or were dead drunk or raving mad. Ironically, the nuclear industry as a whole, including that of the United States, has been slow to adopt automation and informatics, relying instead on procedures and on the selection and training of personnel.

The mistakes led to a runaway of the chain reaction. The design of the reactor was defective, because it should have been made to shut down by itself in this case, without the necessity of human or electronic intervention. As the disaster unfolded, part of the fuel melted and evaporated before being projected into the atmosphere by the explosion. It was then carried toward the northwest by the wind, across Belarus and the Baltic republics, beyond the frontiers of the Soviet Union. The radioactive cloud spread at an altitude of 1000 to 11,000 meters. For a week, airplanes arriving in Moscow were contaminated.

The reactor core was like a volcano, with radiation and neutrons escaping because the 1000-ton slab had been blown into the air before falling back on the reactor and breaking its structures. Fifty tons of the nuclear fuel was thus dispersed. The firemen put out the visible fires, but they were ignorant of the fact that they were walking on debris giving off radiation at a lethal dose for a person exposed for ninety seconds. A fireman was able to work for about half an hour without feeling the effect of the radiation, but he had become a walking corpse, condemned to die in a few weeks as a result of the exposure even if he had worked there for only two minutes.

The operators did not have available detectors for measuring very high radiation intensities; all the routine monitoring devices were blocked as they hit the top of their scales. It is clear that many of the men dispatched to the site died without accomplishing anything useful at all. A minimum of precaution would have avoided most of these sacrifices. The doses received were such that the men tanned in a few hours, as if they had spent two weeks in the sun.

Yet, heroic characters appeared on the scene. Medvedev remarks:

> Aleksandr Grigoryevich Lelechenko went three times to the electrolysis room to disconnect the flow of hydrogen to the emergency generators, thus sparing the younger electricians from spending more time than strictly necessary in the area of intense radioactivity. The fact that the electrolysis room was next to the pile of radioactive rubble, surrounded by fragments of fuel and reactor graphite, with radiation of between 5000 and 15,000 roentgens per hour [a lethal dose of radiation in six minutes or two minutes, respectively], suggests the heroism and high moral fiber of this fifty-year-old man, who deliberately shielded his younger comrades with his own body. And then, while knee-deep in radioactive water, he studied the condition of the electrical switching gear, in an attempt to supply current to the feedwater pumps. His total exposure was enough to kill five people. Yet once he had been given first aid — in the form of an intravenous infusion — Lelechenko rushed back to the unit and worked there for several more hours. He died a terrible, agonizing death in Kiev.[9]

The plant directors poisoned the minds of the Moscow authorities with false information telling them that the reactor had not been destroyed, whereas the entire neighborhood was strewn with pieces of graphite from the reactor core, projected to considerable distances by the hydrogen explosion. Again, Medvedev:

> The authorities refused to believe that the reactor had been destroyed. They also refused to believe the plant's senior radiation protection official,

Vorobyov, who warned them of high radiation levels. Instead, they told him to toss his radiometer in the garbage. However, somewhere deep inside Bryukhanov, a single sober thought had implanted itself. At some subconscious level, he had taken note of the information supplied [by his colleagues] and as a precaution, requested permission from Moscow to evacuate the town of Pripyat. From 2000 miles away in Barnaul, where his advisor L. P. Drach had reached him by phone, an important Soviet official, Boris Yevdokimovich Shcherbina, issued clear orders: "Don't start a panic! There must be no evacuation until the government commission gets there!"

Then Shcherbina arrived at Chernobyl. He continued to refuse to order the evacuation of the 48,000 civilians, although the situation was frightful. Medvedev explains:

When Shcherbina arrived, I went to see him privately before the meeting, told him about the situation, and said that the town should be evacuated immediately. He calmly replied that this would cause panic, and that panic was even worse than radiation.[10]

Medvedev describes Shcherbina, now the man in charge:

This rather puny man, of medium build, now a little paler than usual, his tight-lipped face showing its age, his thin cheeks deeply marked with the lines of authority, was calm, collected, and focused. He still did not realize that the air around him—in the street and inside the room—was saturated with radioactivity, and emitting gamma and beta rays which penetrated whoever happened to be in their way—ordinary mortals, Shcherbina, or the devil himself. As for ordinary mortals, there were about 48,000 of them in the town that night, including senior citizens, women, and children. . . . Shcherbina . . . alone was empowered to decide whether to evacuate, and whether to classify what had happened as a nuclear disaster.

His behavior was typical of the man. Initially he seemed quiet, modest, and even a little apathetic. This small and frail individual evidently savored the colossal power he wielded, which was so hard to keep in check; he thought of himself as a godlike figure, with the ability to punish, or forgive, as he pleased. However, Shcherbina was just human, as his later behavior showed. At first he appeared outwardly composed, but gradually a storm was gathering strength within him, and by the time he had understood the disaster and decided on a policy for coping with it, he burst forth in a frenzy of impatient energy, driving everyone relentlessly to work harder and faster.[11]

He was furious when he learned that the only rational solution was to bury the plant under tons of sand dropped from helicopters.

Medvedev goes on, quoting a participant, G. A. Shasharin:

Shcherbina was extremely impatient. With the helicopter engines roaring outside, he yelled at the top of his voice that we were lousy workers, that we were no good. He drove us like cattle—all of us, ministers, deputy ministers, generals, not to mention the others—telling us that we were very good at blowing up reactors, but useless when it came to filling sandbags.

Eventually the first batch of six bags was loaded aboard an MI-6. Antonshchuk, Deigraf, and Tokarenko, who had assembled the reactor, took turns flying on "bombing runs," so as to guide the pilots precisely to their target.[12]

The risks that the "liquidators"—the name given to the people who had the task of liquidating the consequences of the accident—had to take are illustrated by the conditions under which the pilots had to work to bury the reactor in sand:

Colonel Nestorov, a top air force pilot, flew the first sortie, approaching the damaged unit in a straight line at 87 miles per hour (140 km/hr), and taking his bearings from the two 360-foot (110-m) ventilation shafts on the left.

He arrived over the crater of the nuclear reactor at 360 feet (110 m). The radiometer [at the helicopter] read 500 roentgens per hour. The helicopter hovered over the target—the opening, formed by the tilted, white-hot biological shield and the reactor vault. As they opened the door, they felt a surge of heat from below; it was carrying radioactive gas, neutrons, and gamma rays. No one was wearing a respirator, nor was the helicopter protected underneath by lead plating. This was added later, after hundreds of tons of sand had already been dumped. The crew looked out of the open door and, staring down into the nuclear volcano and aiming with the naked eye, dropped the bag. And countless bags thereafter. There was no other way to do it.[13]

One of the astonishing features of this description was the fact that Moscow sent an endless stream of high-level political dignitaries, many of whom were to shovel sand and receive massive radiation doses.

Evgenii Velikhov was one of the competent scientific authorities present.

He is noted for his work on plasma physics and well known in the West also for his part in discussions on the control of nuclear weapons. Velikhov, having already received a tenth of the lethal dose of radiation, explained the situation to the nucleocrats and Academicians:

> You've got to be aware of what happened. The Chernobyl explosion was worse than any other nuclear explosion. Worse than Hiroshima. That was only one bomb, whereas here the amount of radioactive substances released was ten times greater, plus half a ton of plutonium. Today, Anatoly Ivanovich, you have to count people, and lives.[14]

Characteristic of the way people, "our most valuable capital" according to Stalin, were treated was the decision of a director to close a crucial water valve located in a reactor pool. He managed to get hold of two watertight diving suits, sent two men into the highly radioactive pool, and promised them that if they died, their families would get an apartment, a dacha, and an automobile. They both died.

What followed in the lives of the high-ranking participants is partially described in the testimony of M. S. Tsvirko:

> When we arrived in Moscow, my blood pressure was very high; both my eyes were bloodshot. While everyone was assembling to take the bus to No. 6 clinic, I sent for my official car and went to my usual No. 4 medical unit (the Kremlin hospital). The doctor asked me why my eyes were red. I told him that high blood pressure was the probable cause. He took my blood pressure, which was 220 over 110. Later on I discovered that radiation drives up the blood pressure. I told him I had been exposed to radiation in Chernobyl and asked him to check my condition. He replied that I was in the wrong place for such an examination, and that I should go to No. 6 clinic. I asked him to examine me anyway; and after a blood and urine test, he discharged me. At home I washed myself thoroughly. I had done the same in Chernobyl and Kiev. I wanted to lie down and rest, but the phone rang. They had been expecting me at No. 6 clinic and wanted me over there immediately. I went most reluctantly, announcing as I entered that I was from Chernobyl and Pripyat.
>
> I was told to go to the waiting room. A dosimetrist passed a sensor over me and told me I appeared to be clean. I had washed carefully before going there, and I have no hair. In No. 6 clinic, I saw Deputy Minister Semyonov, who had already been shaved completely like a typhoid patient. He was

complaining that after he had lain on the hospital bed his head was more contaminated than it had been before. It seems they had been put in beds used previously by the firefighters and operators who had been brought in on 26 April with severe radiation sickness. It turns out that the linen had not been changed, so patients were contaminating each other through the bed-sheets. I demanded to be discharged and soon went home, where I was able to rest.[15]

The evacuation of Pripyat by bus was not a masterpiece of organization. The uprooting of residents played an important role in the stress that over-whelmed the population and seriously affected their health. Medvedev writes:

When the time came to say good-bye to their pets, there were some distress-ing scenes. The cats, with their tails straight up, stared at their masters with an imploring look, meowing pathetically; and dogs of many breeds were whining plaintively, trying to force their way into the buses, yelping franti-cally and growling when they were pulled away. However fond the children were of their pets, there could obviously be no question of taking cats and dogs, as their fur, like human hair, was highly radioactive. After all, the ani-mals spent the whole day outside on the street and must have picked up vast amounts of radioactive particles.

Some dogs, finding themselves left behind by their masters, ran after the buses for a long way, but to no avail. Eventually they fell back and returned to the abandoned town, where they began to roam around in packs.

Archaeologists once read an interesting inscription on some ancient Babylonian tablets: "When the dogs in a city band together in packs, that city will fall and be destroyed." The town of Pripyat was not destroyed by radiation, but it was abandoned and preserved as it had been, for dozens of years to come—a radioactive ghost town.

First [the dog packs] devoured a large number of the radioactive cats and then turned wild and began to growl at humans. They even attacked humans and abandoned farm animals a number of times.

Then, for three days—27, 28, and 29 April (up to the day the government commission was itself evacuated from Pripyat to Chernobyl)—a hastily formed group of hunters with shotguns shot all the feral radioactive dogs. Their breeds included mongrels, Great Danes, sheepdogs, terriers, spaniels, bulldogs, poodles, and lapdogs. On 29 April the hunt was completed, and the abandoned streets of Pripyat were strewn with the corpses of many dif-ferent kinds of dog.[16]

IMPACT OF CHERNOBYL: CONTAMINATION, EXPOSURES, EVACUATIONS

An extensive international study, conducted under the auspices of the World Health Organization ten years after the accident, provides the basis for realistic conclusions regarding the harm done to the population.[17]

The Chernobyl Unit 4 reactor contained about 190 metric tons of uranium dioxide fuel and fission products. Estimates of the amount of this material that escaped range from 13% to 30%.

The total radioactivity released by the reactor was 300 million curies or 12×10^{18} Bq; of this, 6×10^{18} Bq was Xe-133—a largely irrelevant inert gas with a five-day half-life. Only 3 to 4% of the solid fuel in the reactor was ejected, but 100% of the noble gases and 20 to 60% of the volatile radioelements escaped into the atmosphere. The radioactivity, for the cesium-137 alone, was five hundred times greater than that created by the Hiroshima and Nagasaki bombs. For the iodine-131, it was six times greater. These values are higher than those divulged in 1986 by the Soviet authorities who limited themselves to the fallout on the USSR.

The significant radioelements carried away by the radioactive cloud were about 1.7×10^{18} Bq of iodine-131 and 8.5×10^{16} Bq of cesium-137. Rain deposited large quantities in the region close to the reactor. Initially, iodine-131 produced the most important health effects, because it concentrates in the thyroid. But it has a half-life of only eight days, whereas over the long term, cesium-137, whose half-life is thirty years, is more important.

Because of the atmospheric conditions, 70% of the radioactive precipitation occurred in Belarus, most of the rest of the fallout being distributed among the Ukraine, Russia, and the rest of the northern hemisphere. The localization of rainfall led to a very unhomogeneous distribution of contamination. Even at a distance of 400 kilometers from the reactor, in the Woshin region, northwest of Minsk, part of the population had to be evacuated. In certain locations there was radioactivity of very high intensity—thirty hours after the explosion, military personnel measured at Novodicki, 70 kilometers from Chernobyl, a dose of 30mSv/hr outdoors. Altogether, five million people in the Ukraine and Belarus were exposed to significant doses of ionizing radiation coming from the fallout of radioactive substances. Low doses of radioactivity were also distributed among tens or hundreds of millions of people all over the northern hemisphere.

The accompanying table shows the distribution of received doses and numbers of people involved.[18] An area of 10,000 km² was contaminated at the level of 555 kBq/m² to 1480 kBq/m², while 21,000 km² received 185 to 555 kBq/m². In Belarus, 2.2 million inhabitants now live with a contamination of 37 kBq/m²— i.e., 37 billion Bq/km², or 37 GBq/km².

TABLE 7.1: DISTRIBUTION OF DOSE RECEIVED AND POPULATIONS EXPOSED FOLLOWING THE CHERNOBYL ACCIDENT

CONTAMINATION PER M^2	RATE OF ANNUAL AVERAGE EXPOSURE	POPULATION EXPOSED	MEASURES TAKEN BY THE SOVIET AUTHORITIES
>1480 kBq*	> 5mSv	135,000	General evacuation
555–1480 kBq	2–5 mSv	270,000	Help in relocation
			Mandatory medical follow-up
185–555 kBq	1–2 mSv	580,000	Special medical follow-up
37–185 kBq	< 1 mSv	4,000,000	Continued medical follow-up

kBq = 1000 becquerels.

Workers involved in the recovery and cleanup after the accident received high doses of radiation. In most cases, these workers were not equipped with individual dosimeters to measure the amount of radiation received, so experts can only estimate their doses. According to Soviet estimates, between 300,000 and 600,000 were involved in the cleanup of the 30-kilometer evacuation zone around the reactor. Estimates of the number of cleanup workers brought into the area for accident management and recovery work vary; the World Health Organization, for example, puts the figure at about 800,000. In the first year after the accident, the number of cleanup workers in the zone was estimated to be 211,000, and they received a very high dose of radioactivity. Right after the accident, the main health concern, as noted above, involved radio-iodine, which has a half-life of eight days. Today, the problem is the contamination of the soil with cesium-137, with its half-life of about 30 years.

Soviet authorities started evacuating people from the area around Chernobyl within 36 hours of the accident. By May 1986, about a month later, all those living within a 30-kilometer radius of the plant—about 116,000 people—had been relocated. According to reports from Soviet scientists, 28,000 square kilometers were contaminated by cesium-137 to a level of 185 GBq/km^2. Roughly 830,000 persons lived in this area. Another 10,500 square kilometers or so were contaminated by cesium-137 to a level of 555 GBq/km^2 or more. Of this total, roughly seven thousand square kilometers lie in Belarus, two thousand in the Russian Federation, and fifteen hundred in Ukraine. All told, about 250,000 people lived in these areas.

Natural Radiation vs. Chernobyl Radiation

Normal soil has an activity on the order of 1 kBq/kg, mostly from the decay of natural potassium-40. To see how many Bq there are per square meter (for comparison with the entries in the first column of Table 7.1) we need to multiply by the number of kg of soil per m^2 and to a depth of 6 cm or so. Since there are $100 \times 100 \ cm^2/m^2$, with about 1.5 g/cm^3 for the density of soil, we have about 90 kg/m^2. So natural background is about 90 kBq/m^2 for comparison with the values of Table 7.1. Recall that the average American receives some 0.3 mSv per year from rocks, soils, and building materials—compatible with the figures given above. But recall also that granite from Brittany contains some 750 kBq/m^2 in a slab 4 cm thick, as we have previously indicated.

The radioactive substances penetrate into the soil so that humans are not in direct contact with the radioelements. The table shows that 135,000 people received doses somewhat higher than those tolerated for workers in the nuclear industry, whereas 4 million got doses less than the differences observed between various regions in France, and were counseled to undergo "continued medical follow-up."

The doses received, averaged over the entire country, are very much less than the natural background. It is estimated that cesium-137 penetrates to a depth of 3 cm and remains in the soil, and that is the basis of the calculation of the dose received by the inhabitants. In the least-exposed region around Chernobyl, the radioactivity per m^2 is 185,000 disintegrations per second, whereas the cosmic-ray muons irradiating this area number about 50 per second, and the number of gamma rays from the natural radioactivity of the soil is some 90,000 per second. A cosmic ray passing through the entire body produces a hundred times more ionization than a 1-MeV electron produced by a gamma ray coming from the rocks, and only about 30% of the gamma rays from the soil are emitted upward to be converted into ionizing electrons in the body. The 185,000 disintegrations per second per m^2 added by Chernobyl contributes a dose comparable with the natural background due to radioactivity of soil and building materials, plus that due to cosmic rays.

There is no way to assess accurately the hazard to any given individual. We deal here with averages. Fallout is not uniform. There are substantial differences from place to place. In a city with paved streets, the radioelements are not absorbed into the soil. Where dust can be raised by wind or by automobiles, the radioelements can fly about and find their way into the lungs. Radioactive elements, like cesium-137, can be absorbed by plant roots, enter into the food chain, and settle for a long time in the body. These figures for the average dose are useful for an evaluation, averaged over many inhabitants, of the number of them who will contract fatal cancer, but cannot indicate who, specifically, will be a victim of the accident.

Compared with the multiple causes of cancer, the effects mentioned translate to a very small fraction. We illustrate this point using France as an example. During the first year after Chernobyl, the average dose was 0.05 mSv. Assuming a linear effect for low doses and an estimate of 0.04 lethal cancer per sievert, we find one hundred lethal cancers induced per year and about 300 over thirty years, by taking into account the decay of fission products.

In thirty years, nearly three million French men and women—10,000 times the 300 mentioned above—will have contracted a lethal cancer from natural causes or from smoking, and the contribution of Chernobyl is less than that which would result from a 1% increase in the use of tobacco. It is, moreover, twenty-five times less than that due to the natural background. It is not hard to understand, therefore, why authorities responsible for nuclear safety are irritated when, with each case of leukemia, an accusing finger is pointed at Chernobyl, when it is clear that the vast majority of cancers—with the notable exception of thyroid cancer—are due to other causes, even in the areas most affected.

Conclusions of the WHO Study

The study published ten years after Chernobyl by the coordinated teams of the World Health Organization allows us to separate out the real and indisputable effects of the Chernobyl accident in the nearby regions. It sheds some light, sometimes unexpected, on the consequences of the catastrophe and the way in which society has faced up to them. The lack of information and the official lies of the Soviet government led to devastating anxiety about their health among the populations concerned. The conditions of evacuation and relocation aggravated matters considerably and produced symptoms that are never observed as consequences of irradiation. A psychological and clinical investigation after the accident recorded headaches, oppressive feelings in the chest, indigestion, sleep problems, loss of concentration, and a descent into alcoholism. In children, diseases such as endocrine disorders, mental difficulties, and problems of the nervous, digestive, and genital systems were recorded.

The study evaluated the immediate impact of radiation and the longer-term consequences of thyroid cancer and leukemia. The immediate effects were limited to the plant personnel and the firemen called to extinguish the fire of the highly radioactive graphite moderator expelled from the reactor: two of them died. There were 444 people on the site, and they were seriously irradiated. Of the 300 admitted to hospital, 134 suffered acute radiation sickness, and 28 died during the first three months. Those who survived experienced serious physical and mental disorders. There were no acute radiation syndromes in the population evacuated from a 30-kilometer zone around the plant. More than 5 million in the regions neighboring Chernobyl were given iodine pills to saturate their thyroids and prevent the fixation of radioactive iodine that entered their organisms with food, but it is not known how many actually swallowed the pills. Many children did not receive them, which explains the great increase of thyroid cancer among them.

Some children in the contaminated areas received high doses in the thyroid because of the intake of radio-iodine, a relatively short-lived isotope, from contaminated local milk. The incidence of thyroid cancer among children under the age of fifteen in Belarus, Ukraine, and Russia has risen sharply. These childhood cancers are of a large and aggressive type. If detected early they can be treated by surgery followed by iodine-131 therapy to destroy any metastases and then by thyroid hormone replacement, but because of the poor economic conditions this may not be done.

It is to be noted that in the United States, despite repeated demands on the part of scientists for a program of storage and distribution of iodine-containing pills in case of a nuclear reactor accident, the nuclear industry and the government long resisted this action. The French government acknowledged, in 1996, the necessity of distributing potassium iodide pills to everyone living less than five kilometers from a nuclear power plant, for use in case of a major accident. In August 1998 the U.S. Nuclear Regulatory Commission reversed itself and now will promote the availability of prophylactic iodine—a long-delayed triumph of reason over public relations. Several states are stockpiling the pills in evacuation centers within some kilometers of nuclear plants, and the NRC will pay for such initial stocks. We believe that individual households should also be encouraged to stock such pills, which cost less than a dime apiece.

In the region most affected, Belarus, where there are 2 to 3 million children, Fig. 7.1 shows the evolution of the number of cases of thyroid cancer each year since the accident.

The contrast between the obvious clarity of these data and the equivocation expressed in the conclusions on the effects of the catastrophe by a study group including two hundred scientists gathered in 1990 by the International Atomic Energy Agency is striking: "There are health disorders, but not related to radioactivity . . . , but there are no health disorders that can be attributed to the exposure to radiation. . . . The estimated absorbed doses in children's thyroids are such that there may be a statistically convincing increase in the future, but

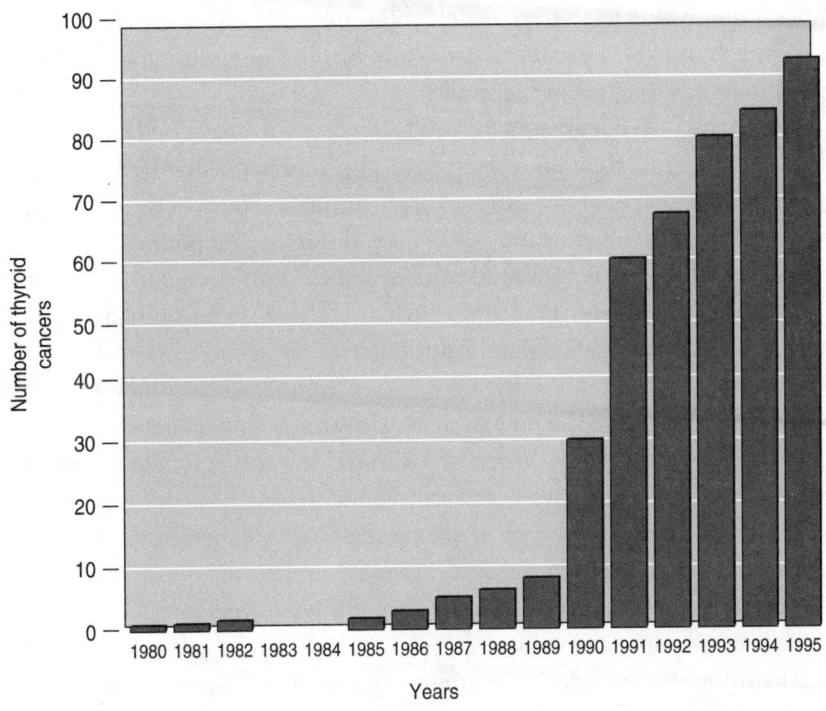

Fig. 7.1. Annual increase in thyroid cancer in Belarus.

that would be difficult to discern." The group surely knew that the very large increase in thyroid cancer would not be "difficult to discern." Our own experience suggests that a clear prediction would have been unpopular in the group and would have been softened in the editorial and publication process; the consequence is a loss of confidence in the organization responsible.

In Ukraine, the number of thyroid cancers has increased by a factor of 8 since the accident. In the contaminated regions, more than 95% of the thyroid cancers were very invasive, that is to say, largely extended to neighboring tissue, which made curative surgical intervention very difficult. The lack of ultrasound equipment for the early detection of cancer was a disaster. A 1999 study ("Radiation and Risk," 1999 special issue 2) shows the annual incidence of thyroid cancer in ten oblasts (regions) of Belarus and Russia increasing from about 220 pre-Chernobyl to more than 1000 annually in 1995–1997, the last year for which data are available.

Cesium, which contaminates animal food and drinking water, presents a longer-term danger because of its longer half-life. Long-term irradiation is almost entirely due to cesium, which enters the body with vegetables and milk and contributes also by external radiation. Strontium and plutonium isotopes account for less than 5% of the irradiation.

The incidence of leukemia was studied in a group of 270,000 children. The observations on its frequency were made in the region that had received an irradiation greater than 555 kBq/m^2 of cesium-137. They show an increase in morbidity due to leukemia and other blood disorders, but persons in neighboring uncontaminated regions also experienced the same trend.

The study concludes that up to now, there have been no changes in the morbidity by leukemia that can be attributed to the Chernobyl accident. That is unlikely to change; it is a disease that does not take a long time to develop. Leukemia cases from the radiation at Hiroshima and Nagasaki peaked in 1953 — eight years after the exposure. Nevertheless, films shown on television screens the world over have portrayed, with considerable emotion, hundreds or thousands of children with leukemia caused by Chernobyl. They could have been filmed ten years before Chernobyl in the same hospitals. The consequences of Chernobyl are serious enough; to exaggerate them with falsehoods is simply not acceptable.

The study also considered injuries to fetuses. It concludes that there was an incidence of mental retardation in certain exposed children. But it is impossible to impute the condition to radiation alone, or even in large part. Stress and the difficult circumstances of life of the parents could have been a factor. It is probable that a study of the lives of the inhabitants affected by the catastrophe would show irreversible damage going well beyond effects that could plausibly be attributed to radiation. The studies show an increased morbidity and mortality among the 800,000 having worked at Chernobyl after the catastrophe, but it is very difficult to come up with the exact number of victims.

Our Own Conclusions on Chernobyl

It is clear that even the major catastrophe of Chernobyl had a minor impact on the health of the average inhabitant of the northern hemisphere. But on the psychological and political level, it has an extraordinary effect, whose consequences on the economy, and even on public health, can be considerable. The problem is serious when people find themselves in a highly contaminated region where there is a severe short-term risk to their health or to their lives. It means very little to them to know that epidemiologists consider that if the radioactivity with which they are afflicted were uniformly distributed, at very low dose, over the entire population of the globe, there would be the same number of victims in total, but the effect would be imperceptible because of other causes of cancer, much more numerous.

The Chernobyl accident had a profound impact on the way nuclear energy is now perceived. It forces us to consider the consequences of ionizing radiation and on the relative dangers of all the sources of radiation by which people are surrounded, natural or manmade. In particular, nuclear wastes force us to contemplate the world that society will leave to its descendants for thousands of years to come.

Each accident in nuclear plants, even an insignificant one, is magnified. The nuclear fuel reprocessing center at La Hague, France, which recovers some 16 tons of plutonium per year from 1600 tons of highly radioactive power-reactor spent fuel, has become a favorite target. Greenpeace and other antinuclear groups have not hesitated to accuse it of irresponsibly releasing dangerous radioactive substances or of causing cancers in the neighborhood. And yet the local or regional radioactive impact of the plant corresponds to 1% of the natural radioactivity. Epidemiological studies on the appearance of cancer in the neighborhood of the plant (or in the South of France following Chernobyl) have shown nothing at all, in contrast to the enormous rise in thyroid cancer among children in several regions of the former Soviet Union.

These attacks that confuse the serious problems of Chernobyl with the minor radioactivity releases from the reprocessing plant at La Hague obscure the real nuclear debate, which should be directed toward the safest choices for

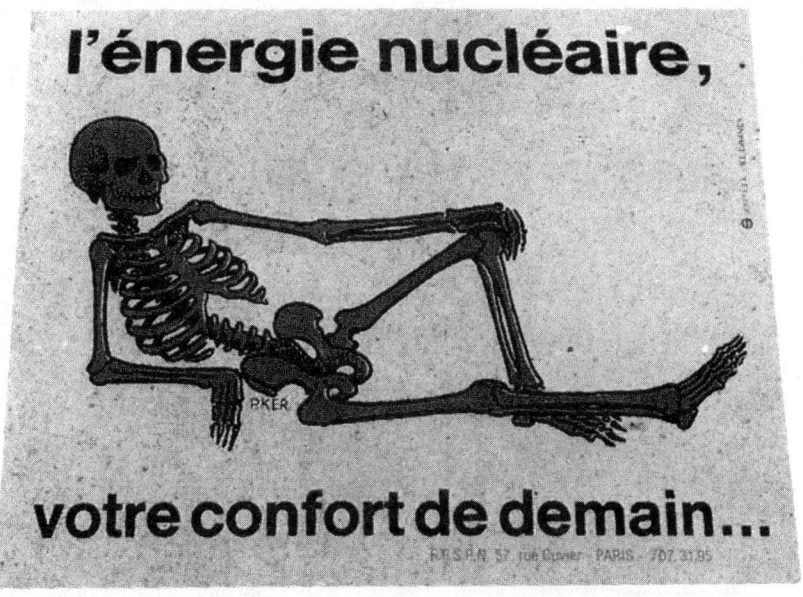

Fig. 7.2. Antinuclear poster. It reads: "Nuclear energy, your comfort for tomorrow."

The will-o'-the-wisp (the mysterious flame occasionally seen fleetingly above a graveyard) is an object lesson in our world fraught with real dangers. The hysterical reactions to will-o'-the-wisps are admirably illustrated by the long commentaries devoted to them in a nineteenth-century edition of the *Dictionnaire Larousse*, the great French dictionary:

"In September of 1849, between Camenz and Koenigsbrück, then again in November near Leipzig, Mr. Vogel saw flames rise over muddy ditches, then go out only to reappear further on, behaving much like a troop of elves gamboling about.

"Usually, because of the air set in motion, the will-o'-the-wisp recedes from an observer who wants to approach it, or follows him when he moves in the other direction. That is doubtless the reason for all of the terror and superstition engendered in the minds of simple people and for all of these mysterious legends which come to us from countries where such phenomena are to be seen.

"Belief in the miraculous is still very current in the countryside; simple natural phenomena have had no difficulty in reinforcing such fantasies. Since it is the way of ignorance to distort the most natural things, it is no surprise that hallucinations and a sense of wonder surround will-o'-the-wisps. That is why mysterious influences are often attributed to them, not only on the temperature and the fruits of the earth, but also on the very facts of human life. In the eyes of country folk, the will-o'-the-wisp is a light that walks, and that light is none other than a soul in torment performing its nightly peregrinations. As soon as one of these phantasms is seen somewhere, the strange news runs over the countryside, commented on in thousands of ways and magnified from mouth to mouth, instilling the greatest terror in the credulous.

"The usual theme of these legends relates to a struggle between a man and the walking light. The hero is on his way, at night, and has no choice but to pass by a cemetery. Suddenly, a few steps away, from the high grass covering the tombs, a luminous globe springs forth, rises, descends, stops, rocks back and forth. At the sight, the man is taken with a frenzy of fear; he takes off like a madman, the phantasm hot on his heels. After a race of a few desperate minutes, the man, suddenly exasperated, stops and turns: the light also stops, a few steps away; he, in his fright, cries out in terror, threatening the luminous globe and accompanying his words with violent gestures. The phantasm comes closer. The man, to repel his assailant, closes his eyes, strikes right and left, lacerating himself with his own hands until, exhausted, he falls to the ground. When he rises, the light has disappeared. All this is amplified in a thousand ways and turns into a terrifying tale.

"When will sufficient education liberate the country folk from these unfortunate superstitions?"

The fear inspired by these will-o'-the-wisps can be compared to the irrational fear that some have of radiation. An exposure to X-rays is tolerated, even welcomed, if it comes from a doctor or dentist. But much smaller doses seem to produce profound psychological disturbances if they are caused by a much smaller accidental exposure to radioactive materials.

We are not trying to minimize the very real and sometimes horrible dangers of major nuclear accidents. It is, however, distressing to note that the investigations concerning Chernobyl, where in 1986 the destruction of a badly designed and badly operated nuclear reactor took place, have revealed the appearance of heretofore clinically unknown maladies that are probably related to the stress and anxiety people felt about the unknown consequences of the radiation released by this explosion. There are, to be sure, real illnesses induced by radiation, such as leukemia or thyroid cancer, but it is neither correct nor constructive to interpret each symptom—or even feeling of unease—as an effect of radiation.

assuring energy sources as little-polluting as possible, and toward considering the problem as a whole and not just for the short term. The need to reflect on the consequences of our choice for centuries to come impels these decisions to be made with care, and demands that society avoid superficial knee-jerk reactions, such as the equally regrettable tendency to minimize, even to totally deny, the consequences of the Chernobyl accident. The very fact that, more than a decade afterward, major Ukrainian forest fires continue to contaminate the atmosphere should lead nuclear power enthusiasts to proceed with caution. The radioactive substances buried in the soil and absorbed by organic matter can be spread over wide areas by smoke and wind and instill fear, perhaps out of proportion to the real danger, but a reminder of the one accident thus far in a nuclear power plant that has caused widespread harm to the public.

THE CRITICALITY ACCIDENT AT TOKAI-MURA (1999)

At 10:35 a.m. on September 30, 1999, workers in a small facility at Tokai-mura, a town 110 kilometers northeast of Tokyo, were exposed with no warning to a lethal flash of radiation in conducting what they apparently considered an ordinary industrial process. Though the accident caused less death and injury than the typical explosion of methane in a coal mine, it will have dire consequences for the nuclear industry in Japan.[19]

A private company, JCO, has for years operated this facility involved with the preparation of uranium oxide fuel for commercial reactors and for the experimental sodium-cooled fast reactor Joyo at Tokai-mura. The fuel consists of typical metal-clad "pins" that are stacks of uranium oxide ceramic disks containing about 19% U-235, in comparison with the 4% enrichment of fuel for a light-water reactor and the 0.71% U-235 content of natural uranium. Because the liquid sodium coolant is by design a poor moderator, the neutron kinetic energy in the reactor is sufficiently high that the fission cross section of U-235 is much lower than for a reactor operating at thermal neutron energy—hence the greater required enrichment.

The accident took place in a small cylindrical tank 50 cm in diameter and 70 cm high which had a cooling jacket extending halfway up the tank. What was involved was the processing of enriched uranium oxide powder that had never been in the reactor and so had no fission product contamination and no significant radioactivity. No more than 2.4 kg of uranium was allowed in the vessel at any one time, to keep the reactor well below criticality—that is, to maintain the amount of fissile material well below what would suffice for a chain reaction that would make the difference between a few neutrons per second and 10^{15} neutrons per second.

Presumably to increase production, the workers mixed batches of uranium oxide powder (with 18.8% U-235) with nitric acid in a bucket, ultimately pouring 16 kg of uranium equivalent into the mixing tank, exceeding the critical mass for a solution surrounded by a water "reflector" of neutrons. The neutron multiplication is far slower than it is in metallic uranium, both because the neutron goes farther and because it moves more slowly, with the result that there is no explosive disassembly as there is in a nuclear weapon. Instead, such a criticality accident (and there have been twenty-two out-of-reactor criticality accidents in the United States, and some in the Soviet Union) suddenly heats the water, which expands both by boiling and by the generation of hydrogen and oxygen by the intense radiation, thus reducing criticality and terminating the chain reaction.

Because the tank at Tokai-mura had a cooling jacket, the uranium soon became critical again. This slow pulsing continued for about 20 hours, and was terminated only when water was drained from the apparatus and the system permanently became subcritical. Three staff were irradiated severely (two have died), thirty-six less severely, and ten persons not working at the facility received some radiation, including three firemen who rescued the severely irradiated workers.

Since there was no shielding against radiation as there is in a nuclear reactor, not only the gamma rays but neutrons escaped and were detected hundreds of meters away from the site of the accident.

On the assumption of an average power of 1 kilowatt during this time, there were about 2.5×10^{18} fissions. This would have produced about 3×10^{12} Bq of the three short-lived iodine isotopes I-131, I-133, and I-135, of half-lives ranging from six hours to eight days. According to data cited by the U.S. NRC in April 2000, about 1 mg of uranium fissioned, producing a total energy release of 22 kWh and the amount of radioactivity cited here.

For comparison, the accident at Three-Mile Island involved about 2.6×10^{18} Bq of I-131 in the reactor core, of which about 12% ultimately entered the large containment dome around the reactor. Fortunately, only about 5×10^{11} Bq is believed to have been released into the atmosphere. So the iodine released from Tokai-mura may have been comparable to that released to the atmosphere from Three Mile Island.

Solutions of uranium cannot result in nuclear explosions that cause damage by blast or by very large amounts of radiation. On the other hand, careless handling of substantial amounts of high-enriched metal could lead to a large-scale nuclear explosion, since the neutron background from this material is so low that full assembly of a critical mass might occur even without the use of propellant or explosive.

Evidently, careless operation on the part of the plant management, ignorance on the part of the workers, and a lack of industry or governmental monitoring in Japan all contributed to this accident.

The consequences of the Tokai-mura accident may include a salutary improvement in monitoring and transparency of Japanese commercial nuclear operations; they may involve the abandonment of the fast reactor that necessarily uses fuel of greater enrichment than that in the water reactors; and they will increase skepticism toward nuclear power in Japan.

HAZARDS FROM INDUSTRIAL ACTIVITIES

Each of the steps in the nuclear power cycle causes some radiation exposure to workers and also radiation exposure to the public, which are shown in Table 7.2, for a single large nuclear reactor and its fuel cycle.[20]

We have analyzed the stages of the nuclear fuel cycle, both the once-through and the reprocessing-and-recycle approach, together with an approach to nuclear energy based on thorium rather than uranium ore. At this point we note that radiation exposure to the public from nuclear power operations would be less than that from coal plants except for the global contributions from mine and mill tailings, and from reprocessing, as the operations have been practiced.

Mine and mill tailings contribute less in the case of the cycle using reprocessing than in the once-through approach, because of the 20% saving in the use of raw uranium. However, the radiation emitted by reprocessing more than compensates that reduction, in the data of UNSCEAR 1993. Taking the

TABLE 7.2. COLLECTIVE EFFECTIVE DOSE TO THE PUBLIC FROM
EFFLUENTS OF THE NUCLEAR FUEL CYCLE. DOSE COMMITMENT
IN PERSON-Sv PER GWe-YR OF OPERATION.

	Local and Regional Component:		
SOURCE	ONCE-THROUGH*	REPROCESSING[†] AND RECYCLE	COAL[‡]
Mining	1.1	0.9	0.002 Mining
Reactor operations (atmospheric)	1.3	1.3	20 Power plant
Total Local and Regional	2.4	2.2	20
	Solid Waste and Global Component:		
Mine and mill tailings (release over 10,000 years)	150	120	
Reactor operation, disposal of intermediate waste	0.5	0.5	
Reprocessing solid-waste disposal	0	1.2	125 Use of ash[§]
Reprocessing, globally dispersed radionuclides (to 10,000 years)	0	217	
Total for Solid Waste and Global Component	150	339	125
Grand Total	152	341	145

*From UNSCEAR 1993, Table 53, p. 200.
[†]From UNSCEAR 93, Table 53; Table 42 for local and regional; Table 51 for globally dis-
persed. This column is per GWe-yr of nuclear plant using reprocessing and recycle.
[‡]From UNSCEAR 1993, p. 56, para. 143; p. 57, para. 151. The reprocessing-and-recycle
approach is assumed to use 20% less uranium than the once-through option.
[§]Buildings constructed with 5% of the ash from power production.

ICRP cancer incidence of 0.04 lethal cancer per person-Sv, we see that the
typical nuclear power plant in operation contributes perhaps 1.8 person-Sv per
year (0.08 cancer death per year), in comparison with the 20 person-Sv/yr from
a coal plant (0.8 cancer death per year), but that the mine and mill residues for
the fuel required to feed a nuclear plant for a year ultimately contribute some 5
or 6 cancer deaths per year. Depending upon the conditioning and protection
of the tailings after the mine has ceased operation, the number could be larger
by a factor of 6 or smaller by a factor of 150.[21] And the reprocessing activity adds
some 217 × 0.04 = 9 deaths per year per plant—essentially none of them local
or regional, because these are due largely to carbon-14 liberated to the atmosphere and disseminated worldwide. This number of estimated deaths from the
worldwide radioactivity disseminated from reprocessing, as listed in the 1993

UNSCEAR report, calls into question the decision procedure that led to the choice of the reprocessing-and-recycle fuel cycle. But the situation has improved — at least for the reprocessing of light-water reactor (LWR) fuel at the British Nuclear Fuels Limited (BNFL) plant at Sellafield, England.

For 1997, BNFL reprocessed 684 tons of LWR fuel and 275 tons of Magnox fuel.[22] BNFL reports that the 10,000-year dose commitment to Europe is 29 person-Sv, and to the world is 422 person-Sv. Assuming this to be due to 34 GWe-yr of reactor operation, the global dose commitment would be $422/34 = 12$ person-Sv per reprocessing Gwe-yr. This compares with the 1993 UNSCEAR report of some 217 person-Sv per reprocessing GWe-yr. The difference is largely due to THORP's capturing most of the C-14 for disposal as solid waste rather than emitting it as gas. The difference between $217 \times 0.04 = 9$ deaths per GWe-yr and $(12 + 6) \times 0.04 = 0.7$ death per GWe-year is striking.

With the ICRP coefficient of 0.04 cancer death per person-Sv, Table 7.2 shows that deaths per year of operation and fuel cycle for once-through nuclear power are 6; for reprocessing and recycle, 14; and for coal with 5% of the ash going to concrete for buildings, 6. If one asks instead about deaths only in the first 500 years, the exposure from mining and milling is reduced by a factor of 20, and that from C-14 from reprocessing by a factor of 10, so the total deaths per year of operation of a 1-Gwe plant are 0.6 for once-through, and about 1.3 for reprocessing and recycle, and the same 6 from coal. The radiation exposure from future operations of mining and milling could be reduced by a factor of 100 even for abandoned plants by appropriate treatment of the waste piles.[23] As for the global radiation dose due to reprocessing, 99% of that is due to carbon-14 and is delivered over 10,000 years; THORP's operation proves that carbon-14 need not be emitted into the atmosphere.

NUCLEAR INDUSTRY HAZARDS COMPARED WITH OTHER RISKS

We are entering an age when public judgment must be provided a solid base of scientific data, data that have not been crudely filtered by powerful pressure groups or pseudo-ecological sects. A decade ago there was an impressive example with the enormous publicity put out by the large cigarette companies contesting the noxious effects of tobacco.

Besides natural radiation, one should also take into account the radioactive substances brought to the surface of the earth by various mining activities. The concentration of uranium and thorium in certain ores of great commercial importance is much higher than the average over the earth's crust. But except for the most extreme cases, this parasitic radioactivity does not produce annual doses greater than 1 mSv, which is of the order of the variation in natural radiation over the Earth's surface. The mineral "monazite" is rich in the radioactive material thorium. On the Arabian Sea coast of Kerala, in India, and on the Atlantic coast of Espirito Santo in Brazil, the dose rate can reach 36 mSv/year—ten times the total annual radiation dose to the average person in the world population. For uranium ores, despite the very small volume of fissionable nuclear fuel consumed in a power reactor, it is necessary to extract 100,000 tons of ore per year for each electricity-generating reactor. Radon leaking from the pile of mill tailings (the ground rock from which uranium has largely been extracted) contributes globally to the radiation exposure.

According to Richard Peto, an Oxford epidemiologist, tobacco causes three million deaths each year, and in thirty years this figure will reach ten million, mostly in developing countries.[24] One might, perhaps, expect that Greenpeace will henceforth turn away from nuclear concerns and mobilize international brigades that will sink cigarette transport ships and set fire to tobacco fields. We absolutely oppose such violence, but three million deaths per year should be kept front-and-center in our consciousness.

We have seen that widespread disease attributed to the Chernobyl disaster could not in fact have been caused by radiation. On the other hand, the nuclear industry's reluctance to take seriously the 24,000 cancer deaths that we expect as a result of Chernobyl is reminiscent of the tobacco firms in their ludicrous and deceptive charade of maintaining, until 1997, that nicotine was not addictive. The nuclear industry and official bodies would benefit from honesty in this matter. For example, in UNSCEAR 1993 (p. 23) we find this candid statement regarding Chernobyl: "The collective effective dose committed by this accident is estimated to have been about 600,000 man-Sv." But in UNSCEAR 2000 there is no overall collective dose estimated—only (vol. II, p. 486) that the "estimated lifetime effective dose" for Belarus, the Russian Federation, and Ukraine totals about 60,000 man-Sv. Ignoring the dose to the rest of the world is not progress.

The solution is not to pretend that there is no harmful effect of low doses of radiation inflicted on the global population (or, worse, to conceal the radiation exposure), but to discuss openly their causes, to evaluate their consequences, determining whether the overall harm is outweighed by the overall benefit, and to develop a system that is acceptably safe. Society should follow the same route taken by industrial societies in responding to the initial insecurity surrounding the railroads or the terrible accidents in the coal mines—that is, to improve safety. The risk of accident has never been completely eliminated, but has been reduced to such a level that no one today dreams of abolishing the railroads or the use of coal, even though, paradoxically, coal contributes substantially to an increase in the natural radiation background. But that is the least of its defects, and is negligible compared to its contribution to the greenhouse effect, as we have already noted.

Without much fanfare, the European nuclear power industry has instituted a study of an evolutionary large (1500–1800-MWe) water-moderated reactor that would be as safe as if it were buried underground. In case of an accumulation of all of the imaginable catastrophes, the products of a core meltdown would remain confined inside the reactor containment.[25] It is certainly an unpleasant prospect for a nation to have to live, for tens of millennia, with a coffin filled with such a devil's brew. But people get used to it when their country is endowed with localized quicksand or volcanic craters, to say nothing of earthquakes as in Southern California. In this next-generation reactor, accidents will be rather rare—less than one case every hundred years for a thousand reactors. But such reactors have not yet been built, nor even fully designed; nor have they been subject thus far to review by knowledgeable and disinterested parties, such as the U.S. Nuclear Regulatory Commission. Such reviews take time and money, and until recently there was something of a chicken-and-egg problem here: NRC review took place only when a reactor was proposed to be built in the United States. The NRC now has a process for certifying reactor design as well as for "banking" sites for building reactors. Any issue that has been resolved during the process of certification cannot be reopened in the subsequent litigation. Thus, NRC has reviewed and certified the designs of three advanced reactors—the Combustion Engineering System 80+, the Westinghouse AP600, and the General Electric Advanced Boiling Water Reactor (ABWR). But there have been no U.S. orders for these plants, and none is expected for several years. However, two ABWRs similar to the U.S.-certified design are now operating in Japan, with more planned. This is progress.

A 1990 NRC study of "severe accident risks" for five specific U.S. light-water reactors estimated annual core damage from "internally initiated accidents" to range from 4 to 60 per million reactor-yrs (Mry). Earthquakes contributed 3 to 120 per Mry, and fires 11 to 20 per Mry. The NRC has since estimated earthquake damage to be somewhat less probable, so that the probability of core damage is below 100 per Mry. The present U.S. population of about one hundred reactors, if maintained to the current standard, would have one severe damage incident per one hundred years, on the average.

The safety of the nuclear industry may be put in context by comparison with the causes of death in the United States, in the year 1994, as summarized by the Census Bureau in Table 7.3. Obviously, the table does not include the

TABLE 7.3. DEATHS AND DEATH RATES FOR 1994 IN THE U.S.: "ALL RACES, BOTH SEXES"; POPULATION = 261 MILLION

CAUSE OF DEATH	NUMBER TOTAL PER YEAR	PROBABILITY PER PERSON AND PER YEAR, PER 100,000
Cardiovascular diseases	2,286,000	876
Malignancies	534,000	205
Accidents and adverse effects	90,000	34
Motor vehicles	42,524	16
Water transport	723	0.28
Air and space transport	1075	0.41
Railway	635	0.24
Falls	13,450	5.15
Drowning	3404	1.30
Fire and flame	3986	1.53
Firearms (unspecified)	1123	0.43
Handguns	233	0.09
Electric current	561	0.21
Accidental poisoning by Drugs and medicines	7,828	3.00
Other solids and liquids	481	0.18
Gases and vapors	605	0.26
Complications due to medical procedures (approx.)	2,700[*]	1.03
Inhalation and ingestion of objects	3,065	1.17
Added by the authors:		
Normal operation of nuclear fuel cycle (6 per year × 300 plants / 10 billion people).		0.02 (200 per billion)
Three Mile Island, ~3 deaths/20 yr × 220 million people		0.00007 (0.7 per billion)
Chernobyl in USSR, 30,000 deaths/20 years		0.6 (6,000 per billion)

But note that a 1999 study of the Institute of Medicine, "To Err Is Human: Building a Safer Health System," indicates that 49,000 to 98,000 Americans each year die as a result of "medical errors in hospitals"—more than die in motor vehicle accidents.

risks in case of war or extraordinary natural disasters, which are discussed in the chapter on nuclear winter.

Supreme Court Justice Stephen Breyer addresses in a larger context not only some of the risks of nuclear power, but also the organization of government and society to make decisions rationally and effectively to regulate risks in general.[26] According to estimates quoted by Breyer, the cost of averting a premature death ranges from $0.1 million for a ban on unvented space heaters and a protection standard for steering columns in automobiles, to $0.8 million for a ban on children's flammable sleepwear, to $45 million for covering or moving mill tailings at active uranium mines, and to $86 *billion* for limiting exposure to formaldehyde. A rational approach is to "pick the low-hanging fruit" by regulating risks up to some $5 million per life saved in the United States.

Breyer notes the discrepancy between expert analysis of risk and the risks as perceived by the general public. Of 30 activities and technologies, nuclear power is judged the riskiest (i.e., first of all 30) by the League of Women Voters and by college students (in data published in 1987), while it was ranked 20th by experts. In contrast, motor vehicles were ranked 2, 5, and 1 by these three groups, respectively, and power mowers 27, 28, and 28. Breyer recommends government organization and career paths for achieving understanding of problems, opportunities, and options for dealing with risk. To which we add the importance of committees of outside, independent scientists and policy experts for this purpose.

In a world full of Good Samaritans, charitable to one another, and where huge meteorites are kept at a distance by powerful guardian angels, this table seems to tell the whole story. But in the real world where human beings seem to cultivate war as an art, its deficiencies are apparent. Moreover, a single meteorite 10 kilometers in diameter, which can be expected to strike the earth about every fifty million years or so, can cause a billion deaths and is not included in the statistics. To be included in the table, a billion deaths in 50 million years would have to be counted as 20 deaths per year, so that the corresponding probability is only one per 250 million per year per person. About one of these deaths per year is ascribed to the United States, so that the corresponding entry in the last column of the table would be 0.0004 (four per billion persons per year).

The table is incomplete, and many other accidents could be included— most of them of very low probability. For instance, two professors of physics and astronomy at Columbia University estimate that since the beginning of commercial aviation, the probability has been one chance in ten that any single passenger flight would be struck in the air by a meteor.[27] They base their argument on the fact that 3000 sufficiently large meteors enter the earth's atmosphere every day, that 50,000 airplanes take off each day, that an average flight lasts 2 hours, so that there are always 3500 airplanes in the air covering two billionths of the surface of the earth. Other hazards are, of course, more likely; in 1997 the United States agencies investigating the loss of TWA 800 just after takeoff from Kennedy Airport concluded that the cause was an explosion of fuel vapor in the nearly empty centerline fuel tank.

Arsenic in Tap Water

On May 24, 2000, the U.S. Environmental Protection Agency (EPA) proposed lowering the limit on arsenic in tap water from 50 parts per billion to 5 parts per billion. Arsenic is an acute poison, but also a chronic poison that has in the past two decades been demonstrated by epidemiological studies to cause cancer of the bladder, lung, kidney, liver, and other organs.

In Bangladesh this is a particularly critical problem, since 97% of the population of 120 million people rely on well water. In the 1970s, the population of 35 million relied on surface water, and the four million wells were welcomed for providing ready access to drinking water free of contagion and contamination. Unfortunately, minerals in the aquifer provide an excessive amount of arsenic, above the country (and U.S.) standard of 50 parts per billion.

At the 500 ppb level of arsenic, internal cancers appear among about 10% of those exposed. Since there is no indication that arsenic at this level contributes to cancer other than in a linear fashion, a reduction to 5 ppb would reduce the probability of cancer to about 0.1% among those consuming such water; about one-tenth of those cancers would be cancer of the bladder. Only in June 1996 did the scope of the arsenic contamination in Bangladesh become clear—a disaster of monstrous scale. In the United States, shortly after taking office in January 2001, the Bush Administration stayed the EPA action, citing costs to local communities. Perhaps the officials do not believe that arsenic at 50 ppb would condemn 1% of the citizenry to internal cancers; as this book goes to press, the EPA has promised another study to define the precise level at which health benefits and costs are in balance.

In no way are we trying to reduce the perception of risk from nuclear power; we want to portray it as accurately as possible. But there is a peculiarity in the individual perception of such risks, as is evident in the relative concern expressed by the American public over commercial aviation accidents, despite the twenty-times-larger death rate (and additional serious injuries) from automobile accidents.

The treatment and management of the spent fuel from nuclear power plants and the choice of reactor and fuel cycle options that can enable nuclear power to play a major role in world energy supply for centuries are crucial questions to be discussed in detail in a later chapter. The special aspect of nuclear power is the radioactivity produced in enormous amounts, which must be isolated with high reliability and for a long time—despite accidents, mismanagement, bankruptcy, and the like. But the all-fossil-fuel option may be less acceptable even over the next century, because of its contribution to global warming.

CHAPTER 8

Reducing Greenhouse Gas Emissions

ENERGY CONTEXT AT THE BEGINNING OF THE NUCLEAR ENERGY PROGRAM

TO THE EXTENT that nuclear power can take the place of fossil fuels, the emission of carbon dioxide can be reduced, and with it world warming from the enhanced greenhouse effect. But CO_2 reduction cannot wait for the massive expansion of nuclear plants.

Until 1973 the worldwide supply of petroleum fuel was dominated by the multinational producing companies, with a relatively minor and passive role played by the nations whose territory contained oil reserves. At the time of the October 1973 war between Israel and Egypt and Syria, the Arab oil nations decided to use the "oil weapon," cutting the supply of oil, taking over ownership of much of the production, and causing world prices to skyrocket. Oil costing $0.25 per barrel to bring to the surface, which had been delivered on the world market for $2, rose to a $12 price to the refinery.

Between 1970 and 1980, it appeared likely—at least to optimists—that nuclear energy offered the possibility of producing electricity more cheaply than from any of the other available sources. The situation varied from one country to another depending on its coal, gas, or oil supplies and its desire to be independent of international trade for its energy supply; it seemed clear to Japan and France, for example, that nuclear energy would win out in the near future. By contrast, for the United States, large domestic coal, oil, and gas resources provided not only a realistic option but economic and political competition to the nascent nuclear power sector. The question now is not whether nuclear power will prevail in the United States, but whether it will be phased out completely.

In the United States, gasoline prices in early 1999 were at an all-time low in "constant dollars," but in other nations state-imposed taxes may maintain the price at the pump at four or five times the U.S. price—highly visible evidence of their governments' commitment to limiting the import of oil and their dependence on foreign energy sources. The difference in gasoline price to the consumer explains a substantial part of the difference from nation to nation in popular views on energy sources and their futures. However, to the extent that energy prices vary among countries because of internal energy taxes or subsidies, the impact of price on the amount or type of energy use is a consequence and not a cause of a national energy-related decision. Taxes imposed so that the user price reflects the country's view of the real cost (e.g., that of repairing environmental damage, or that of armed forces to maintain access to the energy source) serve to help firms or individuals make decisions on the basis of these very real factors.

By August 2000, the average price paid by U.S. refineries for a barrel of oil had risen from $10 to $30—obviously not because of an increase in the cost of production. Rather, the market power of the Organization of Petroleum Exporting Countries (OPEC) cartel showed itself again, with non-OPEC producers—such as U.S. domestic firms—happily benefiting from $20-per-barrel additional profit.

WORLD ENERGY CONSUMPTION

We shall compare the energy and electricity consumption of the United States, France, and other countries. France gets about 80% of its electrical power from nuclear reactors. For the year 1996, France consumed 378 billion kWh of electricity.[1] Of this, only about 15% came from hydropower and 8% from fossil fuels. Assuming a 30% efficiency of conversion of primary energy (nuclear heat) to electricity, 100 billion kWh of electrical energy corresponds to 1.2 exajoules (10^{18} joules) of primary energy, or 1.137 quads. In 1996, France generated 458 billion kWh of electricity, of which 375 was nuclear, 62 hydro, and 21 fossil-fueled thermal. The excess of generation over consumption is largely accounted for by export of electrical energy.

Table 8.1 indicates, for several countries and for the world, the electricity produced from nuclear energy and the total consumption of electricity. Both are expressed as primary energy—joules of electrical energy divided by the efficiency of production. Also shown are the total energy generated and consumed (coal mined, oil pumped, hydropower fed to turbines, biomass, solar, windpower, etc). Electricity is generated from various sources such as coal, oil, hydropower, and nuclear energy.

The first two columns of data indicate the quantities of nuclear energy gen-

TABLE 8.1. WORLD PRIMARY ENERGY GENERATION
AND CONSUMPTION FOR 1996 (IN QUADS)

One quadrillion BTU (10^{15} BTU, or one quad) is 1.055×10^{18} joules—one exajoule (1 EJ). Electrical energy data presented as TWh converted to primary energy with nominal 39% efficiency; 1 quad = 13.0 GW-yr, or 110 TWh.

	NUCLEAR GENERATED	TOTAL ELECTRICITY CONSUMED	TOTAL ENERGY CONSUMED	TOTAL GENERATED
United States	5.94	28.52	93.36	72.58
France	3.31	3.33	9.87	4.97
Japan	2.49	7.76	21.4	4.05
Russia	0.91	6.43	25.98	39.68
China	0.12	8.14	37.04	37.41
WORLD (incl. the five countries above)	20.1	106.0	375.07	375.37

Data from www.eia.doe.gov/emeu/iea/table xx.html, with "xx" = f1, e1, 16, 62.

erated and the electricity consumed. One can see that nuclear energy represents 18.9% of the electricity consumed in the world (i.e., 20.1/106.0), and 5.3% of the total consumed energy.

We see that the United States uses some 28.5 quads per year of energy for electricity, out of its total energy consumption of 93.36 quads; the corresponding figures for France are 3.33 and 9.87—so that 30% of total energy use is electrical in the United States, versus a slightly larger 33% in France. World total energy production was 375 quads for 1996, balanced by world total energy consumption, of which some 106 quads were in the form of electricity, including 20 quads (19%) of nuclear origin.

The third and fourth columns show very different behavior for the five countries listed. For instance, Japan imported a net 17 quads of the 21 quads consumed, i.e., 81%; France imported half its consumed energy, while the United States imported only 20%. Russia, on the other hand, is a net exporter of energy. Almost all nuclear energy generated is consumed in the country where it is created, as a matter of geography.

WORLD FUEL RESERVES

World fuel reserves for the production of energy are shown in Fig. 8.1, which includes proven reserves, "additional" reserves (very likely), and more hypothetical reserves. "Reserves" is a technical term referring to resources well identified as to location and size and which can be produced commercially at current prices and with current technology.

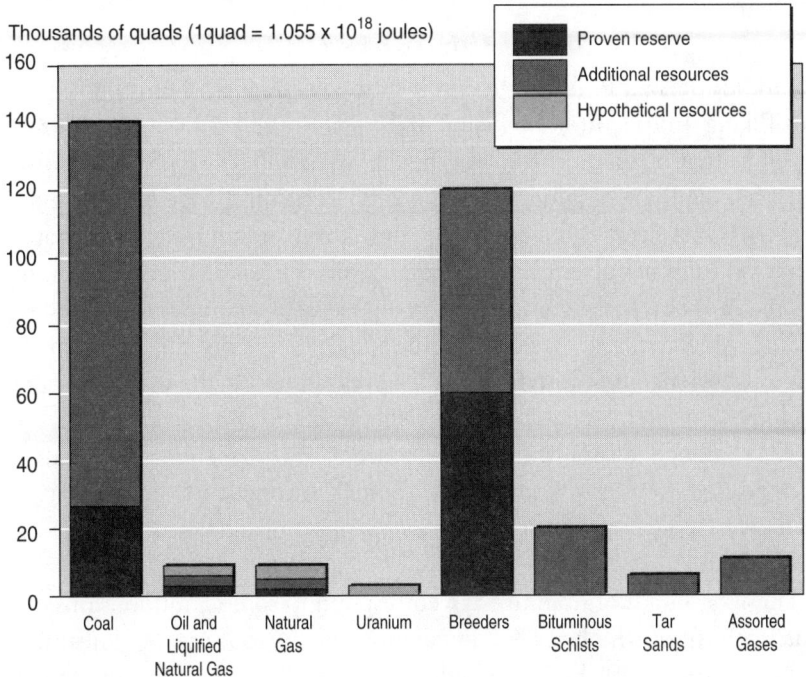

Thousands of quads (1quad = 1.055×10^{18} joules)

Fig. 8.1. Fuel reserves for energy production [Source: Robert Dautray, "Cinquante ans de nucléaire dans le monde," La Vie des sciences, No. 4 (1993), pp. 359–411].

In comparing consumption data in Table 8.1 with Fig. 8.1, the importance of coal is evident; it will be available for several centuries. Coal resources are mainly located in China, India, Australia, South Africa, Siberian Russia, and the United States. France is poor in this respect. The proven reserves of oil can cover the world's needs until 2030–50 and, perhaps, until the end of the present century. Other sources of fuel, bituminous schists (a layered rock altered by heat and pressure inside the earth), and tar sands, from which oil can be extracted, though at higher cost, also appear in the figure. Notable are the "various gases," among which are natural gas hydrates found in the ocean floor and the arctic permafrost. The United States is estimated by the U.S. Geological Survey to have some 340,000 quads of gas hydrates—enough to supply all current U.S. energy consumption for 3,000 years—and the world about one thousand times that amount—enough for one million years. Although the extensive use of this resource and the liberation of the resulting CO_2 would be unacceptable, in the near term the substitution of gas for coal would be highly beneficial in reducing greenhouse gas emissions.

URANIUM FROM THE EARTH'S CRUST

Uranium is present in the earth's crust in the proportion of four parts per million. Taking into account the fission of the uranium-238 in the breeder reactor as well as that of uranium-235, the fission energy that it represents is greater than what could be obtained from the earth's crust if it were entirely made of pure coal. However, if fission energy only from uranium-235 must compete with fossil fuels at current prices, the economical exploitation of uranium in rock requires ores with a concentration of at least one part per thousand.

Since slow-neutron reactors consume essentially only the uranium-235 in their low-enriched uranium fuel, the reserves amount to the very modest number indicated: 1600 quads. Even in the midterm this is a pitifully small reserve—enough to supply the world's current electrical energy demand of 106 quads per year for fifteen years, if all electrical power came from nuclear plants of current design (or about a century for the current level of power from nuclear plants).

The 1600 quads of uranium-235 correspond to some 3 million tons of natural uranium, of which 0.5% (i.e., about 1 part in 200) can be consumed by light-water reactors as low-enriched uranium—although uranium-235 constitutes 0.71% of natural uranium, the "tails" or rejects from the enrichment process still contain 0.2% to 0.3% uranium-235. The uranium-238, which could be burned to near-100% exhaustion by recycle in fast-neutron breeder reactors, corresponds to 200 × 1600, or 320,000 quads.

Of great interest as well are the terrestrial "reasonably assured resources" of uranium, which, according to a leading American expert, John P. Holdren, are likely to amount to 100 to 300 million tons of uranium (corresponding to some 160,000 quads if used in light-water reactors) at a price of \$350 per kg (in comparison with the current spot market price of \$20–30 per kg).[2]

URANIUM FROM SEAWATER

Another, far larger uranium resource exists—the 4.5 billion tons in the water of the world's oceans, of which 2 billion tons could be extracted without the cost increasing much beyond the first exploitation of uranium from seawater. Seawater contains about twice as much uranium as iron—about 3.3 milligrams per cubic meter or ton of water (3.3 parts per billion by weight)—but its extraction would cost today more than obtaining the same quantity from the currently exploited ores. After the alkali metals (lithium, sodium, potassium, rubidium) and the alkaline earths (magnesium, calcium, barium), the next-most-abundant metals in seawater are molybdenum (10 mg/ton) and uranium

(3.3 mg/ton). Iron is present only at 1.3 mg/ton, and this low level of iron limits the fertility of the ocean.

Tadao Seguchi, Director of Material Development at the Japan Atomic Energy Research Institute, has reported a millionfold concentration of sea-water uranium in plastic adsorbents, and has estimated its cost of recovery as some $100 per kg of uranium.[3] He cautions, however, that no systems study has been done to provide a reliable estimate. In field experiments, a mass of 20 kg of modified polyethylene felted fabric is exposed for 20 days in a buoy in the Oyashio current; each kg of plastic then yields 3 g of uranium, 2 g of titanium, 6 g of vanadium, and 6 g of cobalt. The valuable metals can be extracted and the fabric reused many times. In France, Jacques Foos and his team estimate $215–260 per kg of uranium from seawater using the Japanese techniques in the Gulf Stream, which transports about 10 million tons of uranium per year—300 times the amount used per year at present.[4] Foos speculates that this recovery cost could be reduced to $80 per kg by the use of nanofiltration that would preconcentrate the uranium by a factor of 100 and thus allow a similar reduction in the amount of absorbent processed by the ship that is tending the buoys. By using a more selective (and more costly) adsorbent, the French esti-mate that uranium might be obtained from seawater for $18 per kg—by a method less problematical than the exploitation of uranium by mining or leaching. Whether such further advances materialize or not, the availability of uranium from seawater at $100 or even $350 per kilogram is of enormous significance.

The seawater uranium resource of 4.5 billion tons is equal to fifteen hun-dred times the 3 million tons of assured terrestrial reserve; it thus corresponds to some 2.4 million quads if exploited in light-water reactors burning uranium-235, and to 480 million quads if burned in breeder reactors. If the population of the world doubles, and if everyone uses electrical energy at the same rate as do U.S. residents now, the primary energy needed for all electricity would be some 1300 quad per year as compared with the present 106 quad per year; half of the total resource of seawater uranium, if used in pressurized-water reactors, would supply this greatly expanded energy need for 900 years—and if used in breeder reactors the seawater uranium would suffice for 200,000 years.

Therefore, because of the existence of uranium in seawater and in low-grade ores whose exploitation is not profitable at present (because of the low cost of richer ores), the future of nuclear energy will not be limited by the avail-ability of uranium, but rather by the cost of nuclear power plants, and perhaps by the perceived danger of accidents. Other sources of energy, like coal and oil, are more limited than uranium, and their harmful effects can be considerable, as we discuss in Chapter 9, so much so that over the long term uranium will remain one of the great potential energy resources for industrialized societies.

The prospect of a lack of uranium-235 was a major factor impelling the development of breeder reactors—a development that seemed to the nuclear experts at the end of World War II desirable and essential, if nuclear power was to play a significant role in the world's energy supply. The oil shocks of 1973 and of 1979 added to the general instability of Middle East politics and pointed up clearly the potential threat to oil supplies.[5] The sudden increase in prices seemed to validate the potential of nuclear energy, but they subsequently fell. The referenced "A World Oil Price Chronology: 1970–1996" shows the first rise in 1973 from $2–3/bbl to $12/bbl and the second rise in 1979 from $15 to $30–40/bbl, all in current dollars (i.e., the price posted on the day). However, even with the low oil prices of early 1999, the cost of nuclear fuel was less, and it has the additional value of being easy, safe, and cheap to store.

Whether one stores natural uranium, low-enriched uranium, or fabricated uranium fuel, the cost of purchase is small compared with that of oil, coal, or gas. Consider storing as raw uranium ($30/kg) the fuel for one year's operation of a 1-Gwe plant; this is some 200 tons and costs $6 million to purchase. At a 10% interest rate, the interest charge alone for one year's storage is $0.6 million, compared to an annual fuel cost for a coal-fired plant of some $100 million; at a 10% interest rate, to store a year's supply of coal for a single plant would cost an additional $10 million per year in interest charges on the fuel—a charge that would suffice to store nuclear fuel for 16 years. The actual cost of facilities to store 3 million tons of coal, not taken into account in the interest charge on the value of coal, is clearly far more than that of a facility to store 200 tons of uranium.

Nevertheless, the capital cost and time to build a nuclear power plant makes the competition with oil and gas problematic. Because of the relatively small number of reactors now operating—while one prediction had 3000 reactors for the year 2000, the actual number is about 440—uranium ore is more available than foreseen. The price of uranium has collapsed, and even nuclear enthusiasts now agree that the urgent development of breeders is no longer justified in the short run.

NUCLEAR ENERGY CAPACITY: TRENDS AND PROJECTIONS

Among the world's nuclear power plants, ordinary water reactors are in first place. They now represent the most widespread technology, as well as the most tested and the most mature. Pressurized-water reactors are 64% of the total generating capacity, followed by boiling-water reactors at 23%, heavy-water at 5%, gas-cooled graphite at 4%, and water-cooled graphite also at 4%. In 2000,

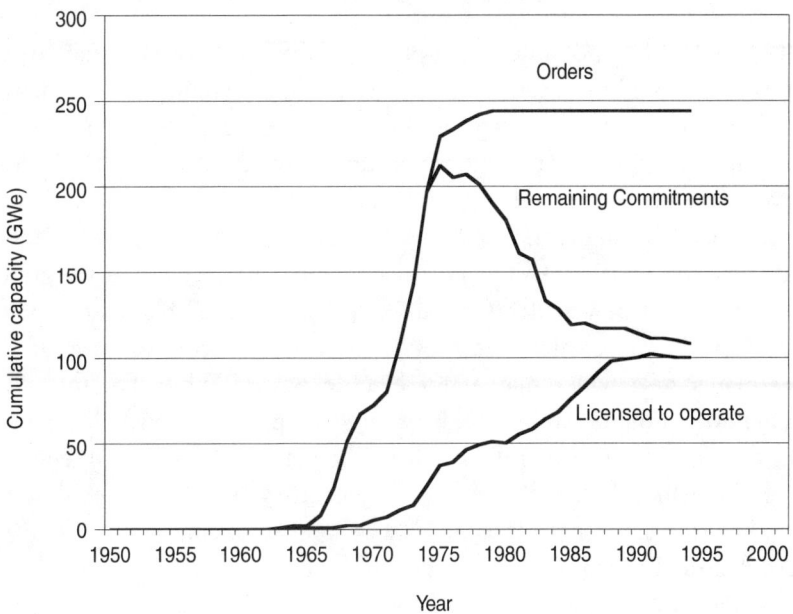

Fig. 8.2. *Total capacity of reactors ordered within the United States, remaining as commitments, and holding operating licenses. (After David Bodansky,* Nuclear Energy: Principles, Practices and Prospects *[American Institute of Physics, 1996], p. 9).*

the world's nuclear generating capacity stands at 349 Gwe, compared with 321 Gwe in 1990 and a projected 420 Gwe for 2010.[6] Thirty countries have nuclear power plants; in terms of nominal 1-Gwe reactors, the United States is first with 95.4, followed by France with 63.2, Japan with 43.5, Germany with 21.3, Russia with 19.8, South Korea with 12.9, the United Kingdom with 12.9, Ukraine with 12.1, and Canada with 10.3.

In the United States, many of the plants ordered were never built. No U.S. reactor ordered after 1973 has been put into service. In large part, this was due to the oil shocks of the early 1970s and the resultant emphasis on the efficient use of energy, which decreased the demand for electricity, from whatever source. Another factor in the United States' turn away from nuclear power was the accident at Three Mile Island. The cessation of nuclear reactor construction in the United States has naturally led to an erosion of competence. In 1978, eighty American universities had departments of nuclear engineering; now only thirty-five do.

This disturbance is not restricted to the nuclear industry alone. The abrupt slowdown struck all large-scale electricity-producing activities. The illusion widely shared in 1975 of a doubling of the consumption of electricity in the industrialized world every ten years evaporated. In the United States in particular, producers had to face up to overcapacity and excessive indebtedness of the kind that have blocked the nuclear power industry, and an increase in competitive market forces will only hurt in this respect.

In Europe, the situation is similar. The Germans estimate that no additional source of electricity will be necessary before 2010. In Europe, the growth of the number of power plants has also slowed because cross-border power transfer has allowed greater efficiency in the exploitation of existing nuclear power plants in a wider market for electric power. While the monopoly nuclear power producer in France, EDF, now supplies London with electrical power through a cable under the English Channel, EDF is also vulnerable to competition within France, because of the new European Union requirement that 25% of the nation's electrical market be opened to competition. The 400 largest electrical consumers in France will be able to choose the supplier of their energy.

In Europe outside France, 14% of the electrical energy is supplied by "cogeneration" plants, in which electrical power is produced in conjunction with the supply of heat to buildings or industry; there has been almost no cogeneration in France. Electrical energy from cogeneration can be inexpensive, since it is thus made with what amounts to near-100% efficiency, and it is produced closer to the point of use, thus saving on electrical transmission and distribution costs, which are often larger than the cost of generation. France is the only country in the European Community with plans for additional nuclear reactors, but the liberalizing of the electrical energy market in Europe is more likely to bring a surplus than an opportunity for sales of nuclear energy.

The economic development of East Asia led to an increasing demand for nuclear energy in that part of the world. South Korea has mastered reactor technology. The cost of nuclear electricity there is regarded as lower than that of all alternative sources and has decreased by a third since 1983. In terms of "capacity factor"—defined as the fraction of the time a plant produces its rated power—the South Koreans have achieved a capacity factor of 87.4%, while, on the average, an American plant operated 62% of the time in 1980 and reached 88% in 1999.

Until the major downturn in early 1998 of most Asian economies, South Korea expected to be a major exporter of nuclear power plants. A large number of such facilities have been quickly built in East Asia in just four or five years. Several are superior in safety and reliability to those which are presently in operation in American power plants. In GWe, the most significant changes from 2000 to 2010 are expected to be: Japan +18.3, South Korea +11.0, China +9.0, Russia +8.4, India +5.2, Ukraine +4.8, Canada +4.1, Taiwan +2.6, and United States −5. However, the U.S. Nuclear Regulatory Commission will probably extend the operating licenses of U.S. nuclear plants from the initial forty years to sixty years; the anticipated 5-GWe reduction may not take place.

UNSUSTAINABLE DEVELOPMENT PATTERNS: THE CASE OF CHINA

In the past, economic development has led to strong increases in energy consumption; widespread development along the paths that have been followed by the industrialized world would greatly increase the combustion of fossil fuels and therefore the risk of climate change from the enhanced greenhouse effect. China is a case in point—the most populous country in the world, with one of the greatest annual growth rates in GDP.

China has experienced a decade of rapid economic growth exceeding 10% annually, while maintaining a heavy dependence on coal for industry, home heating and cooking, and the generation of electrical power. Should an annual growth rate of even 6% be maintained from 1990 to the year 2050 (a factor of 24.7 in GDP) with a continuing dependence on coal, China's emission of carbon dioxide in that year (some 30.55 gigatons of carbon equivalent) would be almost 7 times that of the entire world in 1995—*if* coal consumption per unit of GDP continued unchanged, but it will not.

For instance, the GDP of the entire world increased from $13,190 billion in 1970 to $27,684 billion in 1995 (in "constant 1995 dollars"), while industrial energy use increased from 217 exajoules (EJ)—recall that a quad of energy is 1.055 exajoules—to 370 EJ. Rather than remaining constant, the GDP per unit of industrial energy ratio increased from $60.8/gigajoule (GJ) in 1970 to $74.8/GJ in 1995. Since one ton of coal corresponds to about 31.5 GJ, the GDP per kilogram of carbon was $1.92/kg coal in 1970 and $2.36/kg coal in 1995. Coal at $5, or $10, or $20 per metric ton would cost $0.005, or $0.01, or $0.02 per kilogram. China is on a similar path; GDP quadrupled between 1980 and 1995, with energy use only doubling.

The reduction in energy intensity must be combined with a more environmentally conscious approach to the use of fossil fuels if the 1.3 billion Chinese are to continue their economic development without suffering further serious damage to their health and society from pollution. Table 8.2 shows some actual Chinese projections of energy use.

Of course, 55 years is a long time, but it is notable that these projections (to 240 nominal nuclear plants in China) will still leave nuclear power providing only 13% of China's electricity.

TABLE 8.2. THE USE OF ELECTRICAL ENERGY BY CHINA, 1995 AND 2050

YEAR	POPULATION (BILLIONS)	E(GTCE)	TOTAL	HYDRO	FOSSIL	NUCLEAR	RENEWABLE
			GW(e) of electrical energy				
1995	1.2	1.24	200	49	149	2.1	0
2050	1.5–1.8	4.0	1690	250	1100	240	100
Annual growth:	0.52%	2.13%	3.78%	2.96%	3.63%	8.62%	—

The projections of the evolution of nuclear energy in the world are far from reliable; neither the present slowdown in the Western world nor the early 1990s spurt in East Asia is a sure harbinger of the future, as evidenced by the Asian economic problems of 1998.

ALTERNATIVE NUCLEAR AND NONNUCLEAR ENERGY SOURCES

Because of the enormous development cost and the capital investments involved, only a limited number of concepts were implemented for the exploitation of nuclear energy. Furthermore, at the time, the breeder route seemed inevitable to the French and the Japanese nuclear industries, as well as to the early seers of the U.S. program. Neither graphite-moderated high-temperature gas-cooled reactors nor lead-cooled thorium-based breeder reactors have been in the mainstream of development of nuclear energy.

Nonnuclear, nonfossil (that is, renewable) forms of energy that had been ignored earlier merit an intense research effort, because they are more promising, with respect to cost and pollution, than they appeared to be when the breeder was the only essentially unlimited energy resource within our technical grasp. And cleaner approaches to the use of fossil fuels may help over the next century to satisfy energy needs at acceptable cost and environmental impact, before a greatly expanded nuclear sector can be afforded or built.

The supplies of primary energy in 1995 for the United States and the world are as shown in Table 8.3.[7]

TABLE 8.3. WORLD AND U.S. ENERGY SUPPLY, 1995; PERCENT FROM EACH SOURCE SHOWN

REGION	OIL	COAL	NATURAL GAS	BIOMASS FUEL*	HYDROPOWER	NUCLEAR	OTHER†
U.S.	38	22	24	3	4	8	0.4
World	33	22	20	13	6	6	<0.5

*"Biomass" fuels are wood, charcoal, crop wastes, and manure.
†"Other" in this table are wind, solar, and geothermal.

TABLE 8.4. CONCEIVABLE HARNESSABLE ENERGY FLOWS, WORLDWIDE. TERAWATTS (1 TW = 30 QUADS PER YEAR)

Solar electric	50 (elec)	1% of the land, 20% efficiency
Biomass	20 (chem)	10% of the land, 0.8% efficiency
Ocean thermal	9 (elec)	2% of absorbed sunlight, 2% efficiency
Hydropower	2 (elec)	all practical sites
Wind power	1 (elec)	windiest 3% of the total land area; no at-sea sites
Waves, currents, tides, geothermal	<1 (elec)	

In a 1998 compendium, John Holdren estimates "conceivable harnessable renewable energy flows" for the entire world, as shown in Table 8.4.

There are non-fossil-fuel approaches to energy that are practical and economically competitive in some places—such as the use of solar energy for heating domestic water. Solar energy can heat a central boiler to provide steam for producing electrical power (the so-called "power tower" approach). Furthermore, in the form of "solar cells" made of crystalline or noncrystalline silicon layers it can serve as a source of electricity; nonsilicon systems have been demonstrated as well. After all, the great majority of our artificial earth satellites—broadcasting, weather observing, photoreconnaissance—obtain kilowatts of power from solar cells. Although this is a book on nuclear energy and nuclear weapons, it is important to note that the continued application of science and technology to the solar-cell option—solar photovoltaics—will continue to reduce the cost of this approach. But the sun's light at the earth is a dilute source of power, and one that is absent for much of the day in most places (and even for much of the year in other places), and solar energy is not yet competitive with fossil fuels or nuclear energy.

Much work has also been done on modern windmills for electrical power generation, electrical power from ocean waves, from the tides, from geothermal energy (i.e., either steam vents or hot-dry rock underground), or from ocean temperature differences. Compared with the world's actual industrial use of energy in 1996 of 11.9 terawatts (one TW is 1000 GW) and 1.8 TW of "traditional" energy (fuel wood, crop wastes, dung), these renewable energy flows might conceivably contribute 90 TW. The Department of Energy announced in June 1999 the goal for the United States of obtaining 5% of its electricity from wind energy by the year 2020, up from 0.1% in 1998.[8] In June 1999 the installed U.S. wind-electric capacity was 2.5 GWe, while a year before it had been 1.6 GWe.

One can imagine that with future sources of energy for humanity at stake, the operators of American power plants could be assessed a substantial tax per kilowatt-hour (2%, for example) to finance research and demonstration on energy sources for the long term. With the $4 billion annually in funds thus obtained, certain approaches more expensive in the near term could be financed that might turn out to be indispensable for the future. But we should not minimize the difficulty of obtaining such funds for research or long-term investment in the public interest. In reality, increased taxes of $2 per gallon would still leave gasoline cheaper in the United States than in most of the rest of the world and would make much more money available for research.

A 1997 report of a panel of the President's Council of Advisors on Science and Technology (PCAST) showed the course of government spending for energy research and development over the last thirty years (see Fig. 8.3). It rises from about $1 billion per year in 1966 (80% of it spent on fission energy) to $5–6 billion in 1977–81, and falls to about $1.3 billion in 1997, with about 3% spent on fission energy.

This report recommends also an increase in government-funded energy R&D ("Applied Energy Technology" in the Department of of Energy) from $1.379 billion requested for the fiscal year 1998, to $2.068 billion for FY 2003 (in "as-spent" dollars—that is, dollars corrected for anticipated inflation). More important, however, is the change in allocation of the funding.

The 1997 PCAST report advocates doubling by 2003 the 1997 annual expenditure of $373 million on energy end-use efficiency, more than doubling the $42 million fission research budget, leaving fossil fuel research near $365 million but reorienting it to address the carbon dioxide and oil import problems, increasing research on fusion energy by about 20%, and doubling the $270 million budget for research on renewable energy, such as fuel wood, alcohol from plants, wind energy, and solar energy.

The substantial government expenditure on renewables (which at present supply only 0.4% of U.S. energy) reflects the promise this panel sees in this resource. Notable also is the large ($755 million) annual expenditure recommended for FY 2003 in improved end-use efficiency. The report notes that improvement in energy efficiency from 1975 to 1995 is "now saving U.S. consumers about $170 billion per year in energy expenditures, and is keeping U.S. emissions of air pollutants and carbon dioxide coming from fossil fuel combustion about one-third lower than they would otherwise be."

In a parallel French communiqué, in February 1998, Prime Minister Lionel Jospin commits a new and continuing $90 million annual budget to renewable energy and end-use efficiency, substitution of natural gas for more polluting fuels, and the introduction of cogeneration.

We enthusiastically support the exploration of renewable energy, advances

US Applied Energy Research Expenditures

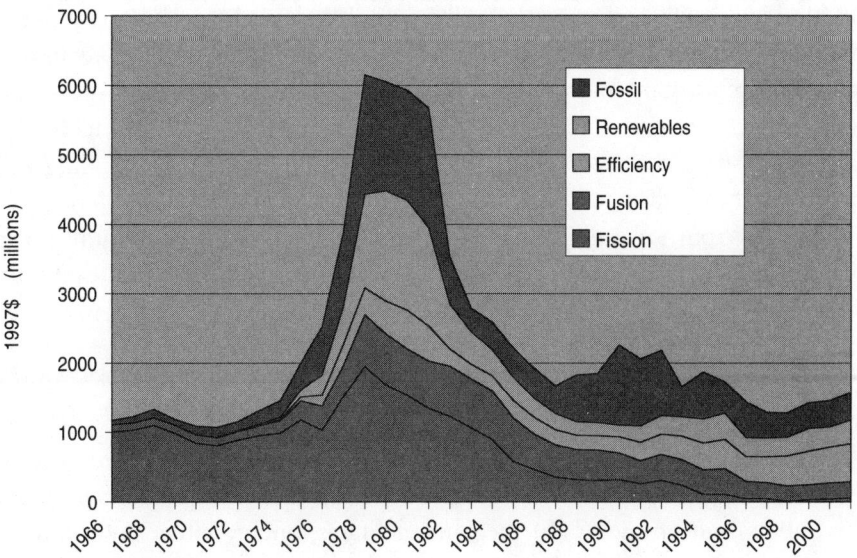

Source: 1997 PCAST report, updated.

Fig. 8.3. Energy Technology R&D budget authority for DOE and predecessor agencies, 1966–1997.

in efficient use of energy, improvements in efficiency and cleanliness of coal-fired power plants, and improved storage and transmission of energy.

Perhaps as a consequence of the PCAST report, the Department of Energy in late 1998 announced a Nuclear Energy Research Initiative (NERI) with initial annual funding of $19 million. This compares with the PCAST panel recommendation of $50 million for the first year, reaching an annual steady state of $100 million in 2003. PCAST advocated also for 1999 $10 million for study of the extension of operating lifetime of existing reactors, and $6 million for

TABLE 8.5. ENERGY-TECHNOLOGY R&D IN THE G-7 COUNTRIES, 1985 AND 1995 *(in millions of 1997 dollars, converted from national currencies at 1995 exchange rates)*

	CANADA	FRANCE	GERMANY	ITALY	JAPAN	U.K.	U.S.
1985	491	NA	1663	110	4558	741	2100
1995	250	704	375	303	4934	87	1282

university research reactors. With a deadline of January 29, 1999, NERI sought to fund innovative technologies for the next decade and "revolutionary" technologies over the next fifty years. It was to focus on reactor and fuel technologies that are resistant to diversion to weapons; new reactor designs with higher efficiency, lower cost, and improved safety; compact reactors; new technologies for onsite and surface storage of nuclear waste; better understanding and performance of advanced nuclear fuels; and fundamental nuclear science.

By May 2001 $79 million had been appropriated, with $35 million in the latest year. Only $6 million is for new awards of two to three years' duration. The total is only 35% of the PCAST recommendation, and not a penny is for vital work on seawater uranium.

ENERGY CONSERVATION

In Western Europe and Japan since World War II, efficiency has been driven by high energy prices to the consumer—largely maintained by taxes imposed by national governments. In the United States, energy efficiency improvements have come largely in the last two decades, in part due to structural changes in utilities, and because of a competitive market. Particularly influential has been a requirement that electrical generating and distribution companies accept power produced by independent suppliers, such as the operators of small hydropower dams and industrial firms that produce electricity from the combustion of waste products like wood chips or sawdust, or from cogeneration. In addition, a restructuring of incentives in the utility industry has allowed profits to be made from providing customers with technology and equipment to use electricity more efficiently.

Beyond lower cost or greater net efficiency of energy sources, experience since the first oil shock of 1973 shows the importance of enhancing the efficient use of energy. We give three examples—residential heating and cooling, cogeneration, and lighting.

In a city like San Diego, not much heating is required in the winter, and the temperature and humidity are low enough in the summer that most homes manage well without air conditioning. The same cannot be said for other regions of the United States, such as Minnesota on the one hand and Florida on the other. In the 1950s many homes were built in cold climates with little insulation to reduce the flow of heat from the interior through the wall to the exterior, and with little weather-stripping to lessen the flow of air. To compound the problem of the cost of heating, some homes used the source of heat that was cheapest to install but most expensive to operate—electrical resistance heating. With such a simple and shortsighted approach, fuel that gives a certain heat of combustion H when burned at the power plant delivers heating to the home of no more than about 0.3 H, given a 30% efficiency of generation of electrical power and transmission to the home. A furnace that safely burns oil or gas or coal requires an investment by the builder and the homeowner several times larger than do electrical heaters—hence the decision of the builder to sell a home more cheaply, but with substantially greater operating cost.

Furthermore, the overall cost to society of the generation and transmission facilities to provide this 0.3 H of electrical energy is far more than the investment that would be required to install in the home a furnace capable of burning about 30% as much of any of these fuels to produce the required 0.3 H for heating the home. Detailed analysis shows that for an owner paying these costs, direct electrical heat would cost $2.90 per hour compared with the $0.86 per hour for gas heat. If assessed the true cost, a homeowner would have to pay much more in capital investment for the electrical heat option, and also much more in operating cost for fuel.

Initially, prospective home buyers made no such detailed calculations, but, shocked at the monthly electrical bills and/or unable to afford them, they either turned down the heat and were uncomfortably cold, or paid an even larger amount than would have been initially necessary to replace the electrical resistance heating with a furnace that burned fossil fuel. Some homes installed a "heat pump" at still larger investment cost, which required only electrical power to run it, but (like a refrigerator run backward) could be used to transfer heat from the cold atmosphere to the warmer house—some 3 times as much heat energy as the electrical energy consumed. This produced about as much heat in the home as if the fuel in the power plant had been burned directly by a furnace in the home, but it incurs the additional cost of both the generating capacity and the heat pump.

One of the results of the oil shock was to make clear the consequences of an increase of the cost of energy and gradually to provide a more efficient market (in the technical sense) for the supply and consumption of energy. The first and most affordable improvement now in the heating and cooling field for new residences is to provide better insulation and weather-stripping. There are even systems called "regenerators" that can allow free ventilation of a home (exchanging air with the atmosphere twice per hour) without wasting the heat that would be expelled if that air went directly from the home to the atmosphere without efficient heat exchange to the incoming fresh air.

In industry as well as in the residence, there is often need for both heat and electric power. Electricity powers the lights and drives motors. Heat derived from fuel is used for space heating and also for "process heat" as in melting

glass, making steel, or cooking. For the most part, a building imports its electricity from a central commercial power plant, but makes its heat locally (we
know that it does not "make heat," but that it facilitates the transformation of
energy from its chemical form to available thermal agitation) from the combustion of gas or oil. Occasionally a facility will have an electrical generator,
but that is fairly rare. We have seen that even large generation facilities convert
only about 30% of the heat value of the fuel to electricity, rejecting 70% to the
environment.

Since most of the applications for heat require fairly low temperatures —
room temperature, for example — it is perfectly reasonable to tap the waste heat
from the electrical generation to warm the building or for process heat (that is,
heat used for an industrial activity such as sterilizing food or drying cloth). This
requires introducing a heat exchanger if hot air is to convey the heat to another
room, or an exchanger to water or other liquid if radiators are to be used. In
either case, the process is called "cogeneration" because the same kilogram of
oil or gas or coal can generate both electrical energy and useful heat.

In the United States until about 1980, public utilities made it difficult or
impossible for independent cogeneration facilities to obtain a reliable, economical supply of electrical power from the utility. And it was almost impossible for such a plant to sell excess power to the utility. In the interest of
efficiency and the public good, these constraints have gradually been overcome so that some facilities have achieved a 50% reduction in fuel cost
through cogeneration, which is now often referred to as combined heat and
power—CHP. A simple gas turbine may produce electrical power equal to
30% of the input fuel energy; using "combined cycle" of gas turbine and steam
turbine provided with heat from the effluent of the gas turbine, an overall efficiency of 55% results. Much of the remaining 45% normally rejected as heat
can be used for building heating, or low-temperature industrial process; if it
displaces the burning of fuel for that purpose, the efficiency of the CHP plant
rises to 85%.

Lighting has long been a major consumer of electricity, and the low-tech
incandescent lamp of Thomas A. Edison, although a marvelous achievement
and far more efficient in producing light from the fuel that generated the electricity than is a candle or kerosene lamp, is still only about 15% as efficient as a
fluorescent lamp. This comes about, fundamentally, because the passage of
electrical current through the gas of the fluorescent lamp heats the electrons to
a temperature far above the melting point of tungsten without conveying
much of the heat to the glass bulb. But the cost of manufacture of an incandescent lamp is lower than that of a fluorescent lamp, together with its fixture, its

ballast, its starting switch, and the like. A 60-watt bulb that can be bought for $0.30 and has a life of one thousand hours consumes over its lifetime 60 kWh of electricity; at an average cost of $0.10 per kWh, this $0.30 lamp uses $6.00 of electricity in its lifetime. A lamp that is on for 4000 hours per year (11 hours per day) thus consumes $24 of electrical energy per year. The "payback" time for a fluorescent fixture that costs about $24 and would consume only $3.60 in electricity per year is thus about one year—a very good investment, because after the first year there is a net annual saving of $20 over the cost of operating an incandescent lamp.

Commercial spaces for decades have used fluorescent lamps, but because small ones with a screw base have only recently been available, it has not been easy for the homeowner to refit to fluorescent instead of incandescent lighting. The compact fluorescent lamp introduced by Philips and now manufactured by many companies does make this possible; one can replace a 60-watt incandescent bulb with a compact 9-watt fluorescent lamp that screws into the same socket, produces the same amount of light, and costs about $15 typically, with a payback period of a year for the (rare) light that is on twelve hours per day. Of course, it is less beneficial to use a fluorescent lamp instead of a light bulb that is on only four hours a day; the payback time for a $15 compact fluorescent is then about 2 years, and each $15 investment then "earns" about $7 per year. Compact fluorescent lamp factories in India and China, created by a pioneering interaction between the Environmental Energy Technologies program at the Lawrence Berkeley Laboratory of the University of California and the World Bank, are saving the equivalent of multiple nuclear power plants. Indeed, in the United States in early 2001, one could buy compact fluorescent lamps made in China.

Refrigerators in the United States are another success story. Federal regulations and technological innovation have reduced the energy consumption of the average refrigerator sold from 1800 kWh/yr in 1974 to 450 kWh/yr in 2001. At $0.10 per kWh, this is a saving to the consumer of $135 per year. Before the profligate 1960s and early 1970s, the average energy use per refrigerator was about 350 kW/yr, but the modern unit is larger and much more satisfactory than that of the 1950s. By the time 150 million U.S. refrigerators have reached the 2001 standard, the annual saving in electrical energy will be some 200 billion kWh from the 1974 level—the output of some thirty-two nuclear plants.

IS NUCLEAR POWER NEEDED BEYOND THE NEXT CENTURY?

Niels Bohr often remarked, "It is very difficult to predict, especially the future." But prediction need not be perfect to be valuable. The modern U.S. economy

TABLE 8.6. POPULATION ESTIMATES 2000–2150, IN BILLIONS.

Year	2000	2025	2050	2075	2100	2150
Population	6.1	8.1	9.6	10.5	11.0	11.4

is not very sensitive to energy prices, as exhibited by continued prosperity in the face of the near doubling of oil prices to more than $30 per barrel in early 2000. But a major shortfall in energy availability is very serious.

Even as recently as 1977, it was widely stated in the energy industry that natural gas resources were too paltry for gas to compete with oil, and certainly not with coal. But in recent years nearly all new U.S. power plants run on natural gas. Forecasting is simplified if one can isolate various factors. In this way, one can write "Energy = GDP × (energy/GDP)." So a society's gross energy needs are obtained by determining the energy intensity (joules per dollar of GDP) and then multiplying by the GDP. In this way, the energy need can be inferred from forecasts of GDP and the trend of energy intensity.

Alternatively, for a population P, one can write "Energy Use = P × (energy/person)," so that a community with three times the number of people using the same technology and the same standard of living would presumably use three times the energy. In turn, for a given activity, (energy/person) = (activity/person) × (energy/activity), so that the total energy used per person is the sum of the energy use in each of the activities, and the amount of activity (transportation, for instance) used by that person.

Various responsible estimates of the world population agree fairly closely on the best estimate vs. time, as shown in Table 8.6, but there is serious disagreement over the uncertainties. For example, the World Bank estimates for the year 2100 ranges from 9.4 to 12.3 billion, while the United Nations provides a broader range of 5.6 to 17.5 billion.

The annual rate of increase of energy use is then the sum of the annual rates of growth in population, in (GDP/person), and in energy intensity. A standard set of assumptions is that population will rise about 1% per year until 2050, the per capita GDP by 1.5% per year, and energy intensity will decrease by about 1% per year. So world energy consumption would grow 1% + 1.5% − 1% = 1.5% per year, to about 900 EJ/yr in 2050. It is unlikely, however, that this estimate is accurate to better than 50%. Per capita consumption of commercial energy in 1995 was 360 GJ in the United States, and only 12 GJ in India. The 900 EJ/yr consumption for 2050, divided by the anticipated world population of 9.6 billion, corresponds to a per capita average consumption of 94 GJ/yr.

In the normal course of events, nuclear power will provide only a small part of this total, since it contributed only about 5% of the primary energy in 2000.

Nevertheless, nuclear energy is one of the few technologically mature and greatly expandable noncarbon sources of energy.

PATHS TO A CARBON-FREE FUTURE

Nuclear energy does not contribute significantly to emission of greenhouse gases or to global warming, but it will be many decades before even a willing world could make the transition to nuclear energy. In the near term, the emission of greenhouse gases from coal combustion can be substantially reduced (even essentially to zero) by carbon "sequestration"—pumping carbon dioxide separated from the combustion gas into exhausted natural gas fields or, after liquefaction, into pools on the floor of the deep ocean.

Natural gas can be exploited more fully, as can unconventional sources of gas, such as those that contain gas bound to water in the cold deep ocean sediment. Automobile fuel economy can be improved by the use of hybrid vehicles, in which a small engine is either off or operating at its most efficient speed while it charges a battery or flywheel to accelerate the vehicle. Energy is not dissipated on slowing, but rather converted back into stored energy by a regenerative system; such hybrids entered into widespread use in 2000 and have about double the mileage of other cars per gallon of gasoline.

Henry R. Lyndon, writing in *The Bridge*, a publication of the National Academy of Engineering, concludes that there has developed a broad consensus that energy systems will move toward electricity for all stationary energy uses, and to hydrogen (compressed, adsorbed, or liquefied) for transportation fuel.[9] The hydrogen would be converted to motive power by onboard fuel cells.

In the same issue of *The Bridge*, Jesse H. Ausubel presents "Five Worthy Ways to Spend Large Amounts of Money for Research on Environment and Resources." His last two recommendations are to build 5-gigawatt zero-emission power plants the size of a locomotive and get magnetically levitated trains (maglevs) shooting through evacuated tubes. These are more in the nature of grand challenges than of immediate practical steps, but they do indicate the potential for change.

The flows of energy and money in the energy sector are enormous. World consumption of oil in 2000 amounts to about 75 million barrels per day, which at a price of $30 per barrel is some $2.25 billion per day, or $0.8 trillion per year. Comparable amounts would have to be spent to reduce substantially the greenhouse gas output of power systems. Even more would need to be spent to replace fossil fuel by nuclear power: to build 15,000 reactors to satisfy the current energy needs of the world would likely cost more than $15 trillion compared with the world's GDP of some $53 trillion annually.

Neither the competing technologies nor the projection of world energy production and consumption is the focus of this book. Nevertheless, these topics provide a context to our concentration on fission energy and nuclear weapons.

We don't know whether the world will ultimately run on large centralized power plants—nuclear or nonnuclear—or on distributed production of electricity from hydrogen pipelines, via the intermediary of fuel cells, or on centralized reactors powered by the nuclear fusion of deuterium from the world's oceans. Fusion energy has its strong proponents and has been receiving substantial research funds.

TWO APPROACHES TO NUCLEAR FUSION POWER

The world mastered the production of energy from nuclear fusion with the first hydrogen bomb explosion by the United States on November 1, 1952. The total energy release of 11 megatons was derived from the fusion of much of a large cylinder of ultracold liquid deuterium, augmented by fission in ordinary uranium surrounding this thermonuclear fuel.

Inertial Confinement Fusion (ICF)

Although this example of burning deuterium by radiation implosion is on a scale impossibly large for electricity production, the same approach is to be applied in inertial confinement fusion (ICF). Here, small pellets of deuterium (or deuterium-tritium ice) of exquisitely careful design and fabrication are to be imploded by radiation pressure in a gold cavity a millimeter or so across. The thermal radiation at a temperature of 300 electron volts (300 eV) would be provided by powerful lasers such as the 192 beams planned for the National Ignition Facility—NIF—at the Lawrence Livermore National Laboratory.

In the ICF approach to fusion, explosive yields on the order of 0.1 ton of high explosive equivalent would be produced by the nuclear fusion in about one milligram of deuterium-tritium (d-t) mixture. NIF became notorious in late 1999 when the program was revealed to have previously unrecognized technical difficulties in damage to its mirrors by the high-power laser beams, along with management problems. Its completion will be retarded by several years, and the cost will rise to more than $2 billion. In NIF, more than a megajoule of laser light is to be converted into thermal X-rays in the gold cavity. As the d-t fuel is compressed by the radiation pressure acting on the outside of the spherical "pusher," it should eventually reach density and temperature so that it can react. The pellet should then explode at speeds comparable with those in a thermonuclear bomb. But because the amount of material is smaller in the ratio of the energy outputs—0.1 ton vs. 1 megaton, or a factor of 10^7—the radius of the system is smaller by the cube root of this number, about a factor of

200. The time of the explosion (and the time required for the implosion) is smaller in this same ratio, hence below the nanosecond range and measured in picoseconds.

Physical problems with this approach are in obtaining the requisite smoothness and precision of the tiny capsule and its contents (which might be a thin shell of frozen d-t ice inside a glass or beryllium pusher) so that the thermonuclear fuel is not mixed with the pusher on implosion. Even if ignition can be demonstrated—i.e., the reaction of a substantial fraction of the d-t fuel—sufficient thermonuclear energy must be produced to repay not only the investment given to the capsule by the X-rays in the cavity, but also the larger amount of energy required to produce the laser light. Recall that one ton of high explosive yields 4 GJ of energy release, so that the 0.1-ton explosion sought in NIF would provide 400 MJ of heat that must be recycled to produce more than 1 MJ of laser light to implode the next capsule.

Beyond the technology and the question of energy efficiency is that of economics. It is by no means clear how ICF could compete with a breeder reactor, in which the heat is produced steadily without a massive investment of energy. Since a conventional nuclear power reactor of 3 GWt produces heat at the rate of 0.8 ton of high explosive per second, an ICF plant would need to have eight explosions per second, each of 0.1 ton, in order to produce the same amount of heat. But only a few percent of the fission reactor's heat is needed to maintain the reaction (i.e., pumping power, and a few percent more to isotopically enrich the fuel). The economic challenge for ICF is not only to produce the fusion but to do so at a cost comparable with the $0.05/kWh or $4 per GJ cost of nuclear heat. Each capsule would thus need to cost a small part of the $1.60 required to pay for a comparable amount of fission heat.

Fusion, however, produces much less long-lived radioactivity than do current fission sources of energy. (See Fig. 6.3.) Although the 14-MeV neutrons copiously transmute nuclei exposed to them, appropriate material choice can minimize the amount of long-lived radioactive materials. A "megajoule laser" is under construction in France, similar to NIF, and will be used in like manner to explore the ignition of thermonuclear fuel. Other approaches to ICF include the pellet implosion by X-rays produced from "Z-pinch" machines that use tens of millions of amperes of current driven by 6 million volts of pulsed power. These currents in a small basket of fine wires already produce megajoules of X-rays at cavity temperature up to 200 eV.

Magnetic Confinement Fusion (MCF)

The other approach to fusion energy is magnetic confinement fusion (MCF), in which a given deuterium or tritium nucleus is burned not in a few picoseconds but over many seconds. As we indicated in Chapter 1, the thermonuclear energy that fuels the sun is obtained by the fusion of light nuclei over billions of years. That in a hydrogen bomb burns in less than a microsecond, and that in the ICF pellet in a few picoseconds. Long preceding the ICF approach was the concept of confining large numbers of deuterium (or deuterium and tritium) nuclei by the action of a magnetic field on the moving charged particles. One cannot put many positively charged nuclei in a given place so that they can react, unless their charge is neutralized by the presence of electrons—not bound to individual nuclei, but present as a kind of background fluid in what is otherwise a vacuum. Such a fully ionized medium (no electrons bound to nuclei) is called a plasma.

In a uniform magnetic field such as that produced in the solenoid of the demonstration laboratory, the magnetic field lines (revealed by the orientation of chains of iron filings on a piece of paper) are parallel to one another in the interior of the solenoid. An ion moving in that field would feel a force that bends the trajectory into a circle if the velocity is in a plane perpendicular to the magnetic field. Ion motion along the magnetic field is not affected by the field. Thus, the general behavior of an electron or nucleus in a magnetic field is to move in a helical path of constant radius. If ions and electrons have similar energy (as would be the case if they had similar temperature), the electron velocity is higher because of its much smaller mass, and the force due to the magnetic field is simply proportional to the field, the charge, and the velocity. The result is that the electrons have a radius of their helical motion about 60 times smaller than that of the deuterium ions. They whiz around their circles at almost 4000 times the deuteron orbit frequency. The electrons whirl at some 2.8 megahertz in the earth's magnetic field, and 100,000 times that quickly (280 GHz) in the field of an MCF device.

Although this concept has existed for a long time, there have been many practical and theoretical problems. Because the particles travel freely along the field lines, they cannot readily be confined to the inside of a solenoid; they would simply emerge from the ends. One immediately thought of winding a wire on a torus (like a doughnut, or the inner tube from a bicycle tire), so that the wire itself forms a succession of little circles around the rod that is the doughnut. This is equivalent to having a long flexible solenoid that is bent into a circle to form a doughnut. The field lines are now confined entirely to the material of the torus, which for a fusion reactor would be a vacuum inside a containment shell on which the wire is wound.

One might imagine that the particle orbit is bent in similar fashion to be a miniature of the wire, although it might be off center from the circular axis of a torus. Not quite true. Orbits of tiny radius are approximately confined to a field line, but because the magnetic field is stronger closer to the linear axis of the torus, the turns of the orbit are not exactly circular; and if the plane of the torus is horizontal, the orbit itself drifts vertically until it reaches the wall. The problem is more complicated because electrons and nuclei drift in opposite directions, and set up an electric field that forces both types of particles to drift outward toward the rim of the torus. When this problem was recognized (it is called the "Fermi drift"), Lyman Spitzer, astrophysicist at Princeton University, suggested taking a long loopy doughnut and bending it into a kind of pretzel. There would be essentially two loops to this bent-up torus, approximately lying in the horizontal plane. Spitzer's idea was that a drift along the axis in one of the loops would be compensated by a drift in the other direction in the other loop, and this is indeed the result that is obtained by detailed calculations and even experiment.

Such ingenuity was exhibited also in Spitzer's initial championship of a large orbiting telescope—a project that eventually became the highly successful Hubble Space Telescope. But for the task of the MCF field, the Stellarator, as Spitzer dubbed his invention, simply allows one to demonstrate the next problem—that of instability of the plasma.

Soviet scientists soon adopted a simpler approach to compensating for the Fermi drift. Instead of twisting the doughnut, they arranged to twist the field lines within an ordinary doughnut. This was done by a few current-carrying "Joffe bars" that lie on the surface of the torus but perform a slow twist as they advance around the surface. For magnetic field lines other than the one at the center of the torus, this provides a twist that increases as one approaches the wall. The device was dubbed a Tokamak by its inventors.

The Tokamak work showed sufficient promise that in 1988 an international consortium of Europe, the Soviet Union, Japan, and the United States began work on the international thermonuclear experimental reactor, ITER—twice as big in scale (eight times in volume, more or less) than earlier Tokamaks. The preliminary design concept was complete in 1990, and work has proceeded since, although the United States dropped out in 1999, because of lack of congressional support. The present design is supposed to operate at a fusion power output of 500 MW, with less than 10% of that required for electric drive to maintain the plasma at operating temperature. The overall construction cost of the machine in 1999 dollars is anticipated to be $4.4 billion.

ITER will use large magnetic fields (on the order of 120,000 gauss—12 tesla) from the main coils that provide the "toroidal field." Eighteen such coils have a total weight of almost 5200 tons and are to be built of high-temperature superconducting wire that will consume no power in maintaining the intense magnetic fields. Keeping the superconductor at its operating temperature near that of liquid nitrogen—77° kelvin (relatively "high" compared with the 4° kelvin temperature of ordinary superconductors such as lead or niobium)— requires a major investment in refrigeration plant and a supply of power for that purpose.

Lighter coils provide the requisite twist in the magnetic field. If it were decided to build ITER, it would take about eight years to complete the construction. During the first ten years of operation, the power output would be 500 MW for a pulse 400 s long, consuming about 0.4 gram of tritium per pulse. Particular problems of the Tokamak involve not only those of stable confinement of the plasma under conditions of burn, but maintaining extraordinarily good vacuum within the confinement chamber, at high temperature and substantial neutron bombardment, which averages about 0.6 MW/m² of wall. For a power economy based on Tokamaks, the machines would need to regenerate their tritium, which would be done for the most part by capturing the d-t neutron in lithium-6.

If uranium from seawater can supply the world's centralized energy needs for thousands of years with light-water reactors, or hundreds of thousands of years with breeders, society will have ample time to demonstrate and exploit practical and affordable approaches to fusion power—beyond the thermonuclear weaponry mastered a half century ago.

CHAPTER 9

Comparing Hazards of Nuclear Power and Other Energy

RISKS OF ELECTRICITY FROM COAL

MOST OF THE WORLD'S electrical energy, consumed in the motors, lights, and heaters of industry, commerce, and our homes, and in transportation and other activities, comes from fossil fuels—that is to say, from coal, oil, or gas. It is possible that there also exist important reserves of primordial gas or oil. Professor Thomas Gold of Cornell University holds this view, but it has not yet been demonstrated unambiguously. There are surely vast additional amounts of carbon-containing fuel in oil shales and in forms of methane below the seabed. The cycle of fossil fuel use begins with the extraction of raw material from mines and wells and includes transport and refining, followed by combustion that produces heat and carbon dioxide and other waste products. The carbon dioxide escapes into the atmosphere, and the ashes resulting from burning coal must be disposed of. Combustion also liberates oxides of sulfur and nitrogen into the air, and the burning of coal releases natural radioactive materials as well. The mechanisms of combustion of fossil fuels are complex, but, in contrast to the potentially catastrophic effects of error in nuclear reactor operation, their understanding and control have been obtained somewhat less rigorously and most of the time without immediate serious consequences if things go wrong. Nevertheless, mining coal used to be one of the most hazardous occupations; there have also been cases where our ignorance has led to local catastrophes, such as devastating fires fed by stored liquefied natural gas.

In recent decades, constraints have been imposed on where and how coal is burned, making the deadly fogs of London a picturesque if murderous memory. But it is well to recall that in December 1952, a four-day temperature inversion there retained the combustion products from the high-sulfur coal used for home heating, and day turned into night. Buses crawled forward only with a person leading the way. The killer fog was responsible for 4,000 to 7,000 premature deaths.

Considerable sums have been spent to limit pollutants other than carbon dioxide (which is not directly hazardous to health and cannot readily be limited—it is part of the burning process), such as sulfur oxides, nitrogen oxides, and particulate emissions (fly ash). There is also now an international effort to reduce carbon dioxide emissions in order to minimize the extent of global warming.

It is well within society's technical ability to collect carbon dioxide from fossil-fueled power plants and, as mentioned in Chapter 8, to dispose of it by injection into exhausted natural gas fields or even into the deep ocean waters. The costs would be substantial—as much as 30% of the cost of electrical energy—but this disposal is technically feasible. One promising approach is to convert coal (and water) to hydrogen and carbon dioxide, inject the CO_2 into unminable coal beds nearby to flush methane gas (CH_4) to collection wells, convert that methane to hydrogen and CO_2 as well, and export hydrogen and electrical power.[1] In this approach, only about one-third as much CO_2 is rejected to the atmosphere as in the normal use of coal for producing electrical energy, and the hydrogen can be used remotely for powering vehicles or for local electrical generation.

Although the dangers in modern coal mines have substantially diminished and the era has passed in which miners suffered black-lung disease and frequent fatal accidents, large quantities of coal must be mined to fuel a power plant. A coal-fired plant, producing as much energy as a typical one-million-kilowatt (1000 MWe or 1 GWe) electric nuclear plant, burns a ton of coal every 12 seconds, which corresponds to 2.2 million tons per year. A nuclear facility of the same power uses less than a ton of uranium-235 per year, which corresponds to about 200 tons of natural uranium contained in 100,000 tons of mined uranium ore.

Exposure to Nuclear Radiation from Coal Combustion

Consideration of the adverse effects on health from the production of energy must take into account, in the pollution resulting from the use of coal and oil, the chemical substances that can cause cancer and other diseases. Hospitals fill up quickly during heavy smog in some cities. The vast fires in Indonesia in 1997, which affected that entire region as far as Malaysia, are an unfortunate case in point. Whole cities were rendered all but uninhabitable for months.

The harmful effects of coal deserve special attention because of its importance in industrialized countries and in developing nations with large populations like India or China. In the United States, 52% of the electricity is produced by coal. The quantity of radioactive material liberated by the burn-ing of coal is considerable, since on average it contains a few parts per million of uranium and thorium. Modern coal-fired electric plants are designed and operated to reduce the emission of particulates from the stack, and also to decrease the emission of sulfur oxide and nitrogen oxide. Older plants, such as the majority of those in China, are far from meeting these standards for fly ash and gaseous emissions. When coal is burned, all the uranium daughters accu-mulated by disintegration—radium, radon, polonium—are also released. The United Nations Scientific Committee on the Effects of Atomic Radiation eval-uates the radiation exposure to the population from this source.[2] Per gigawatt-year (GWe-yr) of electrical energy produced by coal, using the current mix of technology throughout the world, the population exposure is estimated to be about 0.8 lethal cancers per plant-year distributed over the affected population. Table 7.2 summarizes these data. With 400 GWe of coal-fired power plants in the world, this amounts to some 320 deaths per year; in the world at large, some plants have better filters and cause less harm, while others have little stack-gas cleanup and cause far more.

In addition, there is a major exposure to the radioactivity of coal that arises from the use of ash to make concrete. With about 5% of power-plant ash being incorporated into housing, the population dose for the 400 GWe of coal plant leads to an estimated 2000 cancer deaths per year. But if most of the ash went into concrete for dwellings, the annual death toll from radiation from this source would rise to about 40,000.

Some see in accidents a reason to abandon nuclear power in favor of alter-nate ways—so-called soft-energy paths—that they propose to help arrive at a harmonious development of industrial societies. There is much merit in both the more efficient use of energy and in its supply from renewable sources. The world used 375 quads of energy in 1996; the United States used 75. We have noted in Table 8.4 that solar electric power conceivably could amount to about 50 quads per year worldwide; fuel from biomass, 20 quads; and 9 quads from exploiting the temperature difference between the warm surface water of the oceans and the colder water at depth. Biomass, in particular, may develop beyond the 3% of U.S. energy needs that it now meets, as the revolution in biotechnology enables the production of alcohol from cellulose rather than from sugars. It is highly desirable to have small-scale energy sources if they can be achieved at affordable cost and with acceptable environmental impact. It will be necessary, however, to carefully compare these alternatives—including their harmful side effects—to the more traditional ways of producing energy—e.g., fossil-fueled plants burning coal, gas, or oil; hydropower; and nuclear power stations.

THE GREENHOUSE EFFECT

Changes in the earth's climate have a major impact on the possibility of life. According to Fig. 9.1 (S. A. Fetter[3] TWES, page 102), the most recent ice age, 13,000 years ago, is only as distant from the beginnings of recorded Egyptian civilization as is the present from that date. Our ancestors lived in caves in the ice age, when so much water was locked up in continental glaciers that sea level was about 160 meters below its present level. In contrast, if all of the world's glaciers now melted, the sea level would rise by only another 6 meters—though that rise, to be sure, would have severe consequences.

As the earth was gradually emerging from the last ice age, the climate suddenly reverted to the conditions of that age. A mere thousand years or so later, the arctic warmed by 5–10°C over a period of a few decades, leaving today's climate, with occasional warmings of less than 1°C. Life would be very different on earth without the greenhouse effect. The average solar illumination "insolation" is 235 W/m². If the atmosphere were transparent to infrared energy, the average surface temperature of the earth would be a deeply frozen –20°C, with about 30% of the sun's energy directly reflected and the rest radiated by the earth in the infrared. The earth's average surface temperature of 15°C arises because water vapor, carbon dioxide, and other "greenhouse gases" reflect

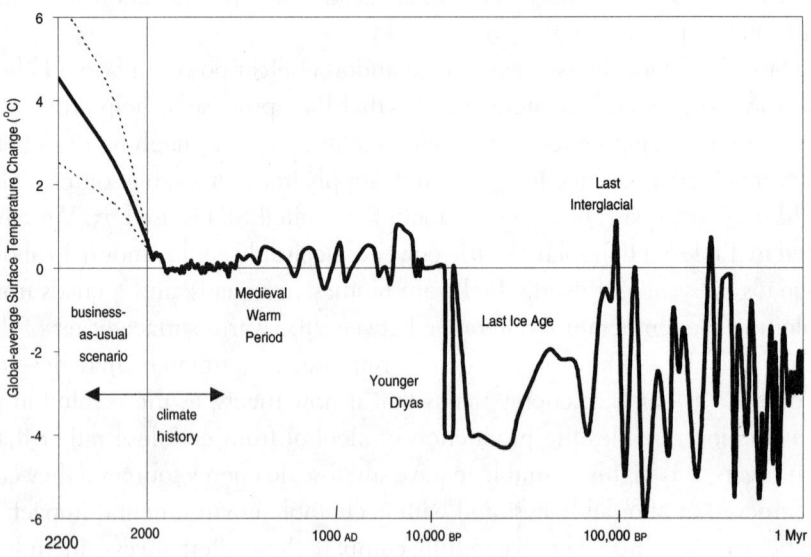

Source: Steven A. Fetter, "Climate Change and the Transformation of World Energy Supply," May 1999. Used by permission.

Fig. 9.1. Variations in average surface temperature over the last million years.

back to earth infrared energy that is radiated out into space—without reducing the sun's illumination of the surface. In preindustrial days, carbon dioxide amounted to about 280 parts per million by volume (ppmv), and it has risen by about 30% to some 360 ppmv today. The concentration of methane has more than doubled. The increase in the earth's temperature due to increase in greenhouse gases is not simple to estimate.

First, there is the "radiative forcing," which is the increase in the amount of net energy falling on a square meter of the earth on the average. A doubling of carbon dioxide from 280 ppmv to 560 ppmv (equivalent) leads to a radiative forcing about which there is no dispute—4.4 W/m². (We follow Fetter's particularly clear exposition.) The radiative forcing of 4.4 W/m² corresponds much less certainly to an estimated average warming of 2.5°C.

Taking into account the increased methane, nitrous oxide (from fermentation in the stomachs of ruminants), and the halocarbons (particularly chlorofluorocarbons), which are potent greenhouse gases and provide radiative forcing on the order of 1.3 W/cm², the limit on carbon dioxide would correspond to radiative forcing of 3.1 W/m² and a concentration of about 460 ppmv.

In 1995, anthropogenic carbon emissions (those due to human activities) amounted to about 7.5 gigatons of carbon per year (GtC/yr). To stabilize at a

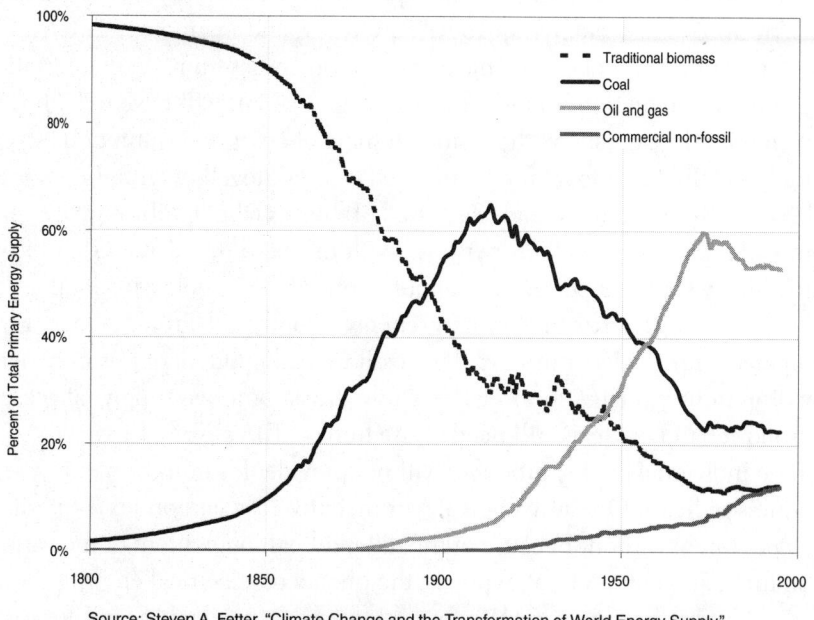

Source: Steven A. Fetter, "Climate Change and the Transformation of World Energy Supply,"
May 1999. Used by permission.

Fig. 9.2. Share of energy consumption by type of fuel.

**TABLE 9.1. HISTORICAL CONSUMPTION AND
RECOVERABLE RESOURCES OF FOSSIL FUELS.**

	CONSUMPTION, 1765–1995		RECOVERABLE RESOURCES	
	(EJ)	(Gtc)	(EJ)	(Gtc)
Oil	4,800	90	$10,000^{+10,000}_{-2,000}$	200^{+200}_{-40}
Gas	2,100	29	$10,000^{+13,000}_{-2,500}$	150^{+200}_{-40}
Coal	5,300	131	$100,000^{+150,000}_{-50,000}$	$2,500^{+4000}_{-1000}$
Methane hydrate and oil shale	—	—	~2,000,000	~40,000
Total	12,200	250	~2,000,000	~40,000

Table reproduced by permission of S. A. Fetter (1999).
Sources: G. Marland, R. J. Andres, T. A. Boden, C. Johnston, and A. Brenkert, "Global,
Regional, and National CO₂ Emission Estimates from Fossil Fuel Burning, Cement Produc-
tion, and Gas Flaring: 1751–1995" (revised January 1998; available at
http://cdiac.esd.ornl.gov/ndps/ndp030.html); C. D. Masters, E. D. Attanasi, and D. H. Root,
"World Petroleum Assessment and Analysis," in Proceedings of the 14th World Petroleum
Congress, Vol. V (Chichester, UK: John Wiley and Sons, 1994), pp. 529–541; World Energy
Council and International Institute for Applied Systems Analysis, Global Energy Perspectives
to 2050 and Beyond (London: World Energy Council, 1995), p. 36 (summary at
http://www.wec.co.uk/energy.htm); and others.

carbon dioxide concentration of 460 ppmv, one could permit in 2025 8.9 GtC/yr; in 2050, 6.0 GtC/yr; in 2075, 4.4 GtC/yr; and in 2100, 3.3 GtC/yr. On the order of 1 GtC/yr of carbon dioxide emission comes from net deforestation and cement production, so fossil fuel carbon emissions will have to be less by that amount. Fig. 9.2 shows the historical share of energy consumption by type of fuel, and Table 9.1 fossil fuel consumed thus far, together with the recoverable resources. We have already discussed (Table 8.5) the stabilization of world population at about 11 billion persons. With on the order of 100 GJ/yr of primary energy supplied per person (about 5 tons of fossil fuel annually per person), by the year 2050 the primary energy supply will grow to 1000 EJ/yr compared with 383 EJ/yr in 1995. The fossil fuel component of this must actually drop from 329 EJ/yr to 270 EJ/yr, while the carbon-free supply (at present hydropower and nuclear) will need to rise from 53 EJ/yr to 730 EJ/yr.

The individual energy producer will not provide this transformation out of goodness of heart. One way to create an incentive is a carbon tax, of perhaps $100 per ton of carbon. Because of the different carbon content of the various fossil fuels, and different initial prices, the impact on electrical energy production costs varies from $0.026/kWh for coal to $0.018 for oil and $0.013 for gas — increases over the average 1997 retail price of $0.085/kWh by 30%, 20%, and 15%, respectively.

TABLE 9.2. OPTIONS FOR SEQUESTERING CARBON DIOXIDE

DISPOSAL OPTION	SEQUESTRATION POTENTIAL (GtC)	SEQUESTRATION PERIOD (YR)	COST ($/tC)
Biomass (included in tables 3–5	100	100+	0–80
Chemical manufacture	0.1/yr	100+	10–60
Underground disposal			
Enhanced oil/gas recovery	20–70	$>10^6$	−40–60
Abandoned oil/gas wells	150–500		
Saline aquifers	100–3000	10^3–10^6	10–60
Ocean disposal	>1000	100–1000	10–60

Reproduced by permission from S. A. Fetter (1999).
Sources: Howard Herzog, Elisabeth Drake, and Eric Adams, **CO$_2$ Capture, Reuse, and Storage Technologies for Mitigating Global Climate Change** (Cambridge, MA: Massachusetts Institute of Technology, January 1997), *http://web.mit.edu/energylab/www/hjherzog/White_Paper.pdf*; International Energy Agency, **Carbon Dioxide Disposal from Power Stations** (Stoke Orchard, UK: IEA Greenhouse Gas R&D Programme, 1995), p. 19, *http://www.ieagreen.org.uk/sr3p.htm*; International Energy Agency, **Carbon Dioxide Utilisation** (Stoke Orchard, UK: IEA Greenhouse Gas R&D Programme, 1995); *http://www.ieagreen.org.uk/sr4p.htm*.

Fetter also provides costs for options for sequestering carbon dioxide (Table 9.2) ranging up to about $60/t. Although he shows a sequestration period of only 100–1000 years for liquid carbon dioxide "lakes" on the floor of the deep ocean, the time could be considerably greater if a durable membrane or other cover were practical. Easier to use for the next decades are abandoned oil and gas wells.

In principle, one should support research to find low-cost ways of lowering emissions of greenhouse gases (or, for that matter, reducing energy received from the sun) in order to make optimum choices in the future. Estimates of the net cost of accepting a doubling of carbon dioxide (that is, more accurately, radiative forcing of 4.4 W/m^2) range from near zero to 5% of gross world product (GWP). For decreasing emissions by 70%, estimates of the net cost range from near zero to some 8% of GWP. These figures, it is clear, are much too rough and imprecise to provide much guidance. In reality, much of the initial reduction of CO$_2$ emissions is profitable rather than costly, although the return on investment may not be so great as might be expected from more conventional opportunities.[4] In 1990, U.S. society emitted into the atmosphere carbon dioxide containing 1300 megatons of carbon (1300 MtC), and by 2010 or so is supposed to reduce its emissions to some 5% below that level. However, since

1990, U.S. society has increased its annual CO_2 emissions at a rate such that they will be some 1700 MtC/yr by 2010. According to extensive experiment and analysis[4] the U.S. building sector can reduce annual emissions by about 70 MtC/yr of carbon in 2010, and save $300 per ton of this carbon (i.e., at a saving of $21 billion/yr). Similarly, U.S. transportation can reduce carbon emissions by some 100 MtC/yr, while saving about $11 billion/yr. Industry can save $4 billion in the process of reducing emissions by 80 MtC/yr. To complete the saving of 400 MtC/yr total, U.S. electrical utilities could save about 130 MtC/yr, but at an annual cost of $6 billion. This would return U.S. emissions to the 1990 level. The United States has been one of the leading advocates of "emissions trading," so that it can fulfill its prescribed reduction under the Kyoto guidelines by paying, for instance, to upgrade electrical plants in China so as to improve their efficiency. Such an approach would be beneficial both to the world's GWP and to the nations involved in such a trade. To reflect a reduction of 5% (65 MtC/yr) the United States wishes to have the Kyoto implementation include the opportunity to prevent the cutting down of forests and thus to "save" emissions of carbon that would otherwise occur. Some other states in the negotiations over the rules, however, doubt that a preserved forest would be as long-lasting as a modification of society's activities through technology and investment so that it reduces further the emission of greenhouse gases.

Except for water vapor, the current proportion of greenhouse gases in the air is very low; the concentration of carbon dioxide is only 0.035% by volume. Scientists now know with precision the evolution of carbon dioxide concentrations over the last 160,000 years, and that the temperature variations during that period are closely related to variations in the concentration of the major greenhouse gas carbon dioxide (Fig. 9.3). Most scientists working in the field believe that a modification of the natural composition of the atmosphere by the excessive emission of carbon dioxide, due to the increased use of fossil fuels, will produce a catastrophic warming of the planet. The energy generated by society is already 40% as great as the totality of energy produced by radioactive decay within the entire volume of the earth, and is responsible for the high temperature in deep mines or wells. In the 250 years since the industrial revolution, the concentration of carbon dioxide in the atmosphere has grown from 270 to 350 parts per million. During the next 100 years, the average temperature at the surface of the earth could increase by 2 to 4 degrees C and be higher than it has been during the last 160,000 years—and especially so in regions farther from the equator. Particulates in the atmosphere have a cooling effect, but according to the Intergovernmental Panel on Climate Change (IPCC) it is very unlikely that global warming can be kept below 1.4°C. That could lead to

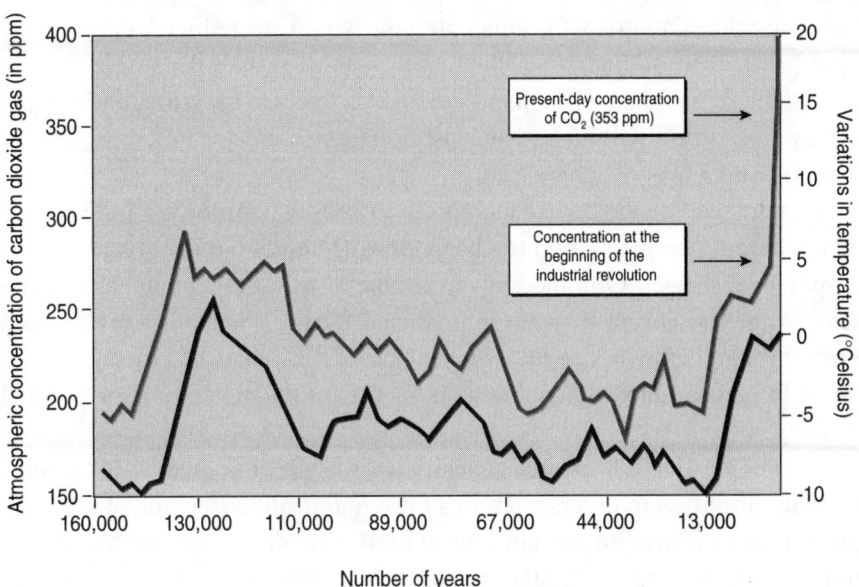

Fig. 9.3. Correlation between the concentration of carbon dioxide in the atmosphere
and the temperature at the surface of the earth.

the melting of the glaciers in the continental areas, the arctic sea ice, and the Greenland glacier and the Antarctic glaciers, a rise in the level of the seas, the flooding of vast continental areas, and an increase in the frequency and magnitude of other important climatic perturbations (typhoons, monsoons).

The IPCC was established in 1988 by the World Meteorological Organization (WMO) and the United Nations Environmental Program (UNEP). Its role is to assess the scientific, technical, and socioeconomic data relevant to proper understanding of the risk of human-induced climate change. Working Group I (WGI) of IPCC assesses the scientific aspects of the climate system and climate change. The IPCC WGI third assessment report of February 2001 contains a readable "Summary for Policy Makers" which is readily available.[5] The 2001 report judges it very likely (that is, a 90–99% probability) that the world will experience during the twenty-first century "higher maximum temperatures and more hot days over nearly all land areas; reduced diurnal temperature range over most land areas and more intense precipitation events." Global mean sea level is projected to rise by 0.09 to 0.88 meter by 2100. Two more of the many substantive conclusions in the WGI report: First, "If, hypothetically, all of the carbon released by historical land-use changes could be restored to the terrestrial biosphere over the course of the century (e.g., by reforestation), CO_2 concentration would be reduced by 40 to 70 ppm," and "The Antarctic ice sheet is likely to gain mass because of greater precipitation, while the Greenland ice sheet is likely to lose mass because the increase in runoff will exceed the precipitation increase." ("Likely" in this report means a 66–90% chance.)

This interval of 160,000 years represents only a tenth of the time during which the human population has been growing, and it has continued to grow despite five ice ages. Our ancestors were able to adjust to climatic changes as serious but less abrupt than those predicted for the twenty-first century. Yet time for adaptation is crucial—time to breed heat-tolerant plants, time to invest in new systems at an achievable fraction of the gross domestic product (GDP).

It is not possible now to predict accurately the circumstances in 2050, and a certain humility is in order, extending even to our understanding of the basic phenomena involved in human impact on the environment and thus on our own civilization. Only in the last two decades has the importance of the greenhouse effect become clear, and the destruction of the ozone layer by specific gases such as refrigerants has led to the totally unexpected local elimination of the stratospheric ozone layer over Antarctica, discovered only in the 1990s. (However, the term "greenhouse effect," with an analysis of its influence on global warming, was used by Alexander Graham Bell in 1914, and Svante Arrhenius (1859–1927) published a very modern analysis of the greenhouse effect even earlier, in 1896.[6]

Some adversaries of nuclear energy contest this argument and suggest that global warming is tolerable or even desirable: if, after one or more centuries, the United States becomes a torrid desert and no longer produces wheat, and if Siberia, on the other hand, enjoys a temperate climate and becomes the breadbasket of the world and grows pineapples like Florida, is that really tragic? Fifteen thousand years ago, much of the United States was covered by a gigantic glacier; these skeptics add that it is human destiny to adapt to climatic variations and it will be easier in a technically developed society. Or, if necessary, can't people just migrate? Perhaps, but there is no way to be sure. In an overpopulated globe there is no "vacuum" into which these mass migrations can fit. The only way is for one population to displace another—a formula for disaster.

The enormous changes in topography since the latest ice age, as are evident on the East Coast of the United States and the west coast of France, clearly demonstrate that it is naive to think that the surface of the planet is forever fixed and that only the harm that is done by society is to be feared. In several million years, our descendants will find cathedrals in geological strata, witness to great civilizations over hundreds of millennia. They will also find burial sites where radioactive waste has decayed, some of which might serve them, in wartime, as mines for uranium-235 resulting from the decay of buried plutonium, unless care is taken to include for this reason some depleted uranium in the waste.

One sees that the temperature of the earth has tracked the level of carbon dioxide: it is undeniable that the temperature rises with the concentration of CO_2. But which is the cause and which the effect? This temperature rise is obviously an argument in favor of the development of nuclear and solar energy. It is also particularly important to improve energy efficiency.

Though the greenhouse effect as a global warming problem has been widely recognized for only about twenty years, the immediate local impact of power generation has long been a separate concern. In this respect, nuclear power plants pour 70% of their energy into the oceans and the atmosphere in the form of heat, as opposed to 55% to 60% for modern fossil fuel plants at the same power level. For the Chinese coal-fed power plants, this fraction is still at 70–80%. The greenhouse effect greatly prolongs and enhances the immediate warming effect of coal-fired plants. Half of the carbon dioxide, although diluted in the atmosphere, remains there for about a hundred years.

It is easy to calculate the increased heat input to the earth due to the enhancement of the greenhouse effect produced by a one-gigawatt coal-fired power plant operated for a single year. One finds that the carbon dioxide produced during that year retains additional solar heat during a hundred years, adding every eight months solar heat equivalent to that liberated by combustion over that single year. During its forty years of continuous operation, a normal coal-fired power plant produces a warming of the planet equivalent to that of forty nuclear power facilities put into continuous operation at the rate of a new one every year (Fig. 9.4). In France, 8% of the CO_2 emitted is from the generation of electricity, while in the world as a whole, 40% of CO_2 emissions accompany the production of electrical energy. Nuclear-electric power would not in itself, however, solve the problem of global warming; it accounts for only 6.3% of the energy consumed in the United States and 33% in France.

The greenhouse effect can be reduced by expanded use of natural gas (methane), which produces only 60% as much carbon dioxide per unit of energy as does coal. But methane itself can contribute powerfully to the greenhouse effect. Methane in the atmosphere is seventy times as effective a greenhouse gas as is carbon dioxide, although methane remains in the atmosphere only about one-tenth as long as does added CO_2. Thus a 10% leak in the natural gas distribution system would make methane about as bad a greenhouse fuel as coal for the same energy production. Commonly accepted estimates for methane losses range from 3% to 9% — in Russia, some 20%.[7]

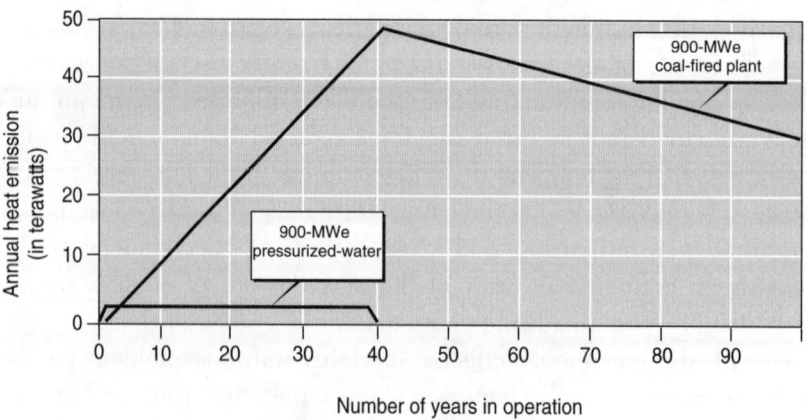

Fig. 9.4. Average heat released to the atmosphere per year by a 900-MWe coal-fired plant and by a 900-MWe pressurized-water nuclear reactor.

In modern economies, the use of any commodity is limited by price. In some cases, a large component of price may be taxes imposed by a state to avoid excessive dependence on foreign sources of energy. Taxes may also be used to reduce consumption and hence environmental impact, or even to compensate to some extent for health or environmental impacts. The industrialized nations are committed to reducing their carbon emissions by the year 2012 to no more than 95% of the 1990 levels. Considering the balance between emission reductions now and decades hence, Steven Fetter's strong conclusion is that reductions in carbon emissions over the next couple of decades are important only as they help to achieve the target in 2050. He concludes that "it is probably better to invest money in future reductions (via energy research and development) than to pay for costly reductions today."[8]

OUR CHANGING EARTH

It is instructive to observe the evolution of natural conditions in Aquitaine, a region in southwest France, as it faced up to the onslaught of the Atlantic Ocean. In the inset maps (Fig. 9.5), the distance from Bayonne to La Rochelle is about 300 km; Bayonne is almost at the Spanish border.

The Aquitaine Eighteen Thousand Years Ago . . .

The growth of the glaciers immobilized an enormous quantity of ocean water so that sea level was more than 100 m below its present level: the Aquitaine coastline was 50 to 100 km farther into the sea. The gulf called the Gironde,

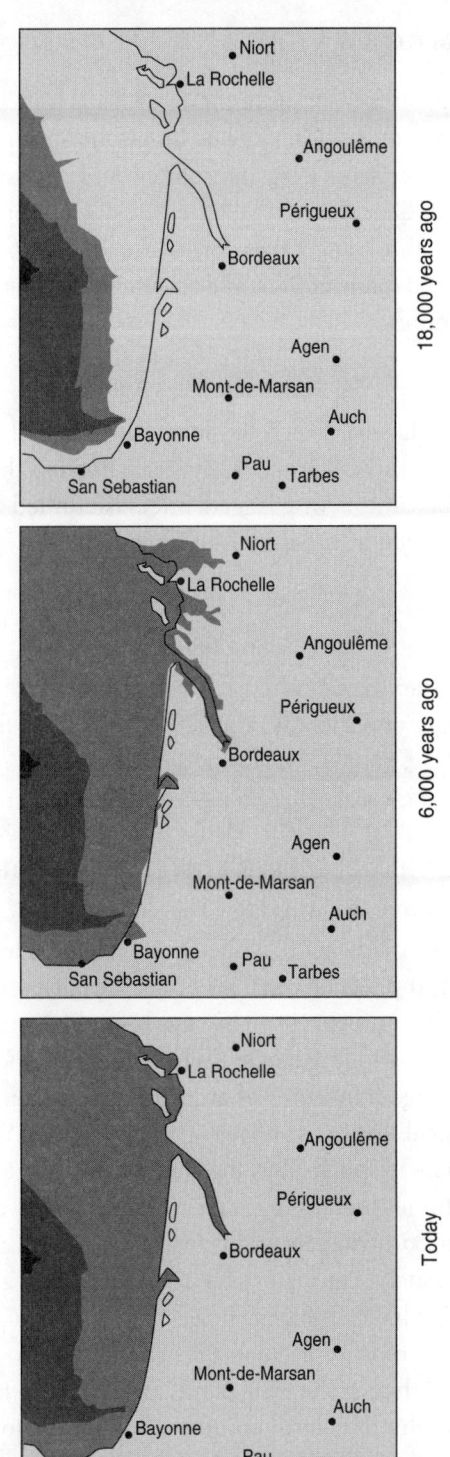

Fig. 9.5. Evolution of the geography of the Aquitaine Basin.

extending northwest from the present city of Bordeaux, was eroded and cut deeply into the Eocene and Cretaceous limestone substratum. Large quantities of gravel and river sand were transported and deposited on the internal continental plateau. Glaciers crowned the Massif Central and the Pyrénées— the east-west mountain range that forms the border between France and Spain. People hunted mammoths and decorated the walls of the Lascaux grottoes with their paintings.

Six Thousand Years Ago . . .

The melting of the glaciers caused by global warming led to a rapid rise in the sea to very nearly its present level. The ocean flooded the river valleys. The Gironde estuary became a vast jagged inlet. Islet-filled rocky bays became shielded from the ocean by the islands of Ré and Oléron. The Aquitaine coast was lined with dunes.

And Today

The sea level has stabilized around the present coast, varying by only a few meters. Sedimentary continental alluvial deposits have filled in the flooded parts of the bays and estuaries. The Gironde today is only 6 meters deep. Heavy swells have eroded the coast, straightening out the shoreline.

THE CALIFORNIA ELECTRIC POWER CRISIS OF 2000–2001

Although for the most part in this book we consider the long term, only a tiny fraction of electrical energy in the United States is stored locally, and so the system is subject to disruption by peak load pressures or even by relatively small reductions in generation capacity. This was brought home to California residents and businesses by rolling blackouts in January and February 2001, and by high prices beginning about April of 2000. In 1996 a typical price paid to the producer of electrical energy was $0.06 per kWh ($6/MWh), but the lure of marketplace efficiencies persuaded the state legislature to deregulate wholesale prices. The utilities that transport electricity and deliver it to the customer were expected to maximize their profits by buying electrical energy at the lowest price and, ultimately, competing for contracts with customers to sell at as high a price as they liked. Initially, however, wholesale prices were deregulated, while rates for most residential customers were frozen until 2002. The utilities sold most of their generating plants, and were, in large part, forbidden to buy electricity under long-term contracts. They were to go to the spot market, where suppliers would bid to sell electricity to the grid, and utilities would pick the lowest-priced commodity.

The companies buying the power plants looked ahead for profit opportunities, and they found them, beyond the normal economies of production. Recently, when demand approached the contracted capacity, it became highly profitable not to be the producer that sold the last gigawatt (GW) of power by putting a plant online quickly that was down for maintenance, but, instead, to work at normal pace or even to retard the return of that plant to generating electricity, in order to drive up the spot price on this highly "inelastic" market. The utilities were in a bind. They were forbidden to buy under long-term contracts; electrical energy cannot be stored, as such; and consumers were not receiving signals in terms of higher prices to cut their electricity consumption at the moment. Annual electricity demand in California had increased from about 250,000 GWh in 1996 to 300,000 GWh in 2000. At the same time, generating capacity had fallen from 56 GWe to 54 GWe. But the monthly spending for electricity generation in the summer months of 1999 averaged about $800 million, while in 2000 it was more like $3.5 billion. For monthly consumption on the order of 21 million MWh in those months, the average monthly spending for electricity generation was $167/MWh. But rates to consumers had been set as if the cost of electricity were $60/MWh, and the utilities lost more than $12 billion and were on the verge of bankruptcy.

The desirability of real-time pricing of electricity to the consumer has long been known—a point made, for instance, in a book published in 1977 of which one of the present writers, Garwin, was a coauthor.[9] In fact, some large commercial customers do have contracts of this type. On the supply side, as well, it is possible for an electricity supplier to the California grid to contract to sell more power than it would normally have available, because it can "buy power back" from large consumers. Exactly this happened in late 2000, when aluminum producers in the Northwest closed their operations for some months, to sell at the spot market rate the power that they had firm contracts to buy from the local producer, who had no shortage.

If frost damages an orange crop, the price goes up and most people eat fewer oranges or none at all. And if there are none on the shelves, then people eat something else. If there is a sudden big demand for airplane seats for a convention or sporting event, low-priced seats in limited amount sell out, and the airlines sell the rest at the highest prices available, but there is typically a cap on the price. And if there are not enough seats, the passengers who have seats fly, and those who do not stay home, fly earlier, or, if it's feasible, drive.

But with an electrical grid, if there is not enough electricity, one cannot simply reduce the amount of electrical power going to each customer. This is attempted to some extent in a "brownout," in which voltage can be reduced by 5% without much impact on the end user. The power used by incandescent bulbs drops, but the power used to perform particular tasks increases because motors, transmission systems, and distribution become less efficient at the lower voltage. Hence exhortations to reduce power consumption, and, ultimately, rolling blackouts. In January 2001, northern California electricity prices at peak hours exceeded $300/MWh ($0.30/kWh). In March 2001, the Federal Energy Regulatory Commission asked generators to justify the prices charged during January and February, or to refund the excess over $273/MWh or $430/MWh, respectively.

The solution to the California problem will be found in bringing the ultimate consumer as well as the generator to the real-time market, which will spur energy conservation as well as new generating capacity and cogeneration (combined heat and power), and will reduce the profit from manipulation of supply of electrical power.

The near-term crisis in California, might be significantly lessened by the adoption of an ingenious scheme of Willard H. Wattenburg, a physicist long associated with the Lawrence Livermore National Laboratory. He proposes to give the authorities a tool to reduce demand more gracefully. It turns out that at peak times in California, almost half the residential demand for electrical power comes from large home air conditioners. These, as well as electric stoves and clothes dryers, operate on 240 V, while lights, refrigerators, medical equipment, and the like operate on 120 V. A home is normally supplied with "3-wire service," with 240-V loads connected between the outer two wires, and 120-V loads connected between the middle (neutral) wire and one or the other of the outer power leads. If one were to equip the power transformers which each serve about ten homes with a remote-control switch, one of the power leads could thus be switched from the transformer to be put in parallel with the other. All 120-V loads would thus be fed normally, while 240-V loads would have no voltage across them at all and would thus draw no energy. This elegant scheme could probably be implemented for about $60 per household, and a total cost statewide of about $1 billion. Even something much less than full implementation would remove the crisis aspect and prevent the charging of extortionist rates for power, while preserving market incentives for greater supply.

The greenhouse effect provoked by gases emitted by the urban and industrial activities of our societies is certainly a threat that warrants great attention. It is partly masked by natural climatic fluctuations. It is only in the long run that irreparable harm may become evident, which might have been avoided by a wider development of nuclear energy. But there are other potent weapons against global warming as well.

We believe that one of the highest duties of society as a whole is to assess and to choose its destiny. In this book our goal is less to prescribe than to inform our readers of the options as we see them. Of course, change is inevitable and often desirable; today's situation is highly unsatisfactory to those whose children die of hunger or disease, or who find themselves pursued and murdered in ethnic or religious conflicts. In considering nuclear energy we do not in any way intend to denigrate other approaches to providing for the needs of society—including renewable energy, improved efficiency to reduce energy needs, and the like. Nevertheless, all these options will have direct and indirect effects on the environment; some will be acceptable, while some will not.

CHAPTER 10

Making Best Use of Scientists

WE HAVE SEEN that seawater, with its resource of four billion tons of uranium, promises an adequate supply of nuclear fuel for hundreds of centuries, and even thousands of centuries if it is combined with breeder reactors. Ores less rich than those presently exploited could also offer a source of uranium at costs higher than the present $25 per kg but still less than the $100–300/kg that might be required to extract it from seawater. There is, then, no chance that the affordable fuel supply for nuclear energy will be exhausted for many millennia to come. Accidents may still happen in the nuclear energy system, to be sure, and their consequences must be compared with the harm to be expected from the accelerated and exhaustive use of fossil fuels like coal and oil.

In considering the negative reactions to nuclear energy, we often find a fear of change, an instinctive mistrust of the unfamiliar. This has been the fate of all great technical innovations. It is important to separate myth from reality.

Nothing illustrates myth better than the way in which railroads were regarded in the early nineteenth century. To quote the Bavarian College of Physicians, dismayed by the unprecedented speed of trains (about 35 miles per hour): "The speed of movement would addle the passengers' brains, causing a kind of Delirium Furiosum. Even if the passengers are willing to face up to this danger, the State must, at least, protect those who watch the trains go by, because they will be affected in the same way. It will be necessary to construct a high wall on either side of the right of way."[1]

One Professor Kips, of the University of Erlangen in Bavaria, Germany, took the railroads to task for the effect they would have on the breeding of horses. He predicted that the army would no longer have cavalry or artillery. In case of an invasion of Bavaria by enemy cavalry, the army would not be able to put up serious resistance except by importing foreign horses at exorbitant prices. The charming book from which we glean these gems also tells us that

The future is dangerous but not necessarily gloomy.

the budding railroad lobby proceeded to purchase and burn the anti-railroad pamphlets that were being distributed so that the population would not be influenced.

These masterpieces of visionary ecology were not limited to Bavaria. The great French scholar François Arago, renowned physicist and member of the Academy of Sciences, was totally opposed to railroads. He saw in the tunnels and underground passages the menace of all kinds of diseases for the passengers, such as inflammation of the lungs, pleurisy, bronchitis, colds, catarrhs, and other kinds of distress brought on by the underground chill following suddenly upon the heat of the sun in open air. And to further blacken the picture, he took to the rostrum and expounded on the terrifying danger to passengers if the boiler were to explode in the narrow darkness of a tunnel.[2]

Arago could not foretell the future, yet he was partly right. Consider the indifference that greets the news from faraway countries of horrendous railroad accidents causing hundreds of deaths. Every year, all over the world, thousands of travelers are killed because of equipment failure or human negligence. Nevertheless, no one suggests that railroads be scrapped. Their benefits are evident and countless, while their dangers are negligible compared with the toll on the roads due to the automobile: 8000 deaths and 100,000 injuries in France alone every year; and in the United States 35,000 annual deaths. Politicians are well aware that efforts to reduce the number of victims by imposing speed limits on drivers could practically lead to riots, or even worse, electoral defeat. Recall the uproar when a 55-mph speed limit was imposed in the United States in the 1970s. In no way do we advocate negligence in operation of railroads, or suggest that investments not be made to maintain and improve safety. But we firmly believe that the elimination of trains because of their toll in death and injury would be far more damaging to society than would be their continuation.

Our descendants will experience major or minor nuclear accidents. They will continue, nevertheless, to use nuclear power, along with accepting the obligation to correct anything that compromises its safety, as society gradually does for the railroads.

The shortsightedness of the violent opposition to the railroads in the nineteenth century finds a parallel with regard to nuclear energy in positions held today by some ecological extremists who thereby imperil the valuable contributions of a reasoned ecological movement. Opposition to developments that show no concern for their long-term consequences is warranted and laudable, provided that judgment is correct. To bring the facts to light, we need the help of scientists and others who have made the study of these issues their life's work. Scientists may disagree, and for reasons that may go beyond the insight of bureaucratic politics that "where you stand depends upon where you sit." Reasoned and explicit disagreement among scientists provides the broader society a valuable basis for informed choice, as is exemplified in the controversy on the back end of the fuel cycle that was discussed in a previous chapter.

URGENTLY NEEDED: STATESMEN WITH SCIENTIFIC EXPERTISE

An isolated scientist may sometimes be wrong; a docile committee advised by experts drawn from and representing narrow interest groups is likely not to give proper balance to benefits and costs as they affect society as a whole. But a group of capable scientists, chosen by a learned academy, is less likely to err — even if the members are not initially fully expert. Governments would be well advised to make use of totally independent committees of experts. By "independent" we mean individuals who have neither a financial nor an ideological stake in a problem. It is difficult to assemble a group that is both independent and expert; experts tend to lack independence, and independents tend to lack expertise. The best compromise seems to be independence and competence — competence certified by accomplishment, even if not in the particular question at issue. We shall give later in this chapter several examples in the fields of nuclear power and nuclear weaponry to illustrate this point.

Scientists can also get swept up in patriotic fervor. Andrei Sakharov, who later at great peril to himself opposed his country's nuclear weapons program, was convinced, after the United States exploded atomic bombs over Japan, that the Soviet Union had to have its own. Perhaps Sakharov felt that the United States should have no monopoly on the weapon. He worked with enthusiasm on nuclear weapons and especially in the development of the Soviet hydrogen bomb. It eventually cost the Soviet Union some $50 billion per year in what have turned out to be wasteful military expenditures after the great bloodbath of the Second World War. The irony is that the Soviet Union was not very hospitable to its own scientists even if they worked on nuclear weapons, as Sakharov eventually found out. Stalin was so determined to get the atomic bomb that he put Lavrenti Beria, the head of the secret police, in charge of the project. In 1949, at the time of the first test, Beria had two lists. The first consisted of the names of the scientists who were to be awarded the highest decoration, Hero of Socialist Labor, if the test succeeded, and the other list was of the names of the scientists who were to be shot if it failed. The names on the two lists were identical.[3]

What the Soviet Union could have done with $50 billion per year in helping its people! But the Stalin regime, a police state internally and bellicose toward the outside world, had emptied Russian socialism of its humanity. The arms race long outlasted Stalin, and internal repression did as well.

Foolish and wasteful programs are not unknown to the United States, some of them so dangerous as to imperil its future; among the foolish was a nuclear-powered aircraft program and among the most dangerous an enormous excess in nuclear weaponry. And the United States turned on J. Robert Oppenheimer, the inspired leader of Los Alamos from its inception in April 1943 to the successful detonation of the first three nuclear weapons in July and August 1945. We hope that this book helps to give citizens the understanding needed to recognize and reject some unwarranted programs related to nuclear weaponry and nuclear power.

PITFALLS TO DECISION-MAKING: PRESSURE GROUPS

In July 1996, the French government decided to undertake an extensive reform of the armed forces leading to the dissolution of thirty-eight regiments and the decommissioning of thirty warships, with more reductions to come by the year 2000. In making these reductions, the government had the good sense to adapt the means of the country to the conditions created by a total change in the strategic situation and the necessity of facing up to economic facts. Political reactions were striking. Some, on the part of the unconditional defenders of the military-industrial complex, came as no surprise. But in the political arena, from the extreme right all the way to the distant left, indignant voices were raised. For some, the government was digging the army's grave; others, while paying lip service to disarmament, found fault with the closing of arsenals that now made unmarketable weapons. These reactions can be explained by the fact that France was two years away from important elections and the reform raised the specter of increased unemployment.

The same reactions occurred in India in May 1998, right after its nuclear tests. Reporters who interviewed citizens from all sectors of the political spectrum—including Gandhiesque pacifists—were surprised by the wide approval of the tests. In many cases the approval had to do not with the defense of the country but with the feeling that India without a nuclear weapon was not being taken seriously. Being taken seriously was more important than addressing the country's overwhelming social or even security needs. After Pakistan responded within two weeks with its own underground nuclear explosions, many in India were now critical of their country's decision to test.

One should not automatically admonish pressure groups. It is a normal and legitimate reaction for people to get together to defend a common interest. When the fate of hundreds of thousands of families depends on the indefinite continuation of an activity, it is only natural for them to exert pressure to protect their future livelihoods. Cattle breeders and beet farmers, coal miners, railroad engineers, workers and directors in the arms industry—all tend to react in the same way, employing lobbyists, public relations experts, and self-serving arguments. But there is a huge difference between the interests of farmers threatened with being put out of business by competition from halfway round the world and those of people employed in weapons plants who will be made redundant because of the outbreak of peace. In the first case, we are dealing with people serving a function that goes beyond the simple production of marketable goods. It is not hard to understand that individuals might want jealously to preserve activities that have fashioned a countryside to which they have become deeply attached and with which they identify. The value of these activities is not measured only by the price of the fruit and vegetables harvested locally.

But the government should not strive to preserve costly state-sponsored activities that have long outlived their usefulness and no longer respond to their initial goals. It is up to political leaders to see to it that these essential changes do not lead to unemployment that cannot be absorbed by new activities. This calls for a good deal of imagination and insight. It is one of the most serious challenges of our times; and particularly in the context of European integration, it is by no means clear that traditional occupations such as farming will be or even should be preserved in a new era of competitiveness.

The U.S. Defense Department has had great difficulty eliminating excess bases, depots, and other facilities. Those employed in these facilities, and in the local economy benefiting from their presence, reflexively appeal to their representatives in Congress, stalling billions of dollars of annual savings that would accrue from the downsizing that the Defense Department desires. A few years ago, Congress and the President agreed to set up a Base Realignment and Closing Commission to make a priority list of facilities to be shut down. Congress could then vote the list up or down, but could not amend it. Several rounds of this procedure were successfully implemented (mostly to the financial benefit of the communities) until President Clinton for his 1996 reelection campaign decided to make an exception for two bases in California. The rules infringed in this way, Congress saw no reason to honor further its part of the bargain. The same influences have been evident in the nuclear weapons field—in keeping open the Nevada Test Site and in maintaining the two weapons design labs, Lawrence Livermore National Laboratory in California and Los Alamos National Laboratory in New Mexico, at a level that may or may not be optimum for our national security.

In his farewell speech in January 1961, President Eisenhower warned of the dangers of the "military-industrial complex":

> In the councils of government, we must guard against the acquisition of unwarranted influence, whether sought or unsought, by the military-industrial complex. The potential for the disastrous rise of misplaced power exists and will persist.
>
> We must never let the weight of this combination endanger our liberties or democratic processes. We should take nothing for granted. Only an alert and knowledgeable citizenry can compel the proper meshing of the huge industrial and military machinery of defense with our peaceful methods and goals, so that security and liberty may prosper together.

We judge the situation to be more perilous today than in Eisenhower's time; too many government officials understandably believe that electoral politics—both congressional and presidential—dominate public decisions. To us, this development is likely to cut short American leadership in world affairs; it imperils the future of all.

In principle, these questions of choosing or maintaining certain activities can be addressed by a simple decision procedure—i.e., by choice based on the tools of economics. What are the benefits to be expected from a certain course of action (and the probability of realizing those benefits), and what are the costs and damages and their probabilities? Too often the public, or, for that matter, the decision-maker, receives from proponents the list of benefits without a similar tally of costs or problems; and opponents to the program may present only the costs. The closing of military bases is a routine question for such analysis.

Another example, which has been a stumbling block on the way to the control of nuclear weaponry, is the so-called "peaceful nuclear explosions." The role of the scientist or technologist is not to decide, but to provide the information that will permit a decision; such choices depend not only on the costs and benefits of a particular program but on the "opportunity cost"—the loss of the benefits from other programs that could not be funded because the necessary resources would be consumed in the program in question. Of course, should a scientist or technologist be elected or appointed to a high office, he or she must actually make and execute the choice of programs, after taking effective measures to obtain unbiased and competitive advice and analysis. A scientist expert in one field must be careful not to disdain advice in other fields, or even in his or her area of expertise.

PEACEFUL NUCLEAR EXPLOSIONS

In the past, Soviet scientists used a number of underground nuclear blasts to pulverize enormous amounts of rock, which would be much more expensive to crush with chemical explosives. It is easy for a state that has mastered thermonuclear weapons, as we have seen, to produce explosions with a power equivalent to 10 million tons of dynamite. The Soviets employed nuclear explosions to put out oil well fires and to stimulate wells approaching exhaustion. The energy of the oil that can thus be extracted, according to Russian scientists, can be five to seven times greater than that of the nuclear explosion. Scores of nuclear explosions in the former Soviet Union have served to stimulate the flow of gas or oil on an experimental basis, to provide underground caverns for the storage of petroleum components, or to serve as reservoirs for the disposal of toxic wastes.

Proposals for the use of nuclear bursts to produce electricity were made during the decades when the United States had a substantial program in peaceful nuclear explosions, which preceded but was smaller than the Soviet effort. In particular, one plan of Los Alamos, Project Pacer, called for the production of electrical power by the explosion of thermonuclear explosives in underground cavities filled with high-pressure steam. Each day, a 60-kiloton nuclear explosive would be detonated in a cavity to keep the steam hot, while a relatively conventional steam-turbine power plant would draw on the steam reservoir to produce electrical energy. Nuclear heat from the explosion would simply replace a day's nuclear heat from a reactor.

In 1975, one of the authors (Garwin) worked on an advisory group to the

U.S. government studying the whole field of peaceful nuclear explosions, and Pacer in particular. Although it had been claimed that Pacer was a cheaper road to nuclear power than the reactors that were mature at the time, side-by-side comparison with a normal nuclear power plant showed otherwise. In addition, the scale of the nuclear explosive manufacturing and transport program was almost unfathomable. Each of the 60-kiloton explosives would have had an explosive yield some 4 times that of the bombs that devastated Hiroshima and Nagasaki. For each of the almost one hundred nuclear power plants operating now in the United States, 365 such explosions per year would be required—36,500 per year in total. It is unreasonable to think that humanity might consider technology of this kind, while it is still searching for satisfactory methods for properly managing nuclear power plant waste and trying to reduce the number of nuclear explosives.

Major projects continue to be set forth—most recently by the Russian nuclear weapon laboratories. In an audacious scheme, scientists have analyzed an enormous steel pressure vessel using a year's output of all of Russian steel mills for the container, to be equipped with multiple fountains of liquid sodium inside, for the purpose of shielding the steel container from the force and radiation of the 20- or 50-kiloton thermonuclear bursts.[4] The possible attraction of such explosives in peaceful use lies in part in the fact that only relatively small quantities of fission products and plutonium are produced. For the Russian 120-kiloton explosive used in rock crushing, it amounts to a mere 300 tons of high-explosive equivalent from fission. This results in a factor of 400 less fission products than in a nuclear reactor; the rest of the yield came from fusion of deuterium. This approach would thus compete with approaches for extending uranium fuel supplies—e.g., a breeder reactor—or obtaining uranium from seawater.

In the mid-1990s, it was the turn of the Chinese to consider the possibility of producing electricity by means of nuclear explosions.[5] They suggested that with underground explosions within the yield range of typical thermonuclear weapons (10 to one hundred kilotons), energy could be produced from uranium-238, and thorium-232 could be burned, multiplying the accessible energy of any particular uranium resource by a factor of 100 or so with respect to what can be obtained with ordinary reactors that burn only the 0.71% of natural uranium that is uranium-235. In these underground explosions, thermonuclear reactions could provide 90%, or even 99%, of the nuclear energy released. According to the authors of the project, there would remain only a few modest technological problems to solve. This was not, of course, a brand-new idea, but it was being taken seriously for the first time in China.

The Chinese also had a proposal to transfer enormous quantities of water from the Yaluzangbu River to northwestern China, where 47% of the Chinese population lives but which has only 7% of the country's water resources. This titanic project remains beyond the scope of chemical explosives. Of course, the problem of radioactivity arises in these projects with nuclear explosives, even more so when they are connected with water flows. It is true that peaceful nuclear explosives could be made less polluting than weapons, because they do not have the same weight and (in some cases) size limitations. For example, the ground can be prevented from contributing to neutron-induced radioactivity by means of judiciously chosen shielding. Nevertheless, residual plutonium and radioactivity of fission products are very serious problems in any thorough analysis of peaceful nuclear explosions.

In the United States, a broader scientific community has at times engaged in a serious dialogue with physicists and engineers who dream of economic benefits from nuclear explosions, and the resulting more sober evaluation has shown costs exceeding benefits. The incompatibility of a program of peaceful nuclear explosions with a total ban on nuclear weapons testing strengthened the policy judgment. Indeed, on such a topic, the primary critical analysis should not come first from an outside group; the nuclear weapons laboratories themselves ought to finance independent studies on these questions, as well as on the potential contributions of techniques other than nuclear explosions in areas like the generation of electrical power. Government decisions on projects initiated by the weapons laboratories clearly would benefit by outside analyses commissioned by the labs themselves.

In June 1996, at the Geneva Disarmament Conference, the Chinese agreed, as had the French a year earlier, to discontinue their nuclear testing after July 29, 1996, and to sign the Comprehensive Nuclear Test Ban Treaty, which outlaws all nuclear explosions, including those called "peaceful." In order not to close off some possible economic benefit from peaceful nuclear explosions, ten years after the treaty goes into effect the question would again be examined whether these explosions should be once more permitted or forever banned. Only if there is consensus (i.e., no dissent) among the states party to the treaty would such explosions again be permitted. It is to be hoped that costs and alternatives would be considered, in addition to the claimed benefits.

THE HERITAGE OF EXCESS NUCLEAR WEAPONS

It is essential that we rationally manage the sixty thousand atomic bombs inherited from the decades of Cold War between the two major blocs. It is pointless to discuss whether Chernobyl will cause thirty thousand victims in forty years, or twice as many, when the nuclear devices designed to kill, with maximum efficiency, hundreds of millions of people remain at the ready and can cause much more deadly accidents than Chernobyl. Are we capable of dealing with this major hazard of annihilation—measured in months or years and not tens of millions of years? In Chapter 11 we'll see why this problem has not yet been solved and what must be done to take care of it.

PUBLIC USE OF SCIENCE AND TECHNOLOGY

No observant person can be oblivious to the pace of scientific advance—in physics and astrophysics, in mathematics, in genetics and molecular biology, in chemistry, and in a host of other fields. The pace of technological innovation, as well, is breathtaking—especially when the ultimate consumer is the purchaser, as in the case of personal computers, skis, cameras, and consumer electronics. Indeed, one of the reasons for the increased pace of scientific discovery is the availability of the technology that empowers the scientist. In most fields of science, the team is rather small, with particle physics the exception in the last decade or so—a field in which hundreds of researchers in tens of institutions may be involved in a single experiment. But even in the case of the individual scientist or small team, the work is made possible, for the most part, by modern tools and sets of tools, such as the personal computer and its software; scientific instrumentation for low-temperature or high-pressure research; or the tools for determining the sequence of the four types of nucleic acids in DNA, or the sequence of the twenty types of amino acids in the long chains that make up proteins of interest.

This portion of the scientific enterprise works very well. The evolution of the market economy has also resulted in the creation of a large number of start-up firms to commercialize these various inventions and innovations. The United States has led in this field, with great benefit to the world's public.

Unfortunately, the rapid pace of advancement in pure science, and in consumer technology such as personal computers and electronics, has not carried over to the public sphere, the satisfaction of public needs, and even less to the formulation of public policy. This failure obtains whether we consider public safety or the efficient collection of taxes, or even national defense. Perhaps we are hoping for more than can be expected in a democratic society; indeed, the world eliminated the naturally occurring scourge of smallpox, although smallpox as a weapon of terrorism or war is more to be feared as a result and could instantly bring on a pandemic. And the "revolution in military affairs," as evidenced by the performance of American armed forces in integrating battlefield intelligence and strike in the 1991 Gulf War—and repeated in Kosovo and Serbia in 1999—was a long time in coming and is far from complete. A firm that does not use cost-effective tools is likely to go bankrupt in a competitive economy and soon disappear. In contrast, governments may fall, but they may well be replaced by less efficient regimes. In democracies, the voters are more likely to be influenced by public relations and acting skills than by a concern for the public good. And the ultimate challenge of military conflict or economic competition with other nations usually takes second place to the matter of the next election in the priorities of our leaders. NATO and the European Union show some signs of recognition that the European weapons programs have suffered greatly from duplication—not competition—and the comparison with American military capabilities is now so obvious that the Europeans resolve to modernize their forces and to make them more effective in joint operations. Time will tell.

If matters such as the income tax, the provision of health care, and the operation of schools are not done well even where public understanding is widespread, it is no surprise that public decisions that depend, in addition, on science and technology are taken with even greater difficulty and in many cases with considerable lack of success. Government is difficult, and people as clients and as masters are doubly difficult.

The national government rarely has had the competence and the incentive to make informed decisions in highly technical fields. It is unusual for the head of a cabinet department to be literate or accomplished in science and technology. One notable recent exception was William J. Perry, U.S. Secretary of Defense from 1994 to 1997. Perry is a trained mathematician with long experience in technology applied to national security and intelligence. In 1999 he received the R.V. Jones Award for scientific intelligence. During his earlier service from 1977 to 1981 as Director of Defense Research and Engineering, Perry helped to steer the Defense Department to make the investments and the commitment to bring into being the revolutionary nonnuclear weapon capability that first came to public attention during the 1991 campaign against Iraq to liberate Kuwait, and again in U.S. air operations against Serbia and Kosovo in 1999.

But even a competent cabinet leader has difficulty formulating the detailed options from which to make an informed choice; a group effort is necessary, but there are two problems. First, specialized bureaucratic personnel are usually busy performing their jobs and not available for the hard work of the required analysis, which might take several months to a year. Second, the government as a whole—like any organization—is shot through with conflicts of interest. This doesn't mean that officials have invested in one or another contractor who might be favored by a particular program choice, and does not refer to contributions to congressional and presidential campaigns. Instead, major problems are the presumed programmatic positions of the bureaucracies; the conservative nature of a system that prefers changes at the margin when confronted with the prospect for radical change; and also the interest of some specialists in technology for its own sake. In numerous instances of the procurement of types of ships, aircraft, or weapons, the fact that the technology was fascinating to those who wanted to work on it was enough to overcome their judgment and in some cases even their scruples. Problems of process are less interesting than those of substance, but both must be addressed. We recall a quip by John W. Gardner:

> An excellent plumber is infinitely more admirable than an incompetent philosopher. The society which scorns excellence in plumbing because plumbing is a humble activity, and tolerates shoddiness in philosophy because it is an exalted activity, will have neither good plumbing nor good philosophy. Neither its pipes nor its theories will hold water.[6]

Advances in science and technology are necessary in the long run, but when the public needs can be satisfied more cheaply with existing means, it is a disservice to prescribe research rather than to adopt an existing solution.

INDEPENDENT ADVISORY GROUPS

Since the end of the Second World War, the United States has possessed an exceptional tool less used in other countries. This is the public service of scientists stemming from the highly successful work of scientists and engineers in the major projects of the war effort—radar development at the MIT Radiation Laboratory and associated industries, and the nuclear weapon Manhattan Project. In addition to benefiting from the mobilization of science and technology for the war effort, which also took place in England, the Soviet Union, and Germany, the postwar U.S. government benefited from the continued involvement of scientists and engineers who had been involved during the war. Most of them returned to their universities, but in typically American style formed committees and organizations to help their government and their country. This substituted for and was probably at that time superior to the Confucian and French tradition of officials who had authority, but neither sought counsel nor accepted it.

These organizations included the Federation of Atomic Scientists, formed by the Manhattan Project scientists in 1945, which attempted to ensure recognition of the destructive power of nuclear weapons and proper control of this force, and also interlocking groups of scientists and engineers who have tried to persuade the government to take various initiatives for national security. An early postwar example was to push the government toward nuclear propulsion of submarines and surface ships, and to build strategic missiles with nuclear warheads, based both on land and on submarines. Another example of momentous initiatives by scientists outside the government was to urge the development of tactical nuclear weaponry, against the preference of the military at the time, which had institutionalized nuclear weapons in the Strategic Air Command. Perhaps the initiative for tactical nuclear weapons was of interest only to a nuclear monopolist, since these weapons for use against military forces are probably blocked from use by the likelihood of strategic nuclear retaliation.

In the late 1940s, the then new Atomic Energy Commission (AEC)—the heir to the wartime nuclear weapons program, and which had also the responsibility for the production of reactors for naval propulsion and for electric power—was supported by a general advisory committee. In those days, a much larger fraction of the activity was secret than is the case at present. At the same time, many of the leading scientists and engineers in the war effort, having returned to academia and industry, maintained contact with the government programs that grew from their involvement during the war. Some of these programs were active at government laboratories and arsenals, while some were centered in defense contractors, which were feeling their way in the postwar era. Several of these scientists served on the General Advisory Committee of the AEC and thus had a broader policy view of these technical questions.

Even informed, competent public officials often did not have the power to stand up against well-funded lobbying and pressure campaigns. President Eisenhower, when he took office in 1953, saw the need for a new mechanism—to act as a counterweight.

Committees and boards such as the Army Scientific Advisory Panel, the Naval Research Advisory Board, and the Air Force Scientific Advisory Committee tended to be captives of their sponsoring organization. It was clear that the President needed to possess the best advice on the programs that would come to him for decision and incorporation into the budget. Many such decisions are made in the Bureau of the Budget (now the Office of Management and Budget), which has substantial power and has always been reluctant to solicit outside advice or even to have its own advisory committee.

THE PRESIDENT'S SCIENCE ADVISORY COMMITTEE, 1956–73

The provision of scientific and technological competence in support of the President was assumed by the so-called President's Science Advisory Committee of the Office of Defense Mobilization, but it had little power until Eisenhower assumed the presidency and soon brought the committee into the White House; he created the position of Presidential Science Advisor, who also chaired the committee. Thus the President's Science Advisory Committee became an official governmental organ with members who (with the exception of the chairman) were not government employees. Knowledgeable about both bureaucracy and the military, Eisenhower did not trust the Pentagon to provide objective proposals and recommendations for its own programs. The 18 members of the Science Advisory Committee, serving four-year terms, met two days every month, with remarkable attendance records. At any time it also had something like 12 panels; each recruited among its 10 to 15 members the best experts who could be found in the country to work on the particular problem, in addition to two or three members of the parent committee. Standing panels included those on military aircraft, naval warfare, and antisubmarine warfare and the Strategic Military Panel, dealing with both offensive and defensive systems, especially nuclear weaponry. Other panels were created for a specific study and report, such as one on insecticides and pesticides.

The committee meetings and those of its panels were private. Its job was explicitly not to be a cheerleader for science, but to understand matters on behalf of the President, and to present the options and choices as clearly and fully as possible—not for the management of science itself but for the use of science and technology for government purposes. Much of the influence of the committee came from its frequent and intense discussions with the Pentagon and its contractors, so that programs were improved or deflected without the necessity of a formal written report.

The committee was instrumental in moving many of the cabinet departments—e.g., Transportation and Agriculture—to create a position of Assistant Secretary for Research and Development, an accomplishment that, paradoxically, was one of the reasons used by President Nixon's staff in 1973 to argue that the Science Advisory Committee was no longer needed or effective. Quite the contrary was true, if the President was to be able to lead the nation and if these technically literate people were to work together in the national interest and not serve primarily departmental goals. Technical competence in the departments would allow the committee to operate on a higher plane.

Because of the great secrecy of the subject, the Strategic Reconnaissance Panel, chaired from 1954 to 1973 by Edwin H. Land, creator of the Polaroid Corporation, was not quite a panel of the committee, but instead reported directly to the Presidential Science Advisor. It had major impact on the strategic reconnaissance capability of the United States; choices made at that time influence the field to the present day. Driven in good part by Land and his colleagues from the President's Science Advisory Committee, strategic reconnaissance began in 1956 with the U-2 aircraft and then the Mach-3 SR-71, and from 1960 with imaging satellites of increasing competence and complexity. One of the present authors (Garwin) was a member of the Land panel from about 1961 until its elimination in the early 1970s.

The Strategic Military Panel repeatedly reviewed proposals for deployment of nuclear-armed systems for defense against ballistic missiles, pointing out that they would be worse than useless. But it failed to prevent the deployment of the Safeguard system in the Nixon administration, on which $8 billion was spent from 1968 to 1978 to develop and to deploy two radars and one hundred nuclear-armed interceptors in North Dakota, to defend a fraction of U.S. ICBMs against attack by Soviet missiles.[7] The $8 billion spent would be $21.3 billion in 1996 dollars. The system was fully operational for only four months before it was shut down in January 1976, and it was largely dismantled soon thereafter.

Reductions and control of nuclear weapons and their delivery means has been a long-proclaimed goal of governments, but so has been the development and deployment of newer and additional weapons. In the effort to reduce the threat of nuclear weapons, groups of scientists have played a prominent role. In a 1945 speech in the British House of Lords, the philosopher Bertrand Russell proposed that Western scientists confer with their Russian counterparts as a beginning "toward genuine cooperation" to "somewhat mitigate the disaster that threatens mankind."

Lord Russell's continuing concern about the future of a world armed with nuclear weapons led to the "Russell-Einstein Manifesto" of 1955, signed by eleven world-famous individuals, among them Albert Einstein (who died two days later). They observed that nuclear weapons would certainly be used in any future world war and therefore called upon governments "to find peaceful means for the settlement of all matters of dispute between them." In 1957, twenty-two scientists from East and West met at Pugwash, Nova Scotia, at the home of Cleveland industrialist and philanthropist Cyrus Eaton, and created the Pugwash Movement—more formally, the Pugwash Conferences on Science and World Affairs.

PUGWASH

By October 2000 there had been more than 257 Pugwash meetings—the annual conferences each attended by 150 to 250 people and the more frequent topical workshops and symposia with 30 to 50 participants. A basic rule is that individuals participate in their private capacities, not as representatives of governments or organizations. A 1998 description sketches the purpose and history of Pugwash:

> The purpose of the Pugwash Conferences is to bring together, from around the world, influential scholars and public figures concerned with reducing the danger of armed conflict and seeking cooperative solutions for global problems. Meeting in private as individuals, rather than as representatives of governments or institutions, Pugwash participants exchange views and explore alternative approaches to arms control and tension reduction with a combination of candor, continuity, and flexibility seldom attained in official East-West and North-South discussions and negotiations. Yet, because of the stature of many of the Pugwash participants in their own countries (as, for example, science and arms-control advisors to governments, key figures in academies of science and universities, and former and future holders of high government office), insights from Pugwash discussions tend to penetrate quickly to the appropriate levels of official policy-making. . . .
>
> The first half of Pugwash's four-decade history coincided with some of the most frigid years of the Cold War, marked by the Berlin Crisis, the Cuban Missile Crisis, the invasion of Czechoslovakia, and the Vietnam War. In this period of strained official relations and few unofficial channels, the fora and lines of communication provided by Pugwash played useful background roles in helping lay the groundwork for the Partial Test Ban Treaty of 1963, the Non-Proliferation Treaty of 1968, the Anti-Ballistic Missile Treaty of 1972, the Biological Weapons Convention of 1972, and the Chemical Weapons Convention of 1993. Subsequent trends of generally improving East-West relations and the emergence of a much wider array of unofficial channels of communication have somewhat reduced Pugwash's visibility while providing alternate pathways to similar ends, but Pugwash meetings have continued until the present to play an important role in bringing together key analysts and policy advisors for sustained, in-depth discussions of the crucial arms-control issues of the day: European nuclear forces, chemical and biological weaponry, space weapons, conventional force reductions and restructuring, and crisis control in the Third World, among others. Pugwash has, moreover, for many years extended its remit (scope) to include problems of development and the environment.[8]

Pugwash was awarded the Nobel Peace Prize in 1995, jointly with Joseph Rotblat—by all odds the most active and dedicated Pugwashite from 1957 to the present day—"for their efforts to diminish the part played by nuclear arms in international politics and in the longer run to eliminate such arms."

THE DOTY GROUP

The threat of nuclear weaponry was universal, but the Soviet Union and the United States were the chief actors, and it was those governments that had to be influenced. Faculty members at Harvard and the Massachusetts Institute of Technology had long asked themselves whether there was something that could be done in the way of disarmament by direct engagement with their Soviet counterparts.

At the first Pugwash Conference to take place in the Soviet Union—in Moscow, November 27 to December 5, 1960—American and Soviet scientists explored what was to become in 1962 the U.S.-Soviet Study Group on Arms Control and Disarmament. This new contact between U.S. and Soviet scientists in the field of nuclear weaponry was extremely important, as evidenced by the high-level meeting that took place in Washington on December 6—the day after the Pugwash Conference in Moscow. John F. Kennedy had been elected president a month before, and two of the participants in the Moscow session, Jerome B. Wiesner of MIT, who was to become Kennedy's Science Advisor, and Walt W. Rostow, the future Deputy National Security Advisor, reported to a large gathering of officials from the Pentagon, the State Department, and the White House.[9]

On the U.S. side, the leader of the Study Group on Arms Control and Disarmament until its demise in 1975 was Paul Doty, professor of biochemistry at Harvard. In the 1950s and early 1960s, Doty was a member of the President's Science Advisory Committee and as such was involved with evaluating the threat of Soviet nuclear weaponry to the United States. The Doty group engaged in informal bilateral (i.e., U.S.-Soviet) scientific diplomacy that continued to be conducted before or after Pugwash sessions. There was no guarantee that the Cold War would not evolve into a thermonuclear war leading to the total destruction of both sides and of much of the rest of the world. For the Soviet scientists—some of them well connected with the nuclear effort in their country—it was by no means without risk to engage in such discussions with their American counterparts. These pioneers in Soviet-American Disarmament Studies ("SADS"—another name for the U.S.–Soviet group) exhibited great courage and tenacity.

Soviet scientists could participate in these bilateral sessions only if they were formalized by some connection with Pugwash. So it became legitimate in the Soviet system for a select group of its scientists to be designated by their organizations and their government to participate "as individuals" in these meetings. The Soviet attendees were a mixed bag—some first-rate individuals (measured by both their scientific contributions and their integrity), others who were official hacks, and still others who were probably KGB agents. To avoid the appearance of dealing directly with the Cold War enemy, the Soviets insisted that the meetings take place as part of "Pugwash," in the one or two days preceding or following a Pugwash annual meeting or workshop.

Wiesner resumed his active participation in the U.S.–Soviet Study Group after leaving the position of President's Science Advisor, following the November 1963 assassination of President Kennedy.

In a paper at the January 1964 Pugwash meeting in Udaipur, India, Jack Ruina, an MIT electrical engineer recently returned from a position in the Pentagon, explained the technical and strategic problems of defense against nuclear-armed strategic ballistic missiles. At the same time, the United States proposed at the Geneva-based Eighteen Nation Disarmament Committee that the United States and the Soviet Union should "explore a verified freeze of the number and characteristics of their strategic offensive and defensive vehicles." This departure from the sterile proposals for "general and complete disarmament" led to a focus on the merits and risks of missile defense and ultimately paved the way for the 1972 U.S.–Soviet treaty that effectively banned a ballistic missile defense system that would protect either nation against a substantial threat, the so-called Anti-Ballistic Missile (ABM) Treaty of 1972. Soviet participants in Pugwash for years maintained that "defense is good," so that ABMs should not be limited. When Secretary of Defense Robert S. McNamara, meeting in 1967 in Glassboro, New Jersey, with Soviet leaders, proposed limitations on defense against ballistic missiles, he received no encouragement from the Soviet side. Ultimately, advocacy of a treaty was not capricious, but the result of reasoned analyses on both sides that nuclear-armed missiles could overcome such defenses, with the result that the building of missile defenses would only lead to a massive growth in the offensive nuclear forces, without providing real security. The argument continues to this day in the United States.

With the détente that permitted the 1972 Moscow agreements between the Soviet Union and the United States—the ABM Treaty and the "Limited Offensive Agreement"—and the closer government-to-government contacts, the unique role of the Soviet-American Disarmament Studies group appeared to be less valuable. After the death of the leader of the Soviet team, Mikhail D. Millionshchikov, in 1973, the joint study group was less well connected on the Soviet side. American philanthropic foundations saw less reason to support the group with travel funds, and the two governments did not recognize the need for continuing unofficial contacts. The joint study group ceased operations in

1975, but the U.S. scientists strove to reestablish meaningful contacts for the control of the nuclear arms race.

THE NAS COMMITTEE ON INTERNATIONAL
SECURITY AND ARMS CONTROL

Another influential and constructive group, currently operating, is the Committee on International Security and Arms Control (CISAC) of the National Academy of Sciences. The NAS is a private organization, chartered by the government in 1863, that exists not only as an honorary academy but primarily to do studies for the government. The NAS, the National Academy of Engineering, and the Institute of Medicine constitute the "National Academy of Sciences" structure of the United States. CISAC was created in 1979 and started its work in 1980, with the primary purpose of a dialog with the Soviet Academy of Sciences about security matters affecting the two countries, as well as the rest of the world. The highest priority was the control of nuclear weapons and the reduction of the likelihood of nuclear war.

Several of the founding members of CISAC had been members of the Doty group—Doty himself, Wiesner, Garwin. These and others had been members of the President's Science Advisory Committee. Over the years CISAC has been chaired by physicist Marvin L. Goldberger, then president of the California Institute of Technology; by physicist W. K. H. Panofsky, builder of the Stanford Linear Accelerator Center; and currently by John P. Holdren, a physicist specializing in energy and environment, for many years at the University of California at Berkeley and now at Harvard. Half of the dozen or so members of CISAC are members of one of the National Academy institutions, while the others are specialists and individuals with expertise in international security, technology, or weapons. None is an employee of the government.

For instance, former CISAC member General David C. Jones had been Chief of Staff of the Air Force and Chairman of the Joint Chiefs of Staff. William J. Perry left the committee to become Secretary of Defense in the Clinton administration. Michael M. May had directed the Lawrence Livermore National Laboratory. General Lee Butler was formerly head of the United States Strategic Command, which has responsibility for the operation in wartime of all of the strategic nuclear systems that would be used in destroying many thousands of targets with nuclear weapons.

Regarding its primary purpose, CISAC clearly accomplished a good deal in helping to educate two generations of Soviet scientists in international security matters. As a matter of great fortune, when Mikhail Gorbachev took office in 1985, with a distrust (like Eisenhower's) of the military and policy establishment, several of the CISAC-counterpart participants became his principal national security advisors—Evgenii P. Velikhov, Roald Z. Sagdeev, and Georgi A. Arbatov. At a Russian–U.S. session marking 20 years of CISAC, Velikhov gave credit to CISAC and its members for Soviet termination of work on the ABM radar near Krasnoyarsk (which would have violated the ABM Treaty) and for the Soviet moratorium on testing of antisatellite (ASAT) weapons.

CISAC members maintained their individual connections with the U.S. government, as consultants and members of advisory bodies. In addition, CISAC played a role in keeping Washington informed of the interactions with the Soviets and occasionally advising government officials on questions of interest to them. For instance, Paul H. Nitze, the grand old man of American nuclear diplomacy and at times a leading proponent of the arms race, was again in the 1980s in the State Department. He sought advice from a CISAC subgroup regarding the Strategic Defense Initiative and other matters of national security policy and technology.

Over the years, CISAC studies have had considerable influence. A 1991 report, "The Future of the U.S.-Soviet Nuclear Relationship," provided a logical foundation for the levels and sequence for nuclear weapon reductions in the evolution of the strategic arms reduction talks (START). This study proposed that without any change in targeting or understanding of the function of nuclear weaponry, Americans and Soviets could at any time reduce the number of deployed strategic nuclear warheads to 3000 on each side, and could go to one thousand by simple improvements in efficiency of targeting and employment of the warheads. For instance, if the requirement is to have 95% confidence of a warhead exploding on target, and each missile has 90% reliability, two missiles will need to be used. Alternatively, if there is a means to know whether a nuclear explosion has occurred at the target, just 1.11 (i.e., 1/0.90) missiles would be enough to have not just 95% but 99% confidence or more. The idea was to reduce the scale of destruction if war came (at least to reduce the destruction outside the United States and the Soviet Union) and to improve the control of nuclear weaponry.

A 1997 CISAC study, "The Future of U.S. Nuclear Weapons Policy," extended the 1991 report, taking into account the dissolution of the Soviet Union. It argued for a reduction in START III from the 3000 or 3500 deployed strategic nuclear warheads of START II to 2000 deployed nuclear warheads, and then a post-START era in which the two sides would possess 2000 warheads each. This apparent lack of progress hides a major difference in measuring warheads: the unit of account would now be the nuclear warhead itself, and instead of limiting only the deployed warheads to 2000 (with no limitation on the number of warheads possessed), the *total* number would be 2000—deployed and reserve.

A natural progression would then lead to one thousand on each side, and as the other official nuclear weapon states under the Non-Proliferation Treaty—Britain, China, and France—were brought into the process, numbers on the order of a few hundred were within sight. The 1997 study contemplated the total prohibition of nuclear weaponry, but only if an international regime could be achieved under which security would be improved by such a step. The May 1998 tests by India and Pakistan simply brought attention to the nuclear weapons that these countries were believed to possess—which together with the much larger stockpiles of the five official nuclear states and Israel are the reason for urgent reductions.

AMERICAN PHYSICAL SOCIETY GROUPS

The President's Science Advisory Committee is an example of a governmental institution of independent scientists, while CISAC represents an institutionalized activity with frequent contacts with government. Pugwash is an international group with no government standing at all, which created itself and has been maintained through the energy, integrity, and dedication of several of its longtime leaders, together with the support of its members. In contrast, there are some committees or commissions or panels created to address a particular problem, some of which have been highly influential, some highly reliable, and some both at the same time. Among these are two panels of the American Physical Society.

The American Physical Society was created in 1899, with the purpose of promoting the advance of the science of physics. It and its sister societies have an extensive program of publishing technical articles and of scientific meetings. The APS wished also to put the expertise of its members to the service of society at large, and has occasionally found ways to do so. One such activity was the 1975 APS panel on light-water reactor safety, chaired by Harold W. Lewis of the University of California at Santa Barbara, of which one of the present authors (Garwin) was a member. The APS selected a group of members to do an independent analysis of nuclear safety, at a time when the Atomic Energy Commission was completing a report on the topic. The APS study was reviewed by a panel of scientists separate from the study group and made significant contributions in advancing a simple analysis (a "model" that could be evaluated readily) of the radiological consequences of a major nuclear reactor accident. The AEC study—"WASH 1400"—was released in draft form for comment.[10] It had major errors, which were corrected by the independent study.

In 1987 another independent group commissioned by APS studied the technical aspects of "directed-energy weapons" on which the performance of the Reagan-era Strategic Defense Initiative (SDI) depended. The SDI organization had dealt with its critics by charging that they didn't know what they were talking about, were biased, or were unaware of secret information. The APS panel could not be criticized in this way, since it was clearly competent; several members were from the weapon laboratories themselves, and all had access to secret material. The group found that most of the SDI's directed energy weaponry could be defeated by intercontinental-range rockets that simply achieved their final speed in 100 seconds instead of 200–300 seconds—an improvement that the panel and defense contractors agreed was feasible at perhaps a 5% loss in payload capability. Such rockets could readily have been deployed by the Soviet Union by the time SDI's weaponry could be available, and this prospect should have resulted (and perhaps did) in a decision not to deploy SDI, since one of the "Nitze criteria" was not satisfied—SDI would not be "cost-effective at the margin" because it could be countered at less cost than it could be expanded.

The APS in early 2001 began a technical study of boost-phase intercept that might help to inform the public discussion of national missile defense.

We mention the valuable centers at universities—e.g., the Belfer Center for Science and International Affairs at Harvard, the Peace Studies Program at Cornell, the Center for International Security and Cooperation at Stanford, and the Program for National Security Studies at MIT—as extraordinarily effective sources of ideas and plans for implementation of elements contributing to national and world security. And there are individuals who have made their own vital and sometimes long-continued contributions in this field. The role of a committee of independent experts is often to critically assess and validate proposals by individuals who could not command the respect or even the hearing accorded to a formal group.

These examples could be multiplied (e.g., the "Ford-Mitre study" of 1978, *Nuclear Power: Issues and Choices*), but the lesson is that outsiders selected by a group whose function is quite apart from the industry or government or program involved can make insightful and dispassionate analyses that cannot be counted on from interested parties or even from the administration as a whole. Support by governments and philanthropic foundations—in the United States and in the world at large—of independent, competent committees for solving the problems of future energy supply and the problems posed by both nuclear power and nuclear weapons could lead to significant progress in these vital areas.

CHAPTER 11

From Arms Race to Arms Control

IN AUGUST 1945, the United States had used its two nuclear weapons in wartime (and had expended one in a test in New Mexico the previous month). The United States was demobilizing; its military forces would undergo great change in the reorganization that followed the end of the Second World War. Even if nuclear weapons had no immediate targets, there was little doubt that they would continue to be built. U.S. military leaders were not about to give up the most powerful weapon on earth; indeed, they expected to have responsibility for its development and production, as they did for all other arms, although the hand-built (i.e., not production-line) nuclear weapons were far from the military standard of robust and safe explosives. Nevertheless, the production of fissile material continued: plutonium was coming from the reactors and reprocessing facility at Hanford, Washington, and uranium-235 from the isotope enrichment plants at Oak Ridge, Tennessee.

The scientists who had built the atomic bomb strongly opposed military control of the weapon that held such potential for the death of tens of millions. Several of them, by interrupting their careers for two years of full-time lobbying and public appearances, won the battle for the civilian control of nuclear weapon development. To this day, the Department of Energy (successor to the Atomic Energy Commission) develops and manufactures nuclear weaponry, and after the war years passed before the military was given custody of operational nuclear arms.

Beyond the goal of civilian control, many American scientists involved with the program also argued strongly for the worldwide control and perhaps elimination of nuclear weaponry. These scientists stated that it would take about four years for the Soviet Union (or Britain or France, if they had the motivation) to obtain nuclear weapons, and so they were not surprised in 1949 when Moscow detonated its first. Fear of the consequences of a failure led the

Soviet team to test a copy of the U.S. implosion design (obtained by espionage), with their indigenous and improved design reserved for their second test.

So long as the Soviet Union did not have nuclear weapons, the United States could plan to use its own—in case it was attacked and involved in a major war—not only for strategic purposes (bombing the homeland of the adversary either to destroy cities and people outright or to destroy the basis for military power) but on the battlefield as well.

At its entry into the Second World War, the United States was horrified at the Nazis' bombardment of civilians in cities and attacks on refugees. In retaliation for the German bombardment of Britain, the Americans and the British eventually mounted enormous raids with high explosives and incendiary bombs against German cities, and the United States did the same against Japanese cities. This was, at the time, claimed to be justifiable, since it was aimed not at the population but at the industries that made it possible to carry on the war. Still, the rules of war forbade the intentional targeting of civilians, and in part for that reason American scientists influential with the War Department (as it was called in those days) after the war urged the development of tactical nuclear weaponry, in contrast to the strategic weapons that, because of their inaccuracy and their destructive power, would be used for the destruction of cities.

Tactical weapons were procured to be used on the battlefield. The United States thus provided itself with nuclear-armed artillery shells to be fired from 8-inch or 6-inch cannons, surface-to-air missiles with nuclear warheads, air-to-air missiles with nuclear warheads, and even a tripod-fired nuclear weapon—the Davy Crockett. Not to be outdone, the U.S. Navy had its own ship-to-air missiles with nuclear warheads, nuclear-armed depth charges, nuclear-armed torpedoes, and submarine-launched short-range, rocket-delivered antisubmarine weapons. There were also tactical nuclear weapons in the form of bombs to be delivered on the battlefield by fighter-bomber aircraft. Nuclear weapons had been integrated into the armed forces—army, navy, and the newly created U.S. Air Force. It was the era of "more bang for a buck"—the nuclear weapon was simply a more powerful explosive.

None of these weapons was ever used in combat, but a huge investment was made in such tactical nuclear weapons, in their delivery systems, and in the required specialized personnel as well as security and protective forces.

After the Second World War, Germany, Japan, and Italy had been disarmed and posed no military threat. The Soviet Union was a different matter, and it soon emerged as a major competitor and an avowed foe, so that U.S. nuclear weaponry became focused on that country. Nuclear weapons were considered "the great equalizer"; millions of Soviet and Soviet-dominated troops without nuclear weapons could not avert their destruction or that of their homeland by atomic attack. But it was equally true that the United States could not prevent its own destruction if another nation acquired weapons and could deliver them to U.S. shores. It was this fact that inspired American efforts to avoid the proliferation of nuclear weapons—a goal that has largely been attained but needs continued effort and leadership. It is a goal shared by almost all the nations of the world.

Winston Churchill observed in his famous speech at Fulton, Missouri, on March 5, 1946, "From Stettin in the Baltic to Trieste in the Adriatic an iron curtain has descended across the Continent." Western Europe, struggling to recover from the ravages of the war, now feared an invasion by hundreds of Soviet divisions, augmented by those of the nations that had been taken into the Soviet camp. To counter this threat, the West established in 1949 the North Atlantic Treaty Organization (NATO), vowing that an attack on one member of the alliance would be considered an attack on every one of them. In response, the Soviet Union created the Warsaw Pact in 1955 as a nominal counterpart to NATO, but the "leadership" roles of the United States and the Soviet Union could not have been more different.

The Iron Curtain not only prevented emigration from East to West, but also impeded a valid assessment of Soviet and Warsaw Pact military capability and intentions. Acknowledging the great superiority in numbers of men under arms in the Warsaw Pact nations as compared with NATO, and taking seriously the bellicose threats of Stalin and his successors, the United States deployed tactical weapons with its forces and moved strategic bombers—i.e., those capable of striking the Soviet homeland—to bases in North Africa, England, Turkey, and Japan and on the continent of Europe. Those European nations on whose soil the battle would be fought to repel the feared invasion from the East, especially Germany, were quite understandably never comfortable with the idea that U.S. nuclear weapons would be used against Soviet troops on German territory.

With the 1949 test of the first Soviet nuclear bomb and Moscow's push to build long-range bomber aircraft, the United States was no longer protected by two oceans. It was a dozen hours away from nuclear destruction of its cities—a time that shrank to 30 minutes when intercontinental ballistic missiles were deployed. NATO troops augmented by thousands of tactical nuclear weapons might be able to stop Warsaw Pact forces on the battlefield, but would that really win or even end the war if the Soviet Union could trump the battlefield conflict with its capability to use nuclear weapons against Paris, London, New York, Chicago, and Washington?

With fresh memories of the Second World War, in which 60 million were killed, the United States would have been foolish to imagine that a Soviet Union led by Joseph Stalin (who in 1946 had characterized the war as an inevitable consequence of "capitalist imperialism" and hinted at another war to come) would not initiate nuclear war against the capitalist imperialists. Nuclear weapons and conventional forces were built on both sides to make clear that a war could not be won, and to counter opposing forces to the extent possible if war should begin. Although Stalin died in 1953, the arms race that had begun after the war was to continue for another thirty-five years.

The United States and Canada together built defenses against Soviet bombers, with lines of radars extending across Canada and thousands of nuclear-armed surface-to-air missiles and fighter aircraft. In 1949 the decision was made by President Truman to build the hydrogen bomb, and on November 1, 1952, the United States detonated its first thermonuclear explosive, with nearly a thousand times the energy release of the bomb dropped on Hiroshima. The Soviet threat to NATO expanded following the Soviet Union's detonation of its fission bomb in 1949 and its thermonuclear explosive in 1954.

In 1950 one of the authors (Garwin) had begun work at Los Alamos on nuclear weapons development and testing—both fission and thermonuclear—and not long after (in 1953) on air defense and other military activities related to the use of nuclear arms or to defense against them. He worked for a year on a project to extend to the ocean areas adjacent to the United States and Canada the Semiautomatic Ground Environment (SAGE) air-defense system that was being deployed against Soviet nuclear-armed bombers. Exercises routinely run by the air force and the army never showed an effectiveness of this defense greater than about 15% against the anticipated Soviet bomber raid, armed with nuclear weapons. Even if the defense were to become 50% or even 90% effective, the cost of a nuclear weapon and the aircraft to carry it was small enough that the United States could be destroyed with or without such defenses.

The effort to study the extension of air defenses—Project LAMP LIGHT—was headed by Jerome B. Wiesner and Jerrold Zacharias, both physicists at the Massachusetts Institute of Technology. As we have noted, Wiesner was to become Science Advisor to President John F. Kennedy in 1961. Garwin asked the study leaders in 1953 why they were wasting their time with air defense when by the time anything designed by LAMP LIGHT could be in place the threat would really be Soviet intercontinental ballistic missiles (ICBMs). Zacharias replied that the United States would deploy a defense against aircraft, and then against the missiles when they emerged. Fortunately, the security of the United States and of the world did not depend upon effective defenses against nuclear-armed aircraft or (later) nuclear-armed missiles. Deterrence of attack through the promise of assured destruction was more effective.

NUCLEAR DETERRENCE

With the advent of the atomic bomb in August 1945 and the campaign for its control by the scientists who built it (warning that other nations could soon produce nuclear weapons), farsighted political scientists such as Bernard Brodie tried to understand this future nuclear-armed world.[1] Brodie was among the theoreticians who recognized that the purpose of the atomic bomb was not to win a war but to prevent it. They saw that even with the deepest animosity, two countries armed with nuclear weapons were not condemned to destroy each other even if they could not defend against that weaponry. A nation with a suitable number of nuclear weapons that could survive an attack on these weapons before they could be launched and had a good chance of reaching their targets would *deter* the other side, if the latter cared more about its own survival than about the destruction of the opponent.

Despite this view of their limited utility, nuclear weapons of a wide variety have been constructed since 1945. Each type initially filled a requirement of influential sectors of the armed forces—air, ground, or naval—none wanting to be deprived of weapons they were certain would revolutionize the art of warfare.

The missions assigned to these weapons have also evolved. After a brief period of nuclear monopoly by the United States, during which weapons were treated as if they were simply more powerful tools of traditional battlefield warfare, the acquisition of nuclear devices by the Soviet Union and the difficulty of effective defense against them led to the necessity of "deterrence" as well. A passive defense is analogous to an impregnable fort—the United States, insulated by two oceans from massive military attack and with friendly nations to the north and south, had been passively defended, as if by a moat. But it had suddenly become vulnerable to having its cities destroyed by Soviet nuclear weapons on long-range bombers and, soon, on missiles. An active defense that might destroy even half of the bombers would not protect the country, given the destructive power that could be carried by a bomber—a single bomb would soon exceed in energy yield all the explosive used in the Second World War. But if the fort was no longer unassailable, the attack could clearly be made too costly to sustain—by the capability and the threat to destroy the major military forces and the industrial capacity of the assailants. A pretty good defense against conventional attack was of no consequence against the destructive power of nuclear weapons, but that same destructive power could be used for protection by dissuading the potential aggressor from carrying out such an attack. Particularly valuable and vulnerable to retaliation were the few ports of the Soviet Union, its long-range aviation on the ground, and its centralized government.

In implementing protection by strategic deterrence, the cities, the industrial centers, and all the potential targets—even minor ones—of the Soviet Union (and soon the United States and its allies) were eventually targeted by more and more precise launch vehicles. Next, the opponent's strategic weapons joined the target list, with the idea that a preemptive strike (i.e., launched when one had information that the other side was about to strike) would disarm him and paralyze his offensive and even his retaliatory capability. The purpose was to assure the survival of the nation even if opposing weapons could not be countered.

Over the years, there was much struggle in the United States to interpret "deterrence" as supporting some particular new weapon or course of action—as in the argument current in the late 1970s that "high-quality deterrence" was essential to prevent war, and that this required that the United States possess the ability to destroy the weapons of the other side and thus to remove its ability to carry out (or to deter) attack.

So-called "extended deterrence" began with the formal recognition that the United States would regard an attack on its NATO partners as an attack on itself. The term is used, with resulting confusion, both for the extension of deterrence to nuclear attack on allies (by threat of retaliation), and for the deterrence of large-scale conventional attack by the promise of a nuclear strategic response—that is, on the homeland of the aggressor.

The threat to the nuclear forces themselves, before they could be launched and before they could reach their targets, contributed to an inflation of strategic nuclear forces in the United States and the Soviet Union that went beyond all military logic.

ROLE AND LIMITATIONS OF NATIONAL INTELLIGENCE

In its understanding of the nuclear threat to the United States, the nation of course used whatever information it could obtain by intelligence—foreign publications, spies (human intelligence, or HUMINT), and signals intelligence, SIGINT (the latter includes communications intelligence, COMINT, and electronic intelligence, ELINT). Some of this information was derived from U.S. reconnaissance arcraft flying close to the borders of the Soviet Union gathering COMINT, or intelligence on Soviet radars—RADINT. In 1956 a remarkable unarmed single-seat union of aircraft and camera, the U-2, began to overfly the

Soviet Union at an altitude of 70,000 feet, where it was immune to attack by fighter aircraft or surface-to-air missiles. In 1960, a U-2 flown by Gary Powers was shot down by the evolving Soviet SA-2 missile system, and Powers (unhurt) and his aircraft (in bits) were exhibited at a press conference in Moscow. President Eisenhower, after some initial denial of the overflights by his administration, took responsibility and agreed that there would be no further aircraft overflights of Soviet territory.

In August 1960, the intelligence gap was more than overcome by the first successful film return satellite flight, CORONA, which by 1972 had more than one hundred successful missions, returning all told 33 million feet of film—a million "snapshots" each presenting a patch about 8 miles along the satellite track and 200 miles across the track. A single daylight flight over Soviet territory thus returned an image 200 miles wide and 2000 miles long, to be augmented by a similar harvest 90 minutes later, as CORONA once again overflew the Soviet landmass. The photographs were returned to earth in a "bucket"—a reentry vehicle equipped with a parachute. The idea was to airsnatch the bucket via waiting C-117 aircraft over the Pacific. Eisenhower and his highest national security officials knew almost immediately that the "missile gap" charged by the John F. Kennedy presidential campaign did not exist; rather, it existed in reverse—the United States was ahead.

Over the years, CORONA was augmented and eventually displaced by ever more capable systems for imagery—IMINT—and SIGINT obtained from space. All that can be stated in this volume is that the United States now possesses a near-real-time system for obtaining imagery of the world—a system that does not return film to earth but instead transmits images of great detail and coverage. We are convinced that such national intelligence capabilities played a major role in preventing nuclear catastrophe.

The decision to build vast numbers of nuclear weapons was never taken in a coherent fashion, but was the result of the interplay of powerful forces—most of them directed toward increasing the stockpile of weaponry.

In the Kennedy administration, Secretary of Defense Robert S. McNamara had defined deterrence to exist when the United States had the capability for "assured destruction" of its opponent, and further defined assured destruction to exist when 40% of the population and 70% of the industrial capacity of the principal opponent could be destroyed. His Defense Department then showed that this could be accomplished with 400 warheads of one megaton each, reaching their targets in the Soviet Union. But McNamara was not about to give up tactical nuclear weapons, and in fact deployed 7000 of them to Europe in 1962 to ensure that the Soviet Union recognized the resolve of the United States to defend Western Europe.

Furthermore, the requirement that 400 warheads reach their target led to a strategic force requirement considerably greater than 400 warheads, in view of the necessity to penetrate potential Soviet ballistic missile defenses. There was also the prospect of having some U.S. weapons destroyed before launch.

McNamara once returned from an appearance in Congress and reported to Kennedy that the administration's plan to deploy 500 silo-based Minuteman ICBMs was unacceptable to the legislature, and that he, as Secretary of Defense, had to promise to build 1000. If he had not, he said, Congress would have forced him to build 5000.

For many years, indeed to this day, the United States has had a "triad" of nuclear forces—strategic nuclear weapons delivered on bomber aircraft, on ICBMs, and on submarine-launched ballistic missiles. One of the authors (Garwin) has participated in scores of secret meetings in the Old Executive Office Building across from the White House, with the Strategic Military Panel of the President's Science Advisory Committee, or its Military Aircraft Panel, or its Naval Warfare Panel, concerned with the details and evolution of the triad. Never was there a lack of after-the-fact justification of the necessity for each branch of the triad to be able to impose assured destruction by itself.

McNamara had introduced in his planning the "greater than expected threat," largely to show that even this danger could be met by the forces in being or projected, but he did not take into account that meeting the greater than expected threat would soon become a routine requirement. Thus was born "worst case analysis," which if performed as well by the Soviet Union would lead to an unending upward spiral of arms. McNamara himself contributed to this spiral by announcing in 1962 that nuclear weapons beyond those needed for simple deterrence (for instance, because the Soviet capability for defense against ballistic missiles had not yet reached the greater-than-expected level) would serve a beneficial cause of "damage reduction" by destroying Soviet nuclear weapons before they could be launched. This bonus function of excess warheads led to irresistible demands for improved accuracy of U.S. strategic weapons and no doubt contributed to the Soviet decision both to build more weapons and to be ready to fire them before they could be destroyed.

Later, in a State of the Union Address, President Nixon defined the U.S. nuclear force requirement as "sufficiency," which cheered those who argued that there was no necessity in building U.S. forces to match the number on the other side. But if the rhetoric was sincere, the follow-through was ineffective.

One presidential decision memorandum defined the requirements for U.S. nuclear forces to include the condition that there be fewer Soviet survivors of a nuclear war than American survivors. This seemed to imply that the United

States would have won a war if one million Americans survived while only 200,000 Soviets (or ten Americans and four Soviets) did. With such a justification, and with a "requirement" for ten thousand or more tactical nuclear weapons to destroy tanks and troops in combat, one can see how the United States built altogether some 70,000 nuclear weapons from 1945, and had in 1967 a maximum of 33,000.

Even before nuclear winter was discovered, the scale of potential destruction in nuclear warfare boggled the mind. To destroy 80% of the industrial capacity of the United States, the Soviets would have had to send 600 one-megaton weapons targeted on a hundred centers. In 2000, Russia still had some 6000 operational strategic nuclear warheads of a total force of 10,000 to 18,000 warheads, whereas in 1986 the Soviet inventory totaled 45,000.[2] At the beginning of 2000, there were some 7200 U.S. operational strategic weapons, of an overall force of perhaps 12,000—including tactical and reserve weapons.

The existence of immense quantities of nuclear weapons and their launchers that can reach any place in the world has completely altered military strategy. When the number of nuclear warheads exceeded the number of all imaginable targets, thinking turned toward attacking the enemy missiles. Their launchers were protected by being buried in underground silos; then immense underground tunnels were planned where the missiles could freely circulate and be fired from a variety of openings, chosen at random. It was also envisaged to have airplanes armed with nuclear rocket launchers in flight day and night. Finally, strategists focused on the construction of antimissile missiles that could destroy weapons in flight directed at either cities or at silos—a fascinating arena for technologists and endless work for defense contractors.

The possibility of launching missiles on warning of attack, to destroy the culpable country, surely adds to deterrence of such an assault; if an attack actually comes, it is not clear what is gained by actually conducting the retaliatory launch. This paradox underlies both the theory and practice of deterrence. Our lives may have been saved by a tolerance for paradox, but they were surely at risk for several decades—the more so because achieving deterrence by a launch-on-warning posture bears the grave risk that the weapons will be launched by accident or on the basis of false warning, thus provoking the total destruction that they were built to deter. It should be clear that this peril is largely self-imposed by the United States; if the accuracy of its ICBMs and SLBMs had not been improved to the point of imperiling the survival of Soviet ICBM silos, Soviet nuclear forces would not have needed to be ready for launch on warning of incoming warheads.

Russian weapons ready for launch account in part for the recurrent demand for an effective defense against nuclear weapons—now argued to be necessary in part to guard against small accidental or unauthorized launch of Russian or Chinese missiles, in addition to a primary role of negating a few warheads on ICBMs launched by any of the so-called "rogue states"—North Korea, Iran, or Iraq. An early technical publication on defense against nuclear armed ballistic missiles appeared in 1968, during the administration of President Lyndon B. Johnson, at the time of an intense controversy over the deployment of a defensive system against Soviet missiles armed with nuclear warheads.[3] The article discusses in detail the nuclear-armed interceptors, radars, and other elements of the proposed defensive system, but judges the proposed deployment to be ineffective, largely because of feasible countermeasures. Such judgments led to the ABM Treaty of 1972 between the United States and the Soviet Union, which limited each party to a deployment of one hundred interceptor rockets for destroying incoming strategic ballistic missiles, and with additional restrictions to prevent a defense of the national territory against those missiles. Much current and bitter controversy in the U.S. Congress involves the extent to which continued compliance with the ABM Treaty impedes effective defense of the nation.

In regard to the limitations of nuclear warheads themselves, in 1995, Presidents Clinton and Yeltsin concurred that an agreement should be negotiated to enable the two countries to exchange information, until then kept secret, on the amount of military-grade nuclear material existing on each side. This agreement was authorized by Congress, but Moscow has refused to negotiate. That reflects the fierce resistance of the Russian nuclear weapons lobby and also of the politicians, playing to the electorate, under the pretext that it is possible that the United States might cheat and camouflage large quantities of weapons. Furthermore, Russian politicians resent the process that led to the enlargement of NATO. Americans can hardly scorn partisan politics in the Russian Duma when the appointment of Richard Holbrooke to be U.S. ambassador to the United Nations was held hostage for months by several senators on grounds totally irrelevant to Holbrooke's qualifications for the job.

Today, when Russians and Americans agree that their weapon stockpiles are excessive, the nations' leaders should come to an agreement on a reasonable figure and impose it on their subordinates. When Clemenceau was French head of state, he declared, during a critical period of the First World War, "War is too serious a business to be left to the military." We paraphrase—disarmament is too serious a business to be left to defense departments or nuclear weapon establishments.

We need statesmen capable of reaching a reasoned judgment and saying to their subordinates: "This is the level that we find to be reasonable to reach rapidly, considering the real dangers that threaten us in the foreseeable future.

Get down to that level now." Given that leaders will never be able to please everyone, they might try doing what is right and important.

WHO DECIDED HOW MANY?

The enormous stock of nuclear weapons—33,000 or 45,000, or even 10,000—cannot be justified for any military reason; it was the interaction of domestic politics in the United States with the feeling both there and in the Soviet Union that a large stock of nuclear weapons commanded respect, and also might be used largely to eliminate the nuclear forces on the other side, that led to a nearly unlimited arms buildup.

The view that the nuclear threat from the other side might better be reduced by self-imposed moderation or by treaty or other agreement between the two nations occasionally had some success. For example, in the United States during the administration of President Richard M. Nixon, Senator Edward Brooke of Massachusetts had won the acceptance of a short-lived amendment stipulating that the precision of the nuclear warheads carried by rockets mounted in U.S. submarines was not to be improved to the extent where they would threaten the survival of the launching silos of the adversary. The intent was to arrive at balanced security with fewer weapons, but in reality the commitment was to avoid such a capability only during the Nixon administration, while pursuing its development. Deployment came soon after. With 24 Trident-II missiles per ship, each carrying typically 8 warheads, the U.S. submarine force now has a powerful silo-killing capability; it is just this capability that could motivate Russian commanders to launch the nuclear force "on warning" before it is destroyed.

From the size of the weapons stockpile it is evident that the day was carried by those who paralyzed all efforts toward realistic analyses, by stressing the evil nature of the adversary. In the United States, substantial political power derives from influence over a large part of the nation's expenditures, and the Defense Department's budget in 1998 was $256 billion—16% of total federal expenditures, but nearly 50% of so-called discretionary spending. Henry Kissinger writes of his experience as National Security Advisor to President Nixon, "But our military establishment resists intrusion into strategic doctrine even when it comes from a White House trying to be helpful."[4]

While the military is a powerful entity, the goals of special interest groups are quite divergent. Part of the military establishment considered, with good reason, that the resources invested in nuclear weapons were obtained at the sacrifice of developments more useful for national defense, such as conventional arms. During the Cold War years, the confrontation of the enormous political, financial, and military machines of the Soviet Union and the United

States was dominated by nuclear weapons. Each country could make use of a substantial pool of scientific and technical talent. A very special scientific and strategic culture developed within this circle of interest. Documents reserved for the eyes of the initiated, those with access to data related to national defense, dealt with making use of nuclear weapons and promoting their development. Thousands of technical strategic articles were written, many with a secrecy classification that restricted their reading to security-cleared specialists confined to their own hermetic world, which contained the usual proportions of balanced and of neurotic, ambitious, and unscrupulous individuals.

This secret world had nonetheless a need to communicate with the outside, with the military and political figures on whom it depended for the financing of its projects. In the United States, the relationship with the military was complicated by the rivalry among the army, air force, and navy. The relationship with politicians received a colossal public relations effort, a sort of intellectual intoxication, whose effect depended strongly on the personalities of the leaders. President Reagan was a dream come true for the American weapons lobby: he immediately increased its budget (even though President Carter, running for reelection, had preemptively augmented the budget so as not to seem "weak on defense"—to such an extent that the Reagan administration had difficulty finding a good argument for additional funds). The Reagan administration had initially indicated that it would raise the allocations for conventional weapons, but once in office rapidly decided to favor nuclear weapons. It claimed to emphasize defense against Soviet nuclear weaponry, but in fact attempted to expand the offensive strategic nuclear forces as well.

THE EUROPEAN MISSILE CRISIS IN THE 1970S

In the 1970s, the Soviet Union replaced its old SS-4 and SS-5 nuclear-armed ballistic missiles trained on Western Europe ("SS-" is the Western designation for a surface-to-surface ballistic missile, as contrasted with "SA-" for surface-to-air and "AS-" for air-to-surface) with the newly developed SS-20—a modern, mobile missile with three independently targetable warheads. As one of the sixteen members of NATO, the United States proposed to respond by the deployment of its nuclear-armed, land-based cruise missiles in Europe, and by extending the range of the Pershing intermediate-range ballistic missile (IRBM) to the Pershing II, also armed with a nuclear warhead. Some NATO allies were not anxious to have more nuclear weapons deployed in Europe, particularly ballistic missiles with a very short flight time. Ultimately, the United States and the Soviet Union not only agreed to the removal of the American land-based cruise missile in Europe and of the Pershing II, but agreed to destroy these weapons and to get rid of all Soviet intermediate-range land-based nuclear weapons, which included similar land-based cruise missiles and IRBMs. The prohibition extended to all U.S. and Soviet land-based missiles of range greater than 500 km and less than 5500 km. It therefore left untouched the ICBMs in either country, and also air-launched or sea-launched missiles of any range. The Intermediate-range Nuclear Forces (INF) Treaty was signed in December 1987, during the Reagan administration. It required the declaration and the destruction of these weapons worldwide — whether armed with nuclear warheads or not.

One of the present authors (Garwin) had made it clear in a speech at Cambridge University in 1982 that the SS-20 was not a new nuclear threat to Europe, but was a modernization of the hundreds of Soviet SS-4s and SS-5s that had long held Europe at risk. Almost uncommented on by others was the fact that the SS-20, being mobile, was immune to a first strike from the West, while the SS-4 and SS-5 in their fixed locations were vulnerable to destruction by a U.S. nuclear strike before they could be used. That was the real distinction, not any increased destruction that could have been wrought by the use of the Soviet SS-20 missiles.

The last missile covered by the INF Treaty was destroyed in May 1991. All told, 2692 ground-launched ballistic and cruise missiles have been demolished under the INF Treaty — 2692 nuclear weapons that can no longer be delivered to their targets.

The nuclear warheads for the INF were in no way regulated by the treaty. When Senator Jesse Helms asked at the INF ratification hearings why the warheads were not covered, officials of the Bush administration assured him that warhead dismantlement could not be verified. The authors judge this to have been a lack of desire rather than an inability to verify.

STAR WARS

President Ronald Reagan initiated the Strategic Defense Initiative — "Star Wars" — the chimera in which the United States, thanks to its technical superiority, would supposedly forge an arsenal that would allow it to reach and destroy all the Soviet rockets as they emerged from the atmosphere. Absolute protection for Americans was an absolute nightmare for the Soviets, raising the prospect that as soon as such an American defense was ready, the Soviet Union could be attacked by a powerful preemptive salvo, destroying the majority of its rocket-launching silos, and any retaliatory strike would be nullified by the Star Wars defense. For the United States, the expenditure of some hundreds of billions of dollars was supposed to put an end to the nightmare of a balance of terror with an "evil empire" — a country with mediocre economic and industrial power, but capable of posing an unacceptable threat. For the Soviet Union, this impregnable shield for the Americans was the missing element in the realization of the disarming strike promised by some of the 1980 Reagan campaign literature.

The Star Wars program originated with a television address to the nation on national security by President Reagan on March 23, 1983. The last paragraphs of the speech initiated one of the more bizarre episodes of modern American politics:

> I clearly recognize that defensive systems have limitations and raise certain problems and ambiguities. If paired with offensive systems, they can be viewed as fostering an aggressive policy, and no one wants that. But with these considerations firmly in mind, I call upon the scientific community in our country, those who gave us nuclear weapons, to turn their great talents now to the cause of mankind and world peace, to give us the means of rendering these nuclear weapons impotent and obsolete.
>
> Tonight, consistent with our obligations of the ABM treaty and recognizing the need for closer consultation with our allies, I am taking an important first step. I am directing a comprehensive and intensive effort to define a long-term research and development program to begin to achieve our ultimate goal of eliminating the threat posed by strategic nuclear missiles. This could pave the way for arms control measures to eliminate the weapons themselves. We seek neither military superiority nor political advantage. Our only purpose—one all people share—is to search for ways to reduce the danger of nuclear war.
>
> My fellow Americans, tonight we're launching an effort which holds the promise of changing the course of human history. There will be risks, and results take time. But I believe we can do it. As we cross this threshold, I ask for your prayers and your support."[5]

These paragraphs caught the nation by surprise, not only the public but almost the entire administration. Secretary of Defense Caspar W. Weinberger and Secretary of State George P. Shultz had only hours of warning and no possibility for considered comment on this new policy initiative. The press, with the enthusiastic support of the White House publicists, immediately dubbed the initiative "Star Wars," in view of the role that space-based weapons such as powerful lasers and machines for producing destructive beams of neutral particles were expected to play in this defensive system.

During the summer of 1983 a set of technical committees led by James C. Fletcher, former NASA administrator, studied the prospects for effective defense against the entire force of Soviet ICBMs and submarine-launched ballistic missiles armed with nuclear warheads. The Fletcher Commission outlined a research and development program to spend some $25 billion in five years and $75 billion in ten years to create and demonstrate the technology for such a capability (and probably much larger costs to deploy and operate the actual system). The commission and its panels were exhorted not to determine whether the program was feasible, but to define the achievements and milestones that would have to be met if the program was to succeed. Accompanying the seven-volume report was an optimistic "Executive Summary." Fletcher was later to state that this had been added by someone unknown—"probably in the White House." The result: a largely unread massive report, with an excessively optimistic summary over the name of James Fletcher, who much later disavowed its substance. The report provided a set of tasks and milestones—each rather unlikely of accomplishment, and incredible in combination—which if achieved could lead to a working defensive system. It conferred a cloak of specious legitimacy on the SDI program—specious because it did not take into account the impossibility of achieving the multiple goals and because it largely ignored feasible countermeasures to the defensive system.

An estimate, by an official of the American government, of 1800 per year as the number of clandestine airplane landings on United States territory, each containing a ton of drugs, has been published. Even an amateur could work out a simple and affordable response to the grandiose Star Wars plan. If the KGB could not do it on its own, it should be possible for it to infiltrate the drug underworld and bring into American cities camouflaged thermonuclear bombs, not necessarily small or light, in order to respond to a preemptive strike by a salvo of local explosions if the situation really became unbalanced. There was no need for these amateur strategists, because Soviet scientists were able to convince their leaders, Mikhail Gorbachev in particular, that Star Wars was science fiction, and that if such defenses were actually to be deployed, they could be countered at a small fraction of their cost—not in kind, but by means that were asymmetric.

"Asymmetric warfare" has by now become a commonly used term in the United States, referring to the means that might be used by other nations to respond to the overwhelmingly superior U.S. nonnuclear forces; the Soviet response to Star Wars was an important and early example. One of us (Charpak) spoke personally to Gorbachev, who stated that with 1% of the expenditures the Americans would commit to Star Wars the system could be defeated—if by some miracle it worked at all.

In the Stanley Kubrick film *Dr. Strangelove*, an American military leader participating in a desperate attempt to turn back a U.S. Strategic Air Command plane, sent to bomb Moscow by an Air Force officer gone mad, makes a significant observation. When he learns that the Soviets have built a bomb that can destroy the planet and that is automatically activated in case of a nuclear attack on Moscow, he becomes furious—not because the planet will be destroyed, but because American scientists had proposed the same weapon and the President of the United States had refused to accept it, under the influence of intellectuals who had stuck their noses into things that were none of their business. The general is ready to be blown up with the rest of the planet, but would feel better before dying if he had the same weapon in his own arsenal. This is precisely the type of dangerous defense analyst or military officer who can thrive in the dark secret world of weapons and annihilation.

HOW TO CONTROL ESCALATION

The anticipated results of a large-scale nuclear war between the two superpowers were hardly encouraging. It is fascinating to read, for example, the reports published by the Senate Foreign Relations Committee in 1975 on the effects of a limited nuclear war. For most of those involved in this grisly world, it was clear that a total exchange of nuclear weapons between the Americans and the Soviets was all but unthinkable, because it would lead to reciprocal annihilation, perhaps several times over. On the other hand, one of these reports imagined a more polite exchange that involved the Soviets sending, say, ten strategic rockets (out of 7000) aimed, for example, at all U.S. oil refineries.[6] This "game," if one can call it that, would lead to the destruction of 70% of U.S. refining capacity and the death, as a side effect, of two to five million citizens living close to the refineries. How then, the report asked, should the United States respond? Obviously, not by bombing Moscow or Leningrad, because the destruction of New York and Los Angeles would follow within the hour. One could, therefore, return the favor by destroying the Soviet refineries, and the report describes the effects expected from this response. Of course, to analyze such a use of nuclear weapons is not necessarily to advocate it, but these analyses have attracted adherents and have on occasion led to the development of weapons that would make a similar exchange less unthinkable.

During the Cuban Missile Crisis in October 1962, Leo Szilard—who in the 1930s was the first person to think about the destructive potential of a chain reaction and had been preoccupied by the dangers of nuclear warfare ever since—and one of us (Garwin)—who was then visiting the European Particle Physics Center (CERN) and was concerned with the problems of nuclear deterrence—were sitting together in a Geneva restaurant. The two got to the point of discussing the list of enemy objectives to destroy, in decreasing order of value—in order most effectively to deter not only nuclear war but also all major conflict without bringing about total annihilation if deterrence failed. A current idea was that in case of attack, one would respond by destroying an equivalent objective, for example a city with the same number of inhabitants. Differences in estimation of value on the two sides, however, would lead to unending escalation.

Szilard and Garwin contemplated a scenario in which the response would be directed toward an objective half as important. If the two nations agreed to this idea (as they should, in their self-interest), it would lead to convergence, that is to say, a tapering-off of value destroyed in each response, rather than divergence, or unlimited runaway. The merit of this approach was that the total destruction would be limited to twice the size of the initial target. Such discussions seemed logical at the time because it appeared necessary to take account of the possibility of a single missile, sent by accident or as an isolated act of madness. It would be absurd to wind up with mutual annihilation in this case.

Unfortunately, it was rare to find such civil discussions between individuals whose views and proposals were not so similar as those of Szilard and Garwin. Bitter disputes arose over programs proposed for defending the U.S. population against the effects of nuclear weapons—"civil defense"—and other policy matters.

The choice of targets for the 10,000 strategic warheads, each thirty times more powerful than the Hiroshima bomb, sometimes stirred the consciences of even professional strategists and raised problems going beyond simple questions of arithmetic. In this regard, there is nobody better informed or better prepared than George Lee Butler, U.S. Air Force general (retired), who for three years commanded all U.S. strategic nuclear weapons. Before that, Butler had been responsible, as Director of Strategic Plans and Policy for the nation's armed forces, for drawing a new global portrait. In his public discourse now, Butler notes that "twenty weapons would suffice to destroy the twelve largest Russian cities with a total population of 25 million people . . . ; and therefore arsenals in the hundreds, much less in the thousands, can serve no meaningful strategic objective." His ". . . most urgent concern . . . is the practice of maintaining thousands of warheads on high states of alert . . . in what amounts to immediate launch postures."[7]

Since the breakup of the Soviet Union in 1991, it has become essential to weigh the costs and dangers of maintaining an extensive worldwide stockpile of nuclear weapons, in comparison with their benefits. It is not only the number of weapons that is of concern, but also, as emphasized by Lee Butler, the effectiveness and hazards of the doctrine governing their use.

In particular, the NATO policy permitting and anticipating first use of nuclear weapons (e.g., in response to conventional military engagements with Russia) has been questioned by NATO members Germany and Canada. Russia, which (like its nuclear predecessor the Soviet Union) had previously professed a no-first-use policy for its nuclear weapons against European states, has explicitly proclaimed its intent to respond with nuclear weapons against the now overwhelming conventional force of NATO, in case of war. It is in NATO's self-interest to ensure that Russia maintains tight control over these weapons and takes effective measures to prevent their use.

BOMBS FOR ALL POCKETBOOKS

The great majority of the weapons presently stockpiled are limited to a few well-defined types. We won't go into the details of arsenals but rather illustrate the situation graphically. Figs. 11.1 and 11.2 show the evolution in the number of nuclear weapons maintained by the five official nuclear powers. A glance at the number of weapons and their destructive power (and, for instance, the decline of total U.S. megatons from about 20,000 in 1980 to about 2000 in 1999) gives the impression that there were far too many for any even remotely reasonable objective.[8] Things evidently got out of control.

If the Soviet Union had collapsed ten years sooner, France would be at 250

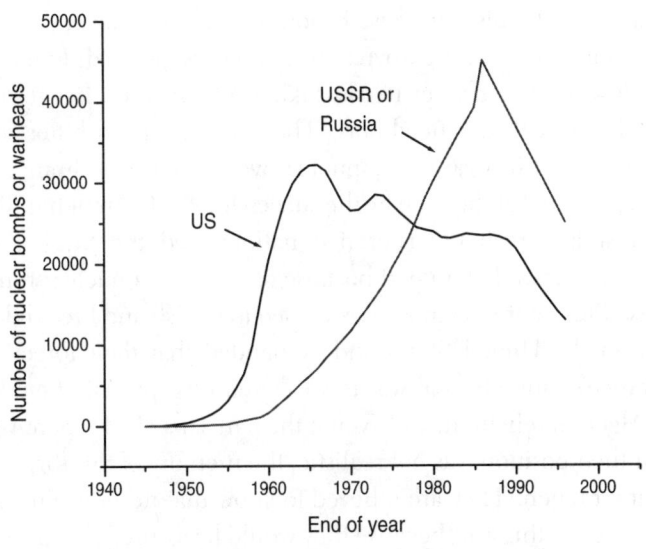

Fig. 11.1. Estimate of the number of American and Soviet/Russian nuclear weapons between 1945 and 1996.

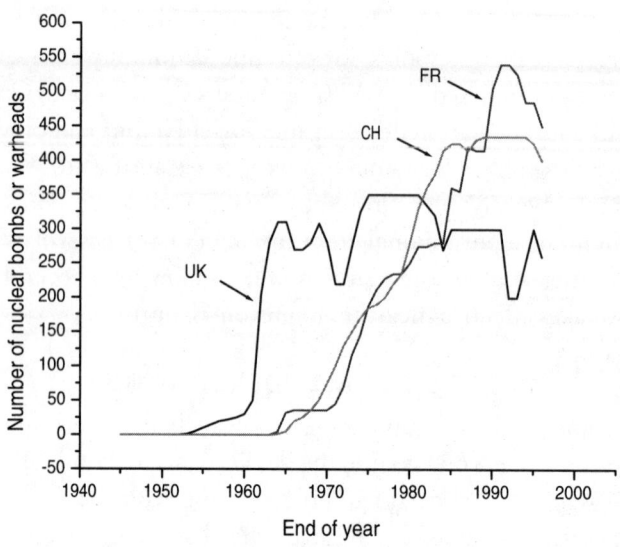

Fig. 11.2. Estimate of the number of British, French, and Chinese nuclear weapons between 1950 and 1996.

nuclear warheads instead of almost 500 (Fig. 11.2). In that case, no statesman would have ventured to say that he wanted to double the stockpile in the ten coming years. No doubts were raised about French security at that time. Now that the principal danger, the Soviet Union, has evaporated, France (like the United States) must assess her needs, taking into account her resources and interdependence with her neighbors. The French have a bitter memory of Khrushchev's threat in 1956 to use nuclear weapons against France and England. It was provoked at the time by the successful Anglo-French military expedition against Egypt after Nasser had nationalized the Suez Canal. The expedition was suspended, in part because of this Soviet nuclear threat and in part because President Eisenhower expressed to British and French leaders the displeasure of the United States and demanded that their forces withdraw. Even if France's ultimate goal was to weaken an Egypt that offered powerful aid to the Algerian rebellion, and even if the partisans of a French Algeria now accept that their position was not realistic, the memory of this threat is intolerable to many French. They are relieved to know that never again will anyone threaten France in this way, because they would be assured of crushing retaliation. This insight into the reasons for the French nuclear weapons program may help in understanding the motivations of other states to build nuclear weapons.

But a nuclear weapons force comes at considerable cost; especially if it is for deterrence, it must be survivable against attack by the potential enemy. If France spends as little as $3 billion per year on a nuclear force, and continues to enjoy fifty years of nuclear peace, it will have cost $150 billion that neither her neighbors, nor other competitors like the Japanese, will have anted up; this $150 billion could be reflected in their technological development, while France would have expended the money on nuclear weapons. In 1997 dollars, French nuclear expenditures in selected years were about $8 billion (1967), $5 billion (1978), $5.5 billion (1982), $4 billion (1995), $4 billion (1996), and $3 billion (1997). The disappearance of the danger of Soviet/Russian imperialism should be taken into account as Europe moves on to a new community of interests and to new political and military thinking. The people of Europe (like the rest of the developed world) must face up to the formidable task of offering an inhabitable world to new generations by exploiting the potential wealth that science has put at our disposal.

Is there any way to estimate the actual cost of the nuclear force? In the United States, the purchasers, that is to say the military, do not pay for nuclear weapons. They are financially responsible for airplanes, tanks, shells, and bombs; but nuclear explosives are provided free of charge—except to the tax-payers—by the Department of Energy, after the approval of the relevant congressional committees and appropriation of funds. For many years, the number of nuclear weapons built and the total "war reserve" stockpile were among the nation's most closely guarded secrets, but most of this information is now freely available—some of it officially released and some published by public interest groups, among them the Natural Resources Defense Council.[9]

It is difficult, and somewhat artificial, to separate the cost of nuclear programs from that of defense in general, which aims to protect the vital interests of a country. The difficulty is compounded by the judgment of almost every government that to survive politically against its domestic rivals, it must never appear, even momentarily, to be in a position of inferiority. It is, nevertheless, useful to evaluate the expenditures relating to nuclear weapons, to help to understand the economic, technological, and human problems involved in any effort to adapt the nuclear arsenal inherited from the Cold War to the new international context. Let's take a look at the American effort, for which more information is available than for that of the Soviet Union; the two programs were more or less of the same size, although the Soviet effort took up a much larger fraction of that country's economic resources.

The report of the U.S. Nuclear Weapons Cost Study Project estimates that the total cost of the military nuclear program from 1940 to 1995, including launching silos, ships, personnel and intelligence systems, as well as the price of their ultimate elimination and destruction, is close to $5.8 trillion, which represents 30% of the total defense budget for the same period.[10] No one can say how much a defense limited to conventional weapons would have amounted to during these fifty-five years, to protect against the same dangers, but it might well have involved some engagements on the battlefield. This outlay must be judged in the light of the deterrence that nuclear weapons may have brought to bear, reducing the likelihood of a major confrontation between the USSR and the United States and their respective allies, although it guaranteed that any large-scale war that did break out would represent the Apocalypse.

The evolution of American military expenditures (Figs. 11.3, 11.4, and 11.5) reflects the major political changes over the last six decades. A comparison of the American defense budget with those of the nine other powers with the highest military budgets in the world (Table 11.1), or with those of a few Middle Eastern trouble spots, is also of interest (Table 11.2).

American Military Expenditures (in billions of constant 1995 dollars)

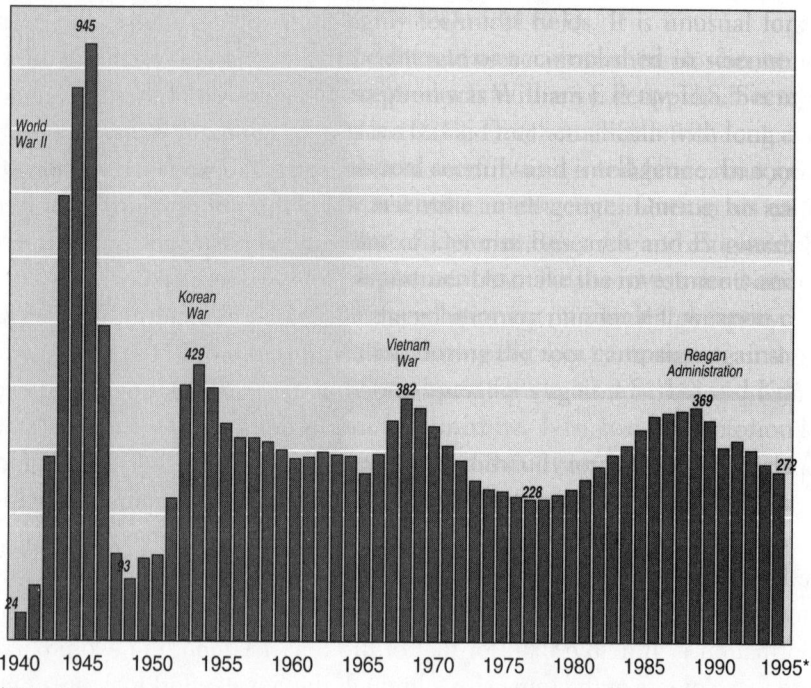

Fig. 11.3. National Defense Budget Authority FY 1946–2005*

TABLE 11.1. THE TEN HIGHEST MILITARY BUDGETS, 1998, IN BILLIONS OF 1998 DOLLARS.

UNITED STATES	RUSSIA	CHINA	UNITED KINGDOM	JAPAN	FRANCE	GERMANY	SAUDI ARABIA	ITALY	SOUTH KOREA
281	64	37	37	35	30	26	18	17	15

Source: Center for Defense Information, 1999; * 1997 expenditures

TABLE 11.2. COMPARISON OF THE MILITARY BUDGET OF THE UNITED STATES WITH THOSE OF THE MIDDLE EAST (1998).

UNITED STATES	ISRAEL	IRAQ	IRAN	LIBYA	SYRIA
281	7	1	6	1	2

Source: Center for Defense Information, 1999

ARMS CONTROL

With his experience as commander of the Allied invasion of Europe on June 6, 1944, President Eisenhower knew too much about war to relish fighting another one. During his eight years in office, 1953–1961, he strove to obtain limits on nuclear weaponry—a difficult task in those days of enmity between the United States and the Soviet Union. His hopes to achieve such aims at a Paris summit with Nikita Khrushchev were dashed when the Soviet Union shot down the U-2 spy plane over Russia in May 1960, and Khrushchev refused to go through with the summit.

Civil society depends much more on contracts than on the use of force, and in the behavior of nations voluntary formal agreements play an important role in trade, in security, in human rights. In the age of nuclear weapons, treaties have been used to limit the number and nature of nuclear arms (bilateral treaties between the United States and the Soviet Union—now with Russia; treaties to limit the spread of nuclear arms, such as the Non-Proliferation Treaty and also the Comprehensive Test Ban Treaty), and even treaties to prevent the outbreak of nuclear war (the various agreements on "hot line" communications between national capitals, a ban on nuclear weapons in space, and the like).

Bilateral U.S.–Soviet Treaties

Beginning in 1972 with the Nixon administration's Strategic Arms Limitation Talks (or Treaties—SALT), U.S. governments often sought to limit the future

A PROPHET

When the authors came upon this parable written in 1919 by Arkady Averchenko, just after the First World War, they were struck by the depth of his insight into the universal nature, and intrinsic logic, of military programs all over the world—the way such programs feed upon themselves quite independently of any objective threat or danger. Averchenko's story bears the title "Practical Common Sense":

The minister of defense of a country, whose name is of no particular importance here, received, one day, a visit from a gentleman with a rather shifty look.

—Take me to someone who can understand things, he said. I have some very important information for him.

—What do you mean by understanding things? he was asked.

—About aviation. I have invented something that I would like to sell. It is an invention that marks a total revolution in the art of warfare. Whoever buys my invention will then have complete superiority over his enemy. From now on, my invention will make the difference between victory or defeat.

Obviously, everyone was delighted to hear that, and the inventor was quickly introduced to a very dignified old general.

The general, as delighted as the others, offered the inventor his most comfortable armchair and inquired with solicitude:

—So what, dear sir, is the nature of your invention?

—I have built a new type of "airplane," said the inventor, which can remain in the air for a week, can transport an entire battalion, and can stand up to any weather. Would you, perhaps, like to buy this air vessel?

After the general had given his word not to take advantage of his visitor's good faith, the inventor took a large packet out of his briefcase and spread out his drawings and plans.

—Yes . . . said the general, after having examined them, that's right. It is exactly as you say. . . . And how much are you willing to sell your invention for?

—For a million.

—That's great! said the general, kissing him on both cheeks. Here is a government check for a million! And the next time you have something, don't hesitate to come to see us. . . .

—I already have something for you, the stranger said slyly. Something quite amazing. . . .

—And what is that thing?

—I have built a missile which can instantly destroy your airplane, so effectively that it will crash to earth like a sack of flour. The airplane has no defense against this missile.

—Where do you get your nerve? said the general, wrinkling his brow. Are you without shame? First you invent a great airplane, and then you destroy it with your own gun?

—I see nothing there to be ashamed of, said the visitor calmly. You will admit that the technique of warfare continually improves and that no one can stop halfway for fear of losing ground and suffering defeat if he is left behind. My airplane is really a formidable weapon! It is obvious, therefore, that a means to defend against it had to be found.

—Hmm . . . in theory that's correct, but in practice . . . I would have thought that, at least, it would be someone else who would have made and offered us the missile. . . . But when it is you, yourself . . .

—My God! said the stranger, wringing his hands. How would that make any difference? Tell me, what would it change if I were to go out now, shave my mustache and change my suit, and then come back by the same door as if I had never seen you before? If that's what you'd like, I'd be happy to oblige.

The general, who, basically, was not that stupid, was, therefore, a bit embarrassed, and realized that he had said something foolish.

—Okay, he said after a moment. We have no choice but to buy your cannon, since we don't want you to sell it to someone else, which you would have every right to do. How much?

—A million.

The general wrote the check, patted the inventor on the back, and said cordially:

—You are really quite a person.

—Ah yes.

—You know, it's quite something to build a missile like that.

—Yes, but things are not as bad as all that. . . . There is always a way out. . . .

—But, said the general, what I mean is . . . after what I saw on the sketches . . .

—Oh yes, this missile is a terrible weapon. But . . .

The inventor sat down once more, looked the general in the eye with an outward show of simplicity, and said cunningly:

—But, what would you say if I were to share a little secret with you, a secret that you will surely find interesting? To protect the airplane from the missile, I have invented a shield so solid that the missiles can't even dent it. . . .

The general took his head in his hands.

—Are you trying to drive me crazy? You have no right to behave like that! It's scandalous, vile, disgraceful . . .

The visitor wrinkled his brow.

—I never behave disgracefully, and don't you forget it. What gives you the right to accuse me that way? Isn't it a good airplane? It is, in fact, superb! And my missile, is there anything wrong with it? It's a masterpiece! What do you want of me? Have I taken advantage of you? Have I lied to you in any way?

—You should have offered me the shield right from the start!

—Now really, said the inventor to the general with a supercilious air. The art of warfare, and the techniques of war in particular, have to develop logically if they are to make any sense. Going by leaps and bounds, as you suggest, just doesn't happen.

The two remained seated in silence for a while, the general deep in thought, the inventor calmly smoking his cigar.

The general would have liked to point out once more that it would have been better if someone else had offered the protective shield, but he was afraid that the stranger would suggest once more that he shave his mustache. . . . No, he didn't want to appear ridiculous on top of everything. Straightening up in his chair, he asked:

—How much?

—A million.

—Maybe you could come down a bit. How about half?

—Out of the question, said the visitor. Others would be willing to pay much more.

—All right . . . sighed the general. You are really insatiable. . . . But if it has to be, it has to be. . . . Here, take your million, even if it ruins us, by God.

The inventor put this check with the others, shook hands with the general, and started toward the door.

—Just a minute, said the general. One more thing. Are you absolutely sure of yourself? Can this shield really stand up to any attack?

The stranger smiled.

—To my cannon? Naturally.

—Then we have nothing to worry about?

—Nothing, except of course if other, more penetrating missiles are invented.

—What? Are you seriously insinuating that they will be invented?

—No question about it.

—Oh God! But when?

—They . . . have already been invented.

—By whom?

—Myself.

—Damn it, now . . . why didn't you say so?

—What do you mean? I was just telling you; these missiles have already been invented.

A derisive smile crossed the general's lips.

—Okay . . . and now you are going to offer the new missiles, right? And when we have bought them, you will come up with a big smile and inform us that you have a shield in reserve, a shield against your own missile. That's it, isn't it?

—Absolutely.

The general shook as if possessed, tore at his hair, and cried out:

—The devil should burn you alive! The earth should swallow you up! You've trapped us and there is no way out! We've been ransacked! You're sucking our blood! You are ruining the country! What's your name? At least give us your name so that we can curse you from the rooftops!

The stranger rose. His face, which, before, had a mocking smile, was now severe, and his lower lip trembled with restrained anger.

—You can insult me as much as you like, he said in an icy tone. That doesn't make you any more intelligent, nor me any more foolish. I won't tell you my name, but if you had a little more sense you would realize that I am logic incarnate, that I am, in fact, Practical Common Sense in person! But don't be too proud of yourself for having caught on. What difference does it make if your country ruins itself in an arms race in ten years or in ten minutes. . . . Genius speaks, and you, like a fool, turn a deaf ear! But that's really none of my business after all; everyone makes his own mistakes and ruins himself the best he knows how. . . . But you, you don't even have the good sense to ruin yourself correctly, once and for all! Good day.

So saying, the stranger slammed the door and walked out of the defense ministry of a country whose name is of no importance to us.

What a prophet!

Soviet nuclear delivery capability by arms control treaties or by less formal unilateral undertakings. SALT consists of the ABM Treaty and the Interim Agreement on Strategic Arms signed by President Nixon and General Secretary Brezhnev in 1972 and approved by the Senate and the House of Representatives. The Interim Agreement was for a five-year span; it posed some limits on the size and numbers of offensive nuclear arms. The Reagan administration criticized such approaches as ineffective and insincere and changed the name to "Strategic Arms Reduction Treaties" (in order to emphasize that reductions are more important than limits or "caps"). The first START agreement between the United States and the Soviet Union was signed by President

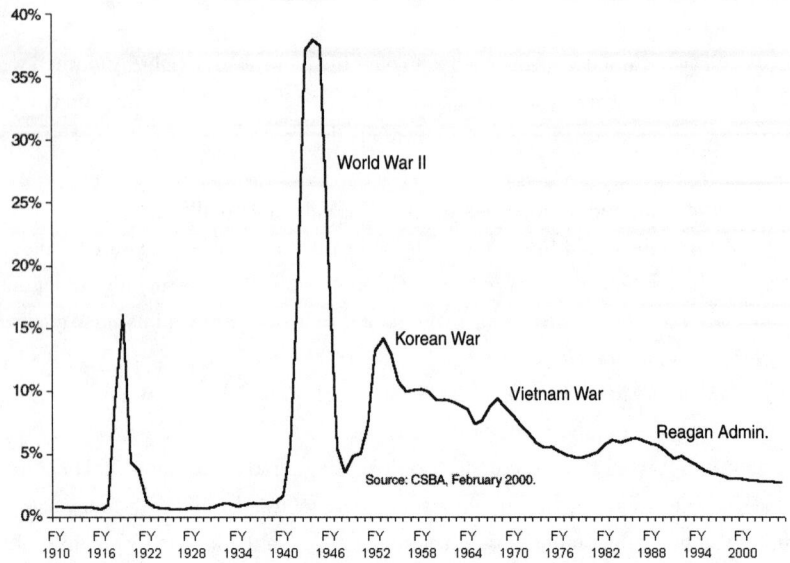

Fig. 11.4. National Defense Outlays as a Share of GDP

George Bush in 1991. START has thus far limited strategic nuclear weapons launchers—bombers, ICBMs, and submarine-based missiles (SLBMs).

In 1992, START I was signed by Belarus, Kazakhstan, and Ukraine, which at that time all still possessed strategic nuclear weapons inherited from the breakup of the Soviet Union. START I entered into force in December 1994. START II was signed in 1993; at the end of the START II phase of the disarmament program scheduled for 2007, there will remain about 3000 strategic nuclear warheads on operational launchers on each side—but about 10,000 or more nuclear warheads—including nonstrategic weapons and reserves. The real number of nuclear weapons to be kept by the United States, as revealed in official documents, is thus on the order of 10,000. Clearly, the Russian military is delighted to be able to justify an equal number. In April 2000, START II was ratified by the Russian legislature.

The January 2001 Memorandum of Understanding provided under START I shows that the United States deploys 7,295 strategic warheads, while Russia has 6,302 treaty-accountable strategic warheads. Russia has expressed its strong desire to limit deployed strategic weapons to 1000 to 1500 in START III, but the U.S. appears to have equally strong views that 2000 to 2500 would be appropriate. START II cannot enter into force until the U.S. Senate approves some protocols, which it may not do, in view of the reservations the Russian Duma has attached to its ratification—particularly a linkage to the

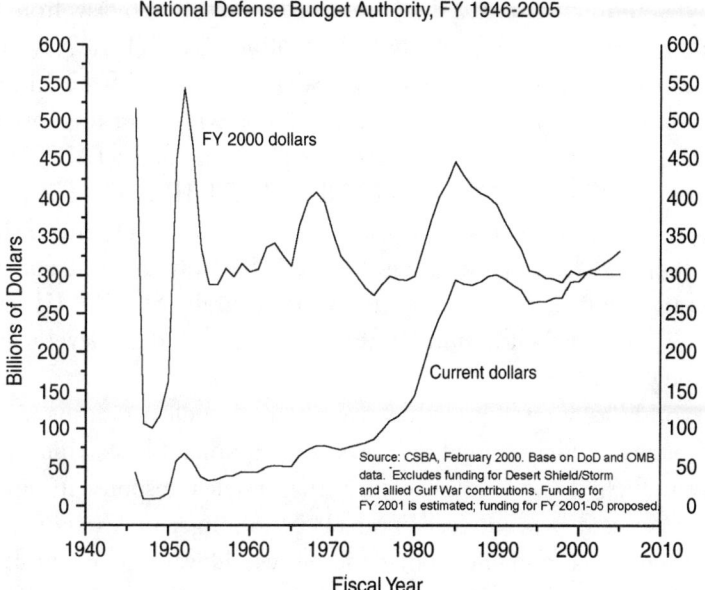

National Defense Budget Authority, FY 1946-2005

Fig. 11.5. American military expenditures in constant FY 2000 dollars (billions).

nondeployment of the National Missile Defense system that the United States has been developing.

The 3000 nuclear warheads deployed on their launch vehicles on each side under START II are still too many. Because START II permits only single warheads for land-based missiles, the Russians don't have enough launch vehicles: their military is, therefore, requesting funds to build new ones, seeking to reestablish an absurd balance. On October 17, 1996, before the Duma, Secretary of Defense William Perry argued for the negotiation of an arms reduction treaty, START III, which would make this race unnecessary. The response was less than enthusiastic, but there resulted nevertheless the Clinton-Yeltsin March 1997 Helsinki agreement that START III would cut the warheads on each side down to 2000 to 2500. The reductions from START II levels of 3000 to 3500 warheads should allow the two camps to save nearly $5 billion.

There seems no possibility that Russia's economic condition will allow it to maintain 3000 strategic warheads in the year 2003; yet its politicians were long unwilling to make an agreement that would thus impose significant constraints only on American forces. They viewed Washington as lacking respect for Russia, as evidenced in the NATO expansion extended to the Czech Republic, Hungary, and Poland in 1998. This is not a trivial matter; it was a similar hypersensitivity to disrespect that drove India to its 1998 nuclear weapons tests.

Congress in 1997 restricted the Clinton administration by law from formal negotiation of START III until Russia had ratified START II. Nevertheless, on July 10, 1998, Russian Marshal Igor Sergeyev—head of the Ministry of Defense—informed the two leaders of the Duma with most responsibility in matters of national security that whether the Duma ratified START II or not, Russian strategic forces would decline to the START III levels. The presidential decree, a government resolution, and instructions for the general staff had all been approved at a July 3 meeting of the UN Security Council, committing Russia to reduce its deployed strategic warheads to the START III level discussed by Presidents Yeltsin and Clinton at Helsinki—to the level of 2000 to 2500 by the year 2010.

The world has entered into an active phase of nuclear disarmament. The United States has discontinued the Airborne Command Post, which was permanently in flight for the purpose of ensuring a nuclear response in case Washington was leveled by a nuclear bomb. It is no longer necessary to respond rapidly, since no nation believes that another would want to, or could, disarm it by a preemptive strike. Nonetheless, both Russia and the United States maintain thousands of strategic warheads capable of being launched within minutes. In addition, in the post–Cold War world, few see our former adversaries as implacable enemies. There nevertheless remains a relatively important minority of military and political authorities who demand absolute protection instead of a simple deterrence, and seem to be unaware of the hazard that steps in this direction might lead to an increase in the danger.

U.S. tactical nuclear weapons have been scrapped or withdrawn, except for a few hundred maintained by the Air Force in Europe, and a similar number of ship-launched cruise missile warheads in the United States. In Russia and in Texas, plants work on the dismantling of nuclear warheads, and U.S. and Russian strategic aircraft with wings and tail cut off are exposed in full view so that their decommissioning can be verified by satellite. It is heartening that these particular vehicles threaten nobody; and sad that so much destructive power was operational for so long, that so many valuable resources have been expended in vain, and that more effective means of destruction persist.

Their mutilation is performed with a six-ton guillotine blade that falls from a height of twenty-five meters on the wings, and then the tail, of the airplane. It shows progress in the decommissioning of an arsenal; it illustrates the enormity of human stupidity given free rein. The fortune gobbled up by their similar bombers would have much better served Soviet defense if it had been used for economic development or the well-being of the population.

The United States and the Soviets were also concerned with preventing the enlargement of the club of nuclear powers, and they worked together with other nations on measures to limit their number. By 1968 the Non-Proliferation Treaty (NPT) had been signed, and it took effect in 1970. In 1999, 185 of the nations of the world were members of this pact; India, Israel, and Pakistan are the notable exceptions, while Iran, Iraq, and North Korea have signed and ratified. The most important elements of the NPT are Articles I and II, under which the five nuclear weapons states (i.e., those that possessed nuclear weapons in 1964—the United States, Soviet Union, Britain, France—and the Peoples Republic of China) promised not to help other states acquire nuclear weapons, and the nonnuclear weapons states promised not to obtain nuclear weapons. France and China did not join the NPT until 1992. In return, the nuclear weapons states committed themselves to provide the non-nuclear weapons states access to the benefits of nuclear power and other non-weapon nuclear technologies. Article VI reads:

> Each of the Parties to the Treaty undertakes to pursue negotiations in good faith on effective measures relating to cessation of the nuclear arms race at an early date and to nuclear disarmament, and on a Treaty on general and complete disarmament under strict and effective international control.

The NPT implied the imposition of restrictions on the sale of certain materials that could be used for manufacturing weapons and the eventual banning of nuclear weapons tests. However, weapons designers in the two superpowers long resisted restrictions on what they consider to be an essential step in building up their own large arsenals: nuclear explosion testing had played a capital role in verifying the validity of evolving concepts. Some also argued that testing was necessary for determining the reliability of weapons already built.

Treaties to Prevent Nuclear Testing

One of the authors (Garwin) was involved in the Eisenhower-era negotiations for a Comprehensive Test Ban Treaty (the CTBT), for the U.S. government, and also in the six-week session in Geneva in 1958–59 for the Conference for Prevention of Surprise Attack—an early recognition that it was not only the numbers and power of nuclear arms that posed a threat to the other side and to the world, but mechanisms, procedures, and mindsets as well.

In the long prelude to the CTBT, finally signed at the UN in 1996, the first stage, the Partial Test Ban Treaty, was achieved in 1963, banning nuclear explosions in the atmosphere, in space, or in the oceans. Groups such as the Pugwash movement had long sought a total ban on nuclear explosion tests as an arms control and nonproliferation measure; the first Pugwash meeting in 1957 was devoted to the effort to stop nuclear tests as a means of limiting the threat to humanity posed by nuclear weapons.

However, the nuclear weapons states had conflicting goals—even internal conflicts over goals. Their nuclear weapon establishments resisted any limitations on their ability to detonate explosives to aid in the development of weapons and to study their effects. The developers were particularly assiduous in arguing that another party signing a comprehensive nuclear test ban might nevertheless continue to conduct underground tests—i.e., that they would cheat and thereby gain a security advantage. The Partial Test Ban Treaty would not have been achieved in 1963 had it not been for the strong public resentment of the radioactive pollution from the tests in the atmosphere. Underground nuclear explosions were permitted, supposedly in order to prevent one side from obtaining an advantage over the other by conducting underground explosions under circumstances in which they could not confidently be detected by the seismometers of the other side—operating outside the territory of the Soviet Union, for instance. The possibility to resolve this would have been to allow inspection of the sites by the other side or by an international team. Agreement could not be reached on the number of on-site inspections that would be permitted annually, in case of suspicious indications on the seismic detection system. In the negotiations to ban all underground nuclear explosions, the United States insisted that the treaty permit seven annual on-site inspections. The Soviets offered three. The logical conclusion is that the two sides did not really want a test ban—or else they would have compromised on a treaty with five on-site inspections annually. In reality, the nuclear weapon states probably valued the progress they might make on nuclear weapons more highly than the limitations on their adversaries, even if no test could have been conducted undetected.

Eisenhower called the failure to achieve a ban on nuclear testing "the greatest disappointment of any administration—of any decade—of any time and any party." Fast forward to the Comprehensive Test Ban Treaty, which was submitted to the U.S. Senate for ratification in September 1997.

The United States supported such an agreement in order to prevent nonnuclear weapon states from developing weapons of a sophisticated design, and to inhibit the other four nuclear weapon states from improving their existing weaponry or obtaining high confidence in weapons that had not yet been tested. But U.S. acceptance of the treaty depended on the assurance that existing nuclear weapon stockpiles could be maintained indefinitely without nuclear explosion testing. As is discussed in the next chapter, this is to be achieved with the support of U.S. nuclear weapons laboratories and a science-based stockpile stewardship program.

Weapon Stockpile Reliability and Safety

At one time during the CTBT negotiations it was asserted by staff of the Los Alamos National Laboratory, but not by their counterparts at Livermore, that it would be essential for the United States to continue to perform hydronuclear tests. Those holding such views were asking for a treaty that would permit tiny nuclear explosions with an energy release one part in ten million of the full yield.

Other nuclear weapon establishments were not interested in these hydronuclear tests. They wanted to be allowed to conduct tests at several hundreds of tons or kilotons. The problem was general mistrust: the Chinese, for example, suspected that the Americans might be learning something that escaped them from these low-power tests. There was, therefore, strong pressure to completely ban them, either by insisting on zero yield or by allowing a much higher threshold like a kiloton.

A review by the JASON group of consultants to the Department of Energy and other U.S. government departments and agencies (of which Garwin is a member) showed that the modifications required to reduce a nuclear weapon's yield to this range made it essentially impossible to obtain information that would be useful for the actual weapon.[11] In contrast, all agree that such tests are useful to ensure that the device is one-point safe; but the nuclear weapons in America's enduring stockpile, to be maintained without nuclear testing under the CTBT, have all been demonstrated to be one-point safe. The other countries, having signed the CTBT, must have reached similar conclusions.

The 1995 JASON report served as support for the decision of the President to order the complete suspension of nuclear testing. Since 1992 the United States had had a moratorium on nuclear explosion testing underground in Nevada; since 1963 all U.S. testing had been underground, in compliance with the Partial Test Ban Treaty.

Given the opportunity in 1992 to conduct as many as 15 tests to resolve any safety or reliability questions (e.g., to redesign the Trident II warhead to use "insensitive high explosive"), the military with responsibility for use of the nuclear weaponry declined the offer. Specifically, the Hatfield-Exon-Mitchell Act of October 1992, provided a one-year moratorium on testing, and then up to five tests per year for three years (strictly to resolve questions of safety or reliability) and then a prohibition of nuclear testing after September 30, 1996, unless another country conducted a test. The United States' signing in September 1996 of the Comprehensive Test Ban Treaty has established now a stronger inhibition of U.S. nuclear testing. The two principal remaining questions addressed by the JASON studies were those of hydronuclear tests and the measures that might be taken to counter unexpected degradation of primary yield due to aging of the primary over many decades. As we have explained, hydronuclear tests had essentially nothing to contribute to knowledge about the health of nuclear weapons, given the great modifications that would need to be made to lower the yield for a perfect implosion by a factor of 5 million or more. The reports also showed that increasing the amount of tritium in the boost gas (by changing the tritium reservoirs in the weapons every few years instead of at longer intervals, or by a different tritium supply system) could compensate for unexpected degradation. This reasoning was persuasive—that the safety and reliability of U.S. nuclear weapons could be maintained under a CTBT. But the President—unlike many in the nuclear weapon laboratories or the Defense Department—must in addition have taken into account the benefit to the United States in the strengthened nonproliferation regime that would result from a total ban on nuclear explosions, in contrast to a limit of one kiloton or even two kilograms.

Another question in ratification of a CTBT is whether potential opponents can perform militarily useful enhancements of their nuclear weapons without underground testing.

India and Pakistan Tests in the International Context

India and Pakistan possess nuclear weapons and both tested them in May 1998, but unless the NPT is amended by its members, neither could join as a nuclear weapon state. In the United States, the Glenn Amendment became law in 1996, its intent being to deter nuclear testing by India or Pakistan, by the imposition of severe penalties—even though such testing violated no applicable international treaty, nor any undertaking by the two countries. The amendment was automatically triggered by the Indian and Pakistani nuclear tests, terminating several government-to-government support programs, U.S. and World Bank lending programs, and all programs but humanitarian aid. Once the Glenn Amendment had failed as a deterrent, its existence was more of a problem than a solution. On the other hand, the lifting of these sanctions (which would require new legislation) would presumably follow only after verifiable assurances by India and Pakistan that would lessen the hazards posed by their nuclear weaponry, and at a minimum, both countries would have to join the Comprehensive Test Ban Treaty. In July 1998, Congress passed legislation permitting commercial food shipments to India and Pakistan—more as a result of pressure by U.S. farmers than as a consequence of an analysis of the effect of this action on restraining or reversing proliferation. The devastating effect of the sanctions on the economy of Pakistan caused further easing of these measures.

Near-monthly discussions between Deputy Secretary of State Strobe Talbott and Indian Foreign Minister Jaswant Singh produced no Indian commitment to sign the CTBT, to stop production of fissile material for weapons, nor to accede to other U.S. requests. India maintains that it has a right to test and acquire nuclear weapons, but various views among the Indian leadership have not thus far resulted in a coherent program. In fact, the tests of May 1998 in the Pokhran desert were urged by and conducted by Indian weapon scientists, with essentially no participation or demand by the military. In May 1999, members of CISAC met in India with a varied group for discussions of the future of Indian nuclear weapons, the meeting having been proposed well before the 1998 nuclear tests. A long interview with Garwin published in the Indian weekly magazine *Frontline* records his judgment that the deployment of nuclear weapons by these two nations in South Asia adds substantially to the security risks for both.[12]

Thresholds for Tests

Another proposal was to have a "threshold" that would allow tests below a yield of one kiloton. Since the purpose of a CTBT is primarily to enhance and support the nonproliferation regime, it would hardly have encouraged those nations who have joined the Non-Proliferation Treaty as nonnuclear weapons states to see the five nuclear states free to test weapons up to one thousand tons of high explosive yield. The indefinite extension of the Non-Proliferation Treaty had been certified in 1995 by 184 of the 190 members of the United Nations; the continued willingness of nations without nuclear weapons to pledge that they would not acquire such arms was primarily a recognition that nuclear arms in the hands of their neighbors would damage their own security to an extent that would not be compensated for by their own possession of such arms. Nevertheless, the nuclear weapon states promised serious motion toward a comprehensive ban on nuclear explosions.

The tests by India and Pakistan in May 1998 demonstrated once again that nuclear weapons designed without testing can give a significant yield. That is no surprise, but we do not in fact know whether the designs worked up to the expectations (and later claims) of the designers.

CTBT Entry into Force

In the last stages of CTBT negotiation in 1996, Britain, China, and Russia apparently introduced the requirement that the CTBT go into effect only after every one of 44 specified nations (those with nuclear reactors or enrichment plants) ratified the treaty. Even without this provision, at least two of the 44, India and Pakistan, might not have signed the CTBT. Israel would have ratified the treaty (despite possessing nuclear weapons and not being a signatory to the Non-Proliferation Treaty), and North Korea would have as well, since it is a member of the NPT and thus is barred from nuclear testing in any case (although how seriously a member of the NPT is another question, since North Korea had announced its intention to leave the NPT but never did so). India would probably not have signed, and—as a result—neither would Pakistan, but it seems unlikely that either would have tested nuclear weapons if the entry into force of the CTBT had not been made dependent on Indian ratification. A CTBT binding all other countries of the world would have imposed greater restraint than is created by signatures on a treaty that will not enter into force. The Indian nuclear weapons scientists would still have wanted to test, but it is likely that the government as a whole would not have approved their request.

Too often, in the past, countries have professed their desire to stop testing while at the same time sabotaging treaty negotiations. One instance of blocking a CTBT, while claiming to support it, occurred in 1978, when the United Kingdom refused limitations of testing because of the cost of surveillance equipment—seismometers to be deployed in Britain—even though the amount was so low that it could have been borne by a charitable organization. And it was the same at the time of the statement, a decade ago, by the director of an American nuclear weapons laboratory, who acknowledged that the lab had not taken up the job of designing durable nuclear weapons, that is to say, weapons requiring neither testing nor frequent replacement, although that had been, for several years, an express Defense Department stipulation.

On August 20, 1996, India used the rule of consensus of the Conference on Disarmament (effectively a right of veto in the CD) to prevent the Comprehensive Test Ban Treaty, negotiated by the sixty-one nations of the Conference on Disarmament, from being presented to the UN General Assembly, although it no longer insisted that the five nuclear powers adopt a clause calling for the elimination of nuclear weapons on a specified date.

India declared that it would never sign the treaty, but that it would not have blocked its presentation by the Disarmament Conference to the General Assembly if the clause relative to its entry in force had not specified that 44 nations, India included, had to ratify it. India thus reserved the right to possess nuclear weapons and to test them, which it did as soon as the nationalistic BJP party came to power in early 1998.

In this regard, less than a week before the test of May 11, 1998, an Indian official indicated to one of the authors (Garwin) that his country had agreed with the United States in the summer of 1996 that it would not ratify the treaty, that it would not test, and that it would not block the transmittal of the CTBT to the United Nations and its eventual entry into force. According to this official, India felt that the United States, in particular, had not fulfilled its side of the bargain, by acceding to a new version of the CTBT that would have as its purpose to end nuclear testing by the weapons states only if every other nation—India, in particular—joined the treaty.

On September 9, 1996, the General Assembly, with a very great majority, adopted this treaty, which the United States and the four other nuclear powers were the first to sign, on September 24, 1996. In the few minutes that followed, many countries followed suit. By October 18, 127 nations had signed. The only ones of the 44 whose signature was necessary, and who had not signed, were North Korea, India, Bangladesh, and Pakistan.

According to the Vienna Treaty on Treaties, signatory states must not undertake any action that would impair the viability of a treaty, even if they have not ratified it or if it has not yet gone into effect. But neither the International Monitoring Organization of the Comprehensive Test Ban Treaty nor the other institutions necessary for the application of the CTBT could be fully implemented before it went into effect, and it was only after three years (i.e., as of September 1999) that the treaty could be modified by the parties that had ratified it by that time. The Clinton administration thus argued that the Senate should give its approval, else the United States would have no voice in the modifications and no way to benefit from on-site inspection of suspected explosions.

The authors believe that it would have been preferable not to have linked the effective date of this treaty with India's signature, so that the rest of the world could have benefitted from a solid test ban regime even if India and one or two other states had not signed. Under those circumstances, international pressure would encourage India to respect the treaty, even if it had not signed it, just as China and France respected the Partial Test Ban Treaty for a long time and tested only underground, though they were not party to the accord.

China and Russia apparently indicated in the summer of 1996 that they would not accept the treaty if the entry-into-force clause did not require its ratification by India. It is up to the reader to consider a preference for our position or that of China and Russia. In the real world, countries take their positions, whether we like it or not. But that is no reason to throw in the sponge and leave one's ideas unstated. Ideas sometimes survive heads of state.

Benefits of CTBT for the United States

Even after the Indian and Pakistani tests, a CTBT remains a good security bargain for the United States. If it can be modified to bring it into force, the world will have greater assurance that others have not tested, by virtue of the CTBT's International Monitoring System (IMS), together with its mechanism for on-site inspection of suspect events—which exists even now in preliminary form. However, the right to on-site inspection will be available only when the treaty enters into force. The United States estimates that, with its own sensors and those of the IMS, it can reliably detect explosions down to about one kiloton, essentially anywhere on earth, and to significantly lower levels at many places, including the national test sites. For instance, in regard to the two 1998 Indian tests that were claimed to have had yields of 200 tons and 800 tons, seismologists estimate that they would have seen an explosion with a yield of as little as 20 tons, from seismometers in Pakistan, which raises obvious questions about the Indian claim. As a case in point, ten seismometers in Kyrgyzstan—dubbed KNET—can be used as a virtual "array" (i.e., a set of seismometers the signals from which can be processed as if the instruments were in a local area) with particularly high sensitivity for the Chinese test site at Lop Nur, with the capability of detecting explosions on the order of ten tons instead of the one kiloton that is the nominal sensitivity of the International Monitoring System.[13]

French Nuclear Tests and the CTBT

Although the United States had long stated its desire to be bound by a Comprehensive Test Ban Treaty, the treaty could be readied for signature only if the other nuclear weapon states and the nonnuclear states found it in their interest to define a single draft. It is worth reviewing the status of French weapons, in view of the fact that France somewhat unexpectedly in 1996, together with the United States, became a vigorous proponent of a CTBT with strictly zero nuclear yield.

French testing was interrupted by President François Mitterrand in 1992 for political reasons. He wanted to draw the nuclear powers into a total and definitive freeze on all testing. He either ignored or rejected objections of the French nuclear weapon labs, which had readied a program of tests. Were they unanimous in their desire to test? Had he gotten wind of discordant voices among the experts, the majority of whom doubtless claimed that the interruption would be detrimental? We have no way of knowing and we shall never know if his associates do not speak up.

There is no question that the majority of the French population was in favor of a nuclear deterrent force, and a poll indicated an 86% favorable opinion at the time of the breakup of the USSR. Opinion has certainly evolved, but a prudent wait-and-see attitude remains.

In November 1994, eight months before the announcement of the resumption of French nuclear tests, Garwin visited Paris with two other American specialists, Ray Kidder and Christopher Paine, in order to understand the objectives of a possible French test series of underground nuclear explosions.[14] Arguments made by those who would be involved in the tests were inconsistent with the proud acceptance of a new warhead for French submarine-launched missiles, without testing.

When Jacques Chirac was elected President of France in May 1995, he terminated Mitterrand's moratorium and ordered the planned nuclear test series to proceed at Mururoa, in the Pacific Ocean. Eight underground explosions were scheduled. Great international protest resulted—something that was totally unwarranted in relation to any hazards to be expected from testing, but also unanticipated by Chirac. In an interview on Australian radio, Garwin asserted that the tests would be not only acceptable but also desirable if they would lead France to support a CTBT; he judged also that Australia had nothing to fear from radioactivity from the test series.

The French leaders decided to pay a high political price despite all obstacles. After six tests the series was declared a success and terminated. And in August 1996, Presidents Chirac and Clinton within a day of each other announced their support for a Comprehensive Test Ban Treaty of strictly zero yield. French citizens have rallied enthusiastically behind the ban. The government has dismantled the Mururoa testing center, eliminating any possibility of backing out of the treaty. In contrast, Washington has not yet decided to close the Nevada Test Site and is explicitly maintaining readiness for nuclear tests. It reserves the right, should national interest so demand, to resume testing. The decision to maintain the Nevada Test Site may have less to do with national security than with job security for the state's two senators.

The CTBT and the Future of the NPT

At the 1995 NPT Review Conference, the five nuclear weapon states committed themselves to implement a total ban on nuclear explosion tests, and that was indeed done in 1996 with the CTBT signed by the five nuclear weapon states and many others. The CTBT has a "zero threshold" in that it bans any nuclear explosion in any medium, of whatever yield.

The Clinton administration delayed submitting the treaty to the U.S. Senate for ratification until September 1997, after which it lay without action and without the usual hearings in the Foreign Relations, Armed Services, and Intelligence Committees. Under pressure from supporters, the Senate Republican leadership called up the treaty for an immediate vote in September 1999, with no warning. Both supporters and opponents agreed that the vote should be delayed for ten days, in order to allow some time for debate. However, the majority had quietly arranged to have senators commit themselves over the previous months to oppose the treaty. No report was issued from the Senate Foreign Relations Committee, in charge of treaties, and on October 13, 1999, the CTBT was ignominiously rejected by a vote of 51 opposed to 48 in favor. (Ratification, of course, as for any treaty, would have required the support of two-thirds of the senators voting.)

The arguments against ratification appeared to be a fear that the United States could not keep its nuclear-weapon stockpile safe and reliable without nuclear tests; that other states would test surreptitiously and successfully and thus gain advantage over the United States; and that a commitment to the CTBT was after all only a signature on the part of some other state, and the nonnuclear weapon states were already bound by the NPT and so the CTBT imposed no further restrictions on them.

Largely missing from the debate was any consideration of the benefit the treaty offered the United States. With its one thousand tests, the United States has nuclear weaponry far superior to that of the other nuclear-weapon states, which would gain relatively more by additional testing. But the primary benefit, even though large, is indirect. The authors believe that nonnuclear-weapon states will not for long forgo the perceived status and (largely illusory) security benefits of possessing nuclear weapons if the nuclear-weapon states continue to test. Therefore, it is wrong to imagine that the nonnuclear-weapon states will remain bound by the NPT. If the United States does not ratify the CTBT and goes back to testing, those countries may reject both the CTBT and the NPT.

As for the potential inability of the United States to maintain safe and reliable nuclear-weapon stockpiles of existing types under a CTBT, the Clinton administration submitted to the Senate six safeguards, worked out with the nation's military leaders and the Department of Energy. Safeguard F is a commitment to resume testing if one of the weapons types critical to the U.S. armory cannot be certified safe and reliable without testing.

In fact, many of the senators opposing the treaty do not accept the desirability of arms control but believe that the United States, as the sole remaining superpower, should be able to set its own course. The extreme or caricature version of this position is that since the United States will obey the law, and outlaw nations will not, why should there be laws? The United States would do the same right thing without laws. Never mind that the same analysis on the domestic scene would lead to the abolition of murder as a crime, and presumably to a substantial burgeoning of murder, theft, and other undesirable behavior. The fact that it is possible to violate a law or a treaty does not mean that it will be widely done. Only when a system of enforcement collapses (as in the collection of taxes) so that a person commonly believes that he or she is foolish to pay taxes (because others don't) does the regime lose most of its value. The year 2001 was when it was expected that so-called "Enhanced Safeguards" would actually begin in signatory countries to the NPT. Following the discovery that Iraq was making a clandestine effort to acquire nuclear weapons, in violation of its undertakings to the NPT, the United States led in the effort to obtain what might be called "anytime, anywhere" inspections of nuclear activities to ensure that they were not oriented toward the acquisition of nuclear weapons. All 187 parties to the NPT agreed to sign contracts specifying the procedure to be applied in that country. Perhaps as a result of the rejection of the CTBT by the Senate, fewer than sixty countries have implemented their agreements with the International Atomic Energy Authority—the body created by the NPT in 1970.

Legally, the United States continues to be bound by signature to the CTBT even though it has not ratified it, until such time as a president formally states that the United States will no longer be bound by the treaty and effectively withdraws its signature.

ARMS CONTROL: A POWERFUL AND ESSENTIAL TOOL

It is appalling to the authors that the literate peoples of the world do not take feasible steps to reduce the threat of the 30,000 or more nuclear weapons still present in the world. Governments find it more popular to rely on weapon programs, many of which are ineffective, to give the impression of protection rather than to negotiate major and verifiable reductions in the threat facing them and the world. Dwight D. Eisenhower stated to Prime Minister Harold Macmillan on London radio and television on August 31, 1959: "Indeed, I think that people want peace so much that one of these days governments had better get out of the way and let them have it."

Forty years after Eisenhower, there are successful arms control and disarmament agreements, and some that have been ineffective. The authors believe that the benefits of arms control justify placing higher priority on and committing more resources to its achievement.

Arms control begins with good defense management, and it is wrong to imagine that military officers or defense department leaders support the growth in arms, independent of costs or consequences. For the most part, these people are trying to do a difficult job under circumstances made more difficult by the bureaucracy and the domestic political pressures felt by the Congress and the President; the drive for reelection so drains the intellectual and physical energies of U.S. representatives (with their two-year term of office) and even senators that many of the most respected have left the Congress at the peak of their influence.

That the world has survived a half century without the use of nuclear weapons is insufficient assurance about the next decades. There is good reason to fear the outcome of nuclear war. To quote the famous remark of Albert Einstein as he contemplated a future with nuclear weapons: "I know not with what weapons World War III will be fought, but World War IV will be fought with sticks and stones."

CHAPTER 12

Current Nuclear Threats to Security

To REDUCE THE THREAT of nuclear destruction, it is important that there be no new nuclear weapon tests and no significant tendency of states to acquire the devices. In preventing the acquisition of nuclear weapons, one must block both a dedicated military program and a nation's ability to obtain materials from the civilian fuel cycle. Reducing the threat to the United States also requires a decrease in the number and destructiveness of nuclear weapons held by potential adversaries and an improvement in the command and control of nuclear forces to eliminate the possibility of an accidental launch or other launches that are not commanded by the leadership. In addition, one has to diminish the possibility of an "inadvertent launch"—that is, one commanded by the leadership on the base of inadequate communication regarding what is taken to be a U.S. attack on an adversary state. Finally, the access of terrorist organizations to existing nuclear weapons or usable materials must be barred as well.

Although the United States is the most powerful nation in the world militarily and economically, it cannot expect to base its security on nuclear weapons while arguing that other countries do not need them—that is, unless it is made persuasively clear that the U.S. nuclear capability augments the security of other nations because of credible security guarantees. In general, the United States can constrain the evolution of other nuclear states only by greatly limiting its own dependence on the weaponry.

We have considered separately civilian and military applications of nuclear energy; but because of the linkage between the two, the imperative to greatly limit nuclear weapons may lead to changes in the strategy of nuclear power production. The neutrons that cause fission and release nuclear energy in electrical plants also produce plutonium that can be used for nuclear weapons; the enrichment facilities that convert natural uranium into low-enriched fuel for these nuclear reactors can also produce uranium enriched to more than 90% and hence ideal for nuclear weapons.

These linkages were evident from the beginning of the era of nuclear energy in 1945, and much effort has been expended to allow the use of nuclear power in a fashion that does not contribute to the proliferation of nuclear weapons. Now vast stocks of weaponry have been accumulated, of which most are now considered an unnecessary threat. Not only must those who seek to acquire nuclear weapon material from the dismantling of these devices be prevented from doing so, but the much larger and increasing amount of weapon-usable plutonium present in spent fuel from power reactors must be similarly protected. Almost 200 tons of plutonium already separated from spent fuel and awaiting fabrication into MOX fuel for power reactors pose a special problem among those that arise in the field of civilian plutonium.

By their adherence to the Non-Proliferation Treaty (NPT), all but three of the countries in the world have expressed their commitment to preventing the multiplication of the number of nuclear weapon powers, and, at the same time, to permitting the extension of the benefits of nuclear electricity to the nations that desire it. The nuclear weapon powers have forsworn a monopoly over the use of nuclear materials and the associated technology in medical and industrial areas.

REQUIREMENTS FOR BUILDING A FEW NUCLEAR WEAPONS

A country, or a powerful group, cannot construct nuclear weapons unless it has available trained men and women. Since the demise of the Soviet Union, thousands of qualified engineers are out of work or not paid regularly, and there will surely be attempts to recruit some of them. Besides, the spread of technological literacy and the commitment under the NPT to share nuclear energy knowledge have aided indigenous weapon competence.

Also necessary are powerful chemical explosives; the ability to make the intense neutron source that, at just the right moment, can trigger the chain reaction; and, finally, plutonium, or uranium enriched to at least 80% in uranium-235. The time required to manufacture nuclear weapons is considerably reduced if there is a source of military-grade, highly enriched uranium, or of plutonium, so that it does not need to be extracted from uranium irradiated in a reactor. Computing power and industrial technology applicable to making nuclear weapons have progressed enormously since 1945. The prevention of proliferation thus depends more and more on physical, political, and legal barriers to deny access to the fissionable materials essential for weapons manufacture.

Highly Enriched Uranium (HEU)

At the dawn of the nuclear era, uranium enrichment was the most difficult task, but at that time all the other steps toward a nuclear weapon were also strewn with major problems. Enriched uranium is the dream material for making bombs. Hundreds of tons of it have been removed, in metallic form, from some of the disassembled nuclear weapons. Uranium can be employed in a relatively primitive device, like the Hiroshima bomb, a gun-assembled explosive, which needed about 60 kg of uranium-235. With the dissolution of the Soviet Union and the great reduction in demand for highly enriched uranium for making weapons, the possibility for theft of a proliferator's bomb—at considerable profit—has substantially increased.

Rather than a gun assembly, the first Chinese bomb used implosion to assemble the uranium, with considerably less of the metal necessary than in a gun device. Comparing the 6 kg of the Nagasaki bomb with the critical mass of 10 kg for naked plutonium not surrounded by a neutron reflector, one can expect, on the assumption that the mechanical properties of the metals are similar, that since the critical mass is 56 kg of naked uranium (enriched to 93.5% in uranium-235), it is possible to manufacture an implosion bomb with 34 kg of uranium or even less.

Because highly enriched uranium is relatively simple to use in a bomb, Saddam Hussein made major investments in the production of uranium-235. His Iraqi regime had tried to produce it with centrifuges and by a very old technique of isotope separation in a magnetic field, using the principle of the so-called "calutrons" at Oak Ridge, Tennessee, that provided some of the weapon uranium for the Hiroshima bomb. He was probably three years from his goal in January 1991. "Report of the Commission to Assess the Ballistic Missile Threat to the United States," with access to all intelligence available to the United States in June 1998, states, "Prior to the invasion of Kuwait in 1990, Iraq could have had a nuclear weapon in the 1993–1995 time frame, although it still had technical hurdles to overcome." Garwin was a member of this commission during its six-month period of operation, January to July 1998.[1]

South Africa had built an armory of six gun-type weapons using uranium-235 enriched by an indigenous process, and, before the transfer of power to the Mandela government, disassembled them and blended the highly enriched uranium with natural uranium to make a low-enriched material that is unusable in nuclear weapons; South Africa then revealed this previously undeclared and largely unsuspected stockpile and joined the Non-Proliferation Treaty as a state without nuclear weapons.

The international fight against proliferation has led to a program to substitute slightly enriched uranium for highly enriched uranium in almost all scientific research reactors. France and the United States have actively participated in this effort (launched in 1977) for reducing enrichment in these research reactors, worldwide. In 1996, the U.S. government announced its decision to take responsibility for the spent enriched uranium fuel removed from American and foreign research reactors and indicated its determination to develop improved performance fuel enriched to 19.9% in uranium-235, rather than the fuel that had been supplied at greater than 90% in uranium-235. Of nineteen university research reactors, nine have already been converted to low-enriched cores and eight are to follow.

The most important exception, on the world scale, is the new research reactor under construction in Munich, Germany; this reactor—FRM-II—is scheduled to begin operation in 2001 at a power level of 20 MW and with a uranium-235 content of 7.5 kg and a fuel cycle length of 50 days. Those substituting low-enrichment fuel in their research reactors take advantage of a new design of uranium fuel element (in the form of a thin metal plate rather than the pencil-like rod of a normal power reactor) conceived in the international reduced-enrichment fuel program specifically so that 20% uranium-235 fuel can be substituted for 93% uranium-235 fuel that could be applied to nuclear weapons; but those building the Munich reactor intend to use this new fuel form with 93% uranium-235 instead, in order to gain an advantage in technical performance of the reactor. That runs counter to the world program, which aims to eliminate, in all nonmilitary reactors, enriched uranium that could be a tempting target for clandestine diversion by proliferators. Besides the Munich facility, a number of existing U.S., French, and Russian research reactors continue to be fueled with highly enriched uranium with no near-term plans for conversion, although a joint U.S.–Russian program has begun to study the use of low-enriched uranium (LEU) in Russian reactors. Evidently, this program is a work in process.

Plutonium

Plutonium presents different problems for the production of a few nuclear weapons. It has the advantage that the necessary mass can be reduced to about 4 kg in perfected bomb models instead of the 60 kg or so for uranium-235 in guns, or something like the 20–30 kg in a uranium-235 implosion design. It was the prospect of an Iraqi plutonium bomb that provoked Israel in June 1981 to undertake the preventive destruction of the Osirak reactor. Plutonium, however, is more difficult to handle than uranium and is less attractive to proliferators if they have the choice—as might be the case if they are able to acquire bomb material.

Each power reactor manufactured and sold today, with electrical output typically in the neighborhood of a gigawatt, produces every week or so the quantity of plutonium sufficient for a bomb. After separation from the fission products and from the uranium in the reactor fuel, the mixture of plutonium isotopes is not optimal for weapons, because the even-numbered isotopes of plutonium—plutonium-240, for example—are somewhat less fissionable. The major problem is that the much larger amount of plutonium-240 than in weapon-grade plutonium makes even the implosion system very likely to preinitiate—and when it does so, it lowers the yield of the simplest system to as little as 2000 tons of explosive, in contrast to a design yield of 20,000 tons (which would still be achieved a portion of the time). Furthermore, more advanced concepts can reduce the penalty.[2]

It would not be difficult, nevertheless, to keep some of the fuel rods in the reactor for only seven months instead of the four years required for optimum economic use. There would then be less than 10% of plutonium-240, which is acceptable for bombs, while the concentration of plutonium-239 in the fuel rods would already be half as great as if the radiation had lasted four years. The economic benefit of being able to fission that plutonium instead of the costly uranium-235 and of keeping the fuel in the pile for three more years are then lost. Frequent refueling was the Soviet practice with graphite-water reactors like Chernobyl, or heavy-water/natural uranium CANDU reactors rather than refueling after years of operation as in light-water reactors. The reactor could thus manufacture plutonium for weapons while producing electricity at the same time.

That is why all these reactors, including those intended only for civilian energy production, must be placed under international supervision, to make sure that the fuel is really unloaded on specified dates and then stored under surveillance and that it is not diverted to illegitimate use.

THE INTERNATIONAL ATOMIC ENERGY AGENCY

The International Atomic Energy Agency (IAEA) was established in 1957 to promote the use of nuclear energy in countries in which the necessary technology did not exist. President Eisenhower, in his Atoms for Peace speech to the UN General Assembly in 1953, had proposed such an agency. According to the Non-Proliferation Treaty, signed in March 1968 and taking effect in 1970, IAEA conducts regular inspections of all *declared* nuclear installations in those signatory countries that do not have nuclear weapons. Recently, inspectors were invited to visit some of the nuclear weapon powers. However, the budget of the agency will have to be considerably increased for it to be able to conduct inspections effectively at the numerous installations in the nuclear weapon states. It is not even sure that the surveillance of power reactors in Russia and the United States is worth the cost, since these states are unlikely to divert civilian material when they have large stores of excess weapon material.

The other signatory countries to the NPT have been able to benefit from essential training for their future experts. Many scientists and engineers have been received in power-plant-manufacturing countries for that purpose; this is obviously an aspect of interaction that can facilitate future proliferation.

The IAEA also has a mission to keep careful records of nuclear materials and to create and monitor "safeguards" against diversion from the civilian nuclear enterprise to military uses. The safeguards are not an aim in themselves; their purpose is to provide "timely warning" of diversion of a "significant quantity" of weapon-usable materials. But the timely warning, as well as the significant quantity, depends on the state of technology available to those acquiring the diverted material. For instance, the IAEA-specified significant quantity of plutonium is 8 kg, but we have seen that the original implosion bomb used 6 kg, and the average U.S. nuclear weapon has on the order of 4 kg of plutonium. Similarly, if a nation or a subnational group has had a years-long program to prepare all of the components of a nuclear weapon except for the fissionable material (as might be the case if it had detailed plans of an existing nuclear weapon, together with on-site help by competent persons experienced with nuclear weapons in the Soviet Union or the United States), it might be only weeks and not months or years from the acquisition of fissionable materials to a nuclear weapons capability. Thus an inspection regime and the decision process of the international community regarding action in response to a possible violation of IAEA safeguards must be tailored to this rather short window.

IAEA cannot punish offenders—punishment is the responsibility of the UN Security Council—but it can trigger an alert so that political or economic pressure can prevent spent reactor fuel from being converted to weapons. Revelations of the state of the Iraqi nuclear weapons program in 1991 pointed up the weaknesses of the agency's policy of inspecting only declared sites. As a result, the agency is now recognized to have the right (and accepts the duty) to request permission for what are called "special" inspections at undeclared but suspected sites. Member states interested in nonproliferation will need to monitor IAEA's performance in this regard.

Reactors whose fuel is continually unloaded present greater difficulties for inspection than those that burn the same fuel for years and are then unloaded on a fixed date, during a shutdown of about a month. In this latter case, counting the fuel elements is easy. In the Soviet reactors, but also in the widely exported Canadian CANDU reactors, on the other hand, where fuel is routinely unloaded without shutting down the reactor, keeping track of the fuel elements is more difficult. As a result, these reactors are better adapted to diversion from their normal activity if purchased by states seeking to acquire nuclear weapons; for this reason, the IAEA inspection regime is more rigorous for the CANDU reactors than for the light-water reactors that are the only ones operating to produce power in the United States and that are the most common type throughout the world.

WEAPON USE OF PLUTONIUM FROM LIGHT-WATER REACTORS

Plutonium has a half-life of 24,000 years compared with 700 million years for uranium. It is, therefore, much more radioactive; it is also much more reactive chemically than is uranium-235. As a product of fuel reprocessing plants, plutonium in oxide form is, in general, packed in welded steel cans containing 2 kg each. After the cans are filled, sealed, and cleaned, they can be safely manipulated with bare hands: the purpose of the cans is to avoid dispersion of the plutonium, which is a severe hazard if inhaled, and to prevent the accidental assembly of a critical mass. These robust packages also facilitate accurate accounting for the fissile material. This plutonium contains about 24% of plutonium-240. In metallic form, pure Pu-240 would have a critical mass of 40 kg, less than that of pure uranium-235, but the mixture of isotopes in civil plutonium has a critical mass of 13 kg as metal, compared to a critical mass of 10 kg for plutonium-239 or military-grade plutonium.

It would be somewhat more difficult to make weapons starting with civilian reactor plutonium, particularly because of the heat liberated by the radioactivity and the larger neutron background produced. To handle this plutonium in large quantities on an industrial scale in conformity with occupational safety and health requirements requires elaborate equipment like that found in plants where plutonium-uranium (MOX) fuel is made. But those who are interested in only a few bombs need handle only a few tens of kilograms or so and can use the old plutonium techniques employed at Los Alamos in 1945.

Of course, the cans of plutonium are not piled on shelves like canned goods in supermarkets. Designing storage and handling technologies that preclude the possibility of diversion by terrorist groups is possible, even against inside accomplices, and such means must be put into effective operation wherever separated plutonium is stored.

Need to Monitor All Plutonium

Excess weapon plutonium, less suitable than uranium-235 for the clandestine production of bombs, is also more difficult to render unusable in nuclear weapons. Unlike enriched uranium, which can be diluted with natural or very-low-enriched uranium for reactors, plutonium cannot be irreversibly diluted, because virtually any isotopic composition of Pu is usable in weapons. Studies at U.S. nuclear weapon laboratories in conjunction with the 1994 CISAC report *Management and Disposition of Excess Weapons Plutonium* showed that it would be much easier to extract Pu from unirradiated MOX—i.e., fresh MOX fuel—than to reprocess spent fuel to extract Pu from it. Therefore, before it is used in a reactor, MOX fuel must be very closely monitored. In countries that already have nuclear weapons, fresh MOX could be protected like their nuclear-weapon material; it would be more difficult if the MOX was routinely used in nations that at present do not have reprocessing plants or MOX fabrication facilities under IAEA safeguards. The UN Security Council and the member states would have to see to it that IAEA's surveillance was adequate. In that regard, more funds from the member states are needed; the IAEA has had flat funding despite its increased responsibilities.

The CISAC study recommends, therefore, that separated plutonium or its compounds, no matter what their origin, be handled according to the same rules that apply to stockpiled nuclear weapons, that plutonium be disposed of in a form in which it is no easier to use for weapons than the spent fuel itself, to which end it should be vitrified with fission products, and a fraction should be used as MOX in existing reactors. Of course, vitrification does not, by itself, solve the problem in the long term, because the intensely radioactive fission products have half-lives much shorter than that of plutonium. At the end of several centuries, the vitrified material would constitute a veritable plutonium mine. The prevention of future proliferation depends on maintaining security and control over the underground burial site, for the nuclear weapons material as well as for the spent reactor fuel.

These problems have also been treated in a study by the American Nuclear Society.[3] It was drafted by an editorial committee that numbered among its members Alexander Haig, former U.S. Secretary of State; Harold Agnew, former director of the Los Alamos Weapons Laboratory; Bertrand Goldschmidt, former director of foreign relations of the French Atomic Energy Commission and once president of the administrative committee of the IAEA; and other eminent individuals from the American, Japanese, and Russian nuclear communities. They recommend the elaboration of foolproof security systems for plutonium from all sources. They don't take a stand on such subjects as the relative merits of reprocessing of spent fuel or direct burial but consider that both methods can be applied with safety if properly conducted.

DISPOSAL PROGRAM FOR EXCESS WEAPON PLUTONIUM

In connection with the program for the disposal of excess weapon plutonium, it is a disappointment to the United States that the suppliers of MOX fuel for European civilian reactors have not agreed to accept weapon-grade plutonium, even at no charge, to be blended with nonmilitary plutonium for fuel in power reactors; it reinforces doubts about the validity of claims that plutonium is valuable for use in power reactors, supporting instead the analysis that a gift of plutonium must be accompanied by a substantial gift of money if its use as nuclear fuel is to occur. Bernard Estève, a former director of EDF's Fuels Department, stated in April 2000 that there was no market for plutonium and that even if there were, the plutonium value would be negative. Hence a program for the United States and Russia to make their own MOX fuel from this material is necessary if this disposal route is to be taken—using it to fuel light-water power reactors in Russia and in the United States, and perhaps elsewhere. The United States would not reprocess spent fuel resulting from this operation, but would subject it to direct disposal; Russia would not reprocess until all the weapon plutonium had been converted into MOX. In addition to this use of MOX derived from excess weapon plutonium, not to be confused with reprocessing of spent fuel (which American industry and government regard as uneconomical and do not favor for others because of the possible diversion or theft of separated plutonium for use in nuclear weapons), some of the excess weapon plutonium will be disposed of in association with fission products in vitrified glass logs. Since it is necessary anyway to vitrify the fission products from the Hanford tanks and the waste from the production of plutonium at Savannah River that ceased long ago, the addition of plutonium costs relatively little. It does not involve reprocessing large quantities of fuel but rather a mass of plutonium that is a hundred times smaller.

It is important to note that, strictly on a technical basis and contrary to a myth propagated by unquestioning advocates of nuclear power, the experts within the nuclear weapon laboratories of the nuclear states agree with the detailed statement of the 1994 CISAC report that civilian plutonium can be used to make powerful and reliable nuclear weapons. While nuclear power has an important role in the future supply of energy to the United States and the world, one of the primary requirements of the industry should be to minimize, monitor, and protect weapon-usable materials from theft, loss, or diversion.

FAST-NEUTRON SYSTEMS AND PROLIFERATION

The present composition of the nuclear industry gives little insight into what will be the best technology fifty years from now, since we can easily foresee fast-neutron reactors, as well as hybrid systems such as Carlo Rubbia's, which are inevitably linked to a reprocessing strategy. We believe that such novel systems, as well as more evolutionary improvements on existing reactors and the supply of uranium from seawater, are important options for energy supply.

It is necessary to consider the proliferation and safeguards implications of proposed nuclear technology. Details matter. For instance, in the case of the system proposed by Rubbia, reprocessing is envisioned, at five-year intervals, of the thorium core containing 10% of uranium-233. Because of the nuclear properties of uranium-233, the necessary mass of the nuclear weapon's fissile core would be half that required for uranium-235, and unlike plutonium, uranium-233 would not produce enough neutrons to cause premature detonation. However, in the thorium/uranium-233 fast-neutron energy amplifier or near-breeder, uranium-233 is accompanied by about 0.1% of uranium-232, from which it cannot be separated chemically. It so happens that thallium-208, derived from uranium-232 after a succession of six disintegrations by alpha particle emission, is itself an intense gamma ray emitter. Even if the uranium were completely purified chemically, a critical mass of uranium-233, two years after chemical separation, would inflict a lethal radiation dose in twenty minutes on a person at a distance of one meter. In addition to the health hazard, the gamma rays make this material easily detectable, unless it is surrounded by heavy shielding—very different from the case of uranium-235, which is almost undetectable. Plutonium-239 is easily shielded from detection because its intense radioactivity is alpha particles and low-energy X-rays, in contrast to the high-energy gamma ray from thallium-208.

This brief comparison shows that monitoring techniques for reprocessed material must take into account the detailed characteristics of the material safeguarded for these future systems, which in some cases can substantially ease the task of protection against diversion.

LOOSE NUCLEAR WEAPONS IN RUSSIA?

In 1989, Mikhail Gorbachev allowed Moscow's Eastern European satellites to become independent; the Berlin Wall fell; and the Soviet empire was well on its way to dissolution. In 1991 the Soviet Union itself disappeared. Its twelve successor states—among them Russia, Belarus, Ukraine, Kazakhstan, and Tajikistan—have been referred to alternately by the term "former Soviet Union" (FSU) or "newly independent states," depending upon the context and the preference of the speaker.

For our purposes, it is important to note that in addition to Russia, the states of Belarus, Ukraine, and Kazakhstan had large numbers of strategic nuclear weapons in 1991. Furthermore, many thousands of tactical nuclear devices were still deployed outside of Russia. It was a matter of urgency for Russia and for the United States to bring this weaponry under centralized control, and Russia did an outstanding job by 1992 in retrieving all of the former Soviet tactical nuclear weapons. By 1996, all of the strategic nuclear warheads had been returned to Russia.

At a time when much in the former Soviet Union that was not nailed down was rather arbitrarily "privatized" by those having immediate possession, there is no indication that even a single one of some 35,000 nuclear warheads suffered this fate. The United States worked with Russia and Ukraine to pave the way for the return of the warheads to Russia, with the latter promising to provide Ukraine the equivalent fissile material in the form of low-enriched uranium for use in the 14 power reactors of Ukraine.

The Soviet Union had been an integrated economy heavily dependent on centralization and scale. For many an item there was only a single plant responsible for its manufacture. Not only was this true for the civilian economy but also for the military, so that the armed forces suffered severe disruption upon the dissolution of the Soviet Union. In particular, Ukraine was the development and manufacturing site for large strategic missiles—the SS-18 and the SS-24, each with ten nuclear warheads. After the removal of the warheads, the missiles still in Ukraine, Kazakhstan, and Belarus posed a substantial problem for these nations, since the fuel is highly toxic.

Russia joined the NPT as a nuclear state and successor to the Soviet Union, while the other former Soviet states have joined the NPT as nonnuclear nations. Any country can adhere to the NPT as a non-nuclear-weapon state if it commits itself not to make or possess nuclear weapons and accepts IAEA inspection and safeguards on its civilian nuclear facilities such as reactors and enrichment and reprocessing plants.

More generally, in 1994, Russia and many of the other former Soviet states, as well as some former Warsaw Pact states, became members of NATO's newly created Partnership for Peace (more precisely, the Euro-Atlantic Partnership Council), which provided a framework for them to actively begin to work with the West on matters of mutual concern in security, environment, education, and the like. The authors regret that NATO expansion—incorporating the Czech Republic, Hungary, and Poland in 1999—has taken precedence over emphasis on the Partnership for Peace, which we believe would be a better foundation for security and peace in Europe.

Surplus Russian HEU

To eliminate some of the danger associated with stocks of Russia's surplus highly enriched uranium, the United States signed a $12 billion contract with

Moscow in February 1993, for the purchase, over twenty years, of 500 tons of military-grade uranium, delivered with an enrichment of 4.4% in uranium-235, adapted to nuclear reactors but unsuitable for weapons. By March 2000, 81 tons of highly enriched Russian uranium had been delivered to the United States in this form. U.S. and world security would be further enhanced if it could be arranged that, as quickly as possible, the entire excess stock of weapons-grade uranium be transformed, by dilution with natural or depleted uranium, into low-enriched uranium, initially to 19.9% uranium-235, so that it could not be used for weapons without a complex enrichment process.

Fissile Material Contraband

The necessity for a nuclear weapons builder to acquire plutonium or highly enriched uranium is the reason for the anxiety over the potential for contraband plutonium from Russia, although this possibly serious source of proliferation thus far has not been shown to have yielded enough material for even one bomb.

In this regard, the Director of Central Intelligence declared before a Senate committee in 1996:

> We have received well over a hundred reports alleging the diversion of nuclear warheads or components during the last few years. The Intelligence Community checks out all reporting of warhead theft and will continue to do so. But to date much of the reporting has been sporadic, unsubstantiated, and unreliable.
>
> Of the numerous reports describing the diversion of weapons-usable material, only a few actually have involved weapons-usable material. And the quantities have been significantly less than that needed for a weapon. In the past two and a half years, European police made the first seizures of weapons-usable material stolen from Russian facilities and smuggled to outside countries. In Germany, police seized about 6 grams of plutonium, a gram sample of highly enriched uranium (HEU), and approximately a half-kilogram sample containing both plutonium and uranium. Czech police seized just under three kilograms of HEU in December 1994, the largest quantity we have encountered.
>
> To date all other reports have been scams, some using low-enriched uranium that is used in reactors. Scams using low-enriched uranium are not surprising because of the tons of this material stored at reactor sites and fuel fabrication facilities, and because security for this material is less stringent than for weapons-usable material.[4]

BUYER: I understand that you're selling the MK28 thermonuclear bomb casing.

SELLER: Yes, there are five thousand in the junkyard. We sell them without accessories, but you won't have any trouble getting hold of them. Here is a list of what you'll need: fast explosives, slow explosives, electric detonators, lithium-6 and deuterium, uranium-238, beryllium, aluminum cones, explosive plug, tritium and deuterium reservoir, uranium plug, neutron generators, and polyurethane.

BUYER: I don't see plutonium on the list.

SELLER: Yes, that's a little hard to find. If you make it yourself, it will cost you ten billion dollars. But you might be able to buy it from Russian military surplus for a billion dollars.

BUYER: And the bomb casing itself, how much?

SELLER: One hundred thousand dollars with the polyurethane thrown in.

As of March 2000, the largest reported seizure is still this three kilograms of highly enriched uranium; some 20 kilograms are required for a uranium implosion weapon. Whether the strengthening of protective measures in the states of the former Soviet Union is responsible for this reasonably favorable result or whether diversions are better hidden is not known. The most recent authoritative report, from the Monterey Institute of International Studies, provides a great deal of detail about these early thefts, particularly those discovered in Russia, but without documenting any more recent than December 1994.[5]

But there continue to be many reports of trafficking in weapon-usable nuclear materials; the absence of confirmed incidents since 1994 does not mean that trafficking has ceased.

ADEQUACY OF CUSTODY OF NUCLEAR WEAPONS

The U.S. scientific community was particularly alarmed with the problem of "loose nukes" (i.e., nuclear weapons not under proper protection) and the dangers posed by access to the weapon material now or soon to be excess. Looking at the incentives for clandestine sale, it was clear that material that had cost on the order of $40,000 per kilogram to produce (plutonium) or $20,000 per kilogram (highly enriched uranium) would be worth millions of dollars per kilogram to a nation such as Iraq, Iran, Libya, or North Korea. These states were spending far more than that for their missile programs and for indigenous schemes for acquiring fissile materials; it was only prudent to assume that they were interested also in foreign sources.

The adequacy of command and control over Russian strategic forces was disputed in June 1998 between the U.S. intelligence community and the head of the U.S. Strategic Command. In 1996 the press had quoted a secret CIA report that concluded that Russian control over its nuclear arsenal was weakening, although the danger of blackmail and unauthorized launching remained slight. "The command and control system of the nuclear arsenal is subject to tensions for which it has never been prepared because of the social change, economic misery and the uneasiness which reigns among the armed forces."[6] The report warned of possible conspiracies in the nuclear forces. The supervision of tactical nuclear weapons and nuclear torpedoes was characterized as poor. There was particular concern for the weapons in the Russian Far East, because living conditions for troops there are particularly deplorable.

In June 1998 there was an authoritative update from General Eugene Habiger, then head of the United States Strategic Command, who visited military sites in Russia that month. At a press briefing in Moscow, Habiger said, "I want to put to bed this concern that there are loose nukes in Russia. My observations are that the Russians are indeed very serious about security." He went on to say that he had been to Russia several times recently and had visited the strategic bases and touched the missiles and the weapons—pointing out that those in the intelligence community responsible for more alarming interpretations had not done so. The intelligence community is not persuaded by this firsthand testimony, perhaps because it has access to varied sources of information.

The Nunn-Lugar Fund

A group at Harvard University—in particular, Professor Ashton B. Carter—was influential in supporting Senators Sam Nunn and Richard Lugar in their efforts to create legislation for what has come to be known as Cooperative Threat Reduction. The resulting Nunn-Lugar legislation in 1991 provided for the expenditure of $400 million from the following year's Defense Department budget for reducing the threat of nuclear weapons located in the former Soviet Union. By now Nunn-Lugar funds totaling some $3.1 billion ($475 million for FY 2000 alone) have been made available for Cooperative Threat Reduction—a far less costly way of destroying Russian nuclear weapons and preventing their use than would have been military action or systems to counter them in war.

Unfortunately, the Defense Department was slow in implementing this legislation, and for several years essentially all the money was spent in the United States on procuring special armored blankets for protecting nuclear weapons in shipment, and on other equipment and services that were to help the transport, storage, and protection of nuclear weaponry in the former Soviet Union. The Armed Services Procurement Regulations prohibited spending funds in that area, where people and organizations had no clue as to how to prepare the required forms; in addition, the Defense Department feared that spending money abroad instead of in congressional districts would imperil support for its budget.

It was clear, however, that there was little incentive for Soviet scientists, engineers, and the military to carry out the measures that were in the interests of both the United States and Russia in securing these weapons, since none of the money was being spent in their country. Even Ashton Carter's service as Assistant Secretary of Defense, 1993–96, did not result in the majority of the Nunn-Lugar money being spent abroad. However, the Department of Energy created a program in which its weapon laboratories work with their counterparts in Russia and with other organizations involved with nuclear materials protection, control, and accountancy—actually spending about one-third of the program funds for Russian salaries, one-third on U.S. laboratory salaries, and one-third for equipment bought in both countries. That program operated at an annual level of about $150 million in 1999. On January 8, 2001, Ted Turner, the billionaire founder of CNN, announced the creation of a new nonprofit organization dedicated to reducing the global threat of nuclear and other weapons of mass destruction. Turner and former senator Sam Nunn co-chair the Nuclear Threat Initiative, based in Washington. These funds are to be spent over five years and should provide a major contribution to leadership in the area pioneered by senators Nunn and Lugar, and supported by Nunn-Lugar funds.

The Moscow Center

Various other programs were created, including the formation of an International Science and Technology Center in Moscow, and a similar establishment in Kiev, Ukraine. The Moscow Center, now funded jointly by the United States, the European Community, and Japan, was instituted for the purpose of providing support for scientific programs proposed by those involved in Russian nuclear weapon labs. The purpose was to prevent unemployment and the subsequent emigration of these technical people to countries looking to produce nuclear weapons. The Moscow Center funds programs at the rate of about $25 million per year. Gradually the Department of Energy program evolved to emphasize material protection, control, and accountancy, resulting in improvements in the poor security of fissile materials in the former Soviet Union.

Much publicity has been given to a cooperative activity conducted by the United States, Russia, and Kazakhstan to remove some 600 kg of highly enriched uranium from Kazakhstan to Oak Ridge in 1994, and a few kilograms of fissile material from Georgia to Britain in 1998. But we note that 600 kg of highly enriched uranium—enough for some 30 nuclear weapons—is less than one part in one thousand of the problem.

The Nuclear Cities Initiative

Nuclear weapons that may be stolen from Russia (or made with knowledge of Russians versed in the art) are more a threat to the rest of the world than they are to Russia, and it is the rest of the world that now has the resources to help protect these arms and this knowledge. It is a matter of providing for the basic human needs of those who have control of the weapons or knowledge useful in their manufacture. The fact that this work in the Soviet Union was confined to ten closed cities is both a problem and a (waning) opportunity. The problem arises from the massive unemployment that would result from the elimination or great reduction of nuclear-weapon-related activity in these single-industry communities; the opportunity from the fact that most of the workers with essential knowledge for the design and manufacture of nuclear weapons are in these cities and not spread across Russia. But a decade after the demise of the Soviet Union, the opportunity to solve this problem is declining, as corruption and poverty have spread throughout the land. President Putin may be able to reverse this decay and build on the strengths of the Russian people and particularly on the competence of those in the nuclear weapon complex. In this, he will need outside support.

The U.S. Department of Energy in 1994 began the Initiative for Proliferation Prevention and in 1998 the Nuclear Cities Initiative, to aid in the downsizing of the Russian nuclear weapon complex by encouraging the transition to other technical nonweapon work.

In his January 1999 State of the Union address, President Clinton announced an Expanded Threat Reduction Initiative, increasing by 60% the expenditures in this area. Such activities included outlays through FY 1998 of $114 million for the Initiatives for Proliferation Prevention, to "stabilize NIS defense institutes and promote long-term employment opportunities for weapons scientists"; $1.4 billion in the Defense Department's Cooperative Threat Reduction ("Nunn-Lugar") Program, to "stabilize NIS defense institutes and promote long-term employment opportunities for weapons scientists"; $67 million in the Defense Enterprise Fund; $428 million in DOE's Lab-to-Lab program, to "bring NIS nuclear materials, protection, control and accounting measures to higher standards"; $98 million for the International Science and Technology Center in Moscow, to "engage NIS weapons scientists in peaceful research to prevent proliferation"; and the intent to spend $600 million on the Nuclear Cities Intiative over the next five years.[7] It is too soon to judge the Bush Administration's performance in these Cooperative Threat Reduction programs, but a cut in the Nuclear Cities Initiative to $7 million this year is not a good omen.

MANAGEMENT OF EXCESS WEAPON MATERIAL

By the year 2010, the reductions in nuclear weapons scheduled in the arms control agreements already under way will result in at least 50 tons of weapon-grade plutonium becoming excess in the United States and a similar or larger amount in Russia, which probably has 1200 tons of highly enriched uranium it does not need. The 52 tons of U.S. plutonium already declared excess includes 14 tons of non-weapon-grade material.

The estimated $12 billion to be paid to Russia over 20 years for 500 tons of HEU works out to some $24,000 per kilogram, far exceeding the substantial cost of blending and delivering the material. This is a case in which it is highly profitable to beat swords into plowshares. Nevertheless, this program was plunged into uncertainty in 1998 with the privatization of the United States Enrichment Corporation, which in July became a publicly held company, with the result that the complicated purchase deal fell apart. It was rescued for the moment by the intervention of Senator Pete Domenici (New Mexico), who spearheaded legislation for an additional $335 million in public funds. In June 2000, USEC directors announced that they will shut down production in 2001 at one of the nation's two uranium enrichment plants despite a provision at USEC's formation in 1998 that it would keep both plants open until 2005. USEC shares sold at $4.50 that day, well off their initial price of about $12. The price of the stock remained around $4 until January 2001, and rose linearly to about $9 in mid-May.

As of March 2001, USEC has purchased 113 tons of Russian warhead HEU, out of the 500-ton total. It has agreed in principle with its Russian partner to modify the agreement to include a discounted, market-based pricing mechanism for purchases of LEU derived from 30 tons of Russian warhead HEU each year. Additionally, LEU will be purchased to make up for previous shortfalls in delivery. This is an important program which is driven by the inherent value of HEU for use as LEU in existing reactors, with little cost involved in its transformation.

The situation with excess weapon plutonium is very different; plutonium has no market value for use as fuel in current reactors, because fuel fabricators would need to be paid to make plutonium-containing fuel from even free plutonium. Although, as we have discussed in our chapters on nuclear reactors, plutonium can be used in a light-water reactor as mixed-oxide fuel (MOX), the cost of making MOX (even given plutonium metal from dismantled nuclear weapons at zero cost) exceeds the full cost of purchase of fuel fabricated from low-enriched uranium. Thus, unlike the highly enriched uranium deal, excess weapon-grade plutonium will not find its way into commercial power reactors without substantial subsidy—on the order of $1–2 billion for 50 tons of excess weapon-grade plutonium.

One of the authors (Garwin) has been involved since 1992 in several joint U.S.-Russian activities to attempt to find a solution to this problem. A January 1994 report of the National Academy of Sciences Committee on International Security and Arms Control, *Management and Disposition of Excess Weapon Plutonium*, reviewed all the proposals that had been advanced as technical means for dealing with the excess weapon plutonium. These included launching the material into space or into the sun; dissolving the plutonium widely in the waters of the world's oceans, and burial in boreholes four to six kilometer deep in rock. Two options were judged by CISAC to be both practical and affordable—the immobilization of plutonium in glass, together with highly radioactive fission products, in order to make it more difficult to steal and isolate the plutonium; and the fabrication of excess weapon plutonium into MOX for use in reactors of existing type.

Despite the fact that the U.S. nuclear industry does not see an economic benefit in reprocessing of spent fuel and the recycle of plutonium, the Department of Energy proposes to dispose of some surplus military plutonium by using it to manufacture MOX fuel for reactors. In order to encourage its use, this MOX will be sold at a price equal to or lower than that of the uranium fuel that it would replace, although its estimated production cost will be higher. After its application to the commercial reactors, the spent MOX fuel will be handled in the same way as the United States handles spent LEU fuel—kept in storage pools for some years, then in dry-cask interim storage, and eventually interred in steel storage casks in the Yucca Mountain repository.

That latter approach was the subject of a second CISAC report of July 1995, "Reactor-Related Options," and the two options—immobilization and burning as MOX in light-water reactors—have been adopted by the DOE for the disposition of up to 50 tons of excess U.S. weapon-grade plutonium. The underlying idea is that this material can be transformed either by partial burning in a nuclear reactor or by mixing with intensely radioactive fission products, so that it is no more accessible than the much larger amount in spent fuel from power reactors worldwide—some 70 tons of newly generated plutonium each year. CISAC suggested that either approach would result in converting excess weapon plutonium to a form that meets the "spent-fuel standard." And CISAC proposed that until final disposition of the material, the excess weapon plutonium should be handled securely so that it would meet the "stored nuclear weapon standard." Russia has been highly resistant to the idea that any of the plutonium that cost so much to produce should be discarded as waste without making use of its energy content; Viktor Mikhailov, the former head of the Soviet Ministry of Atomic Energy, MINATOM, refers to the material as "blood plutonium" in view of the enormous efforts spent and lives lost in its production.

Russia initially wanted to reserve weapon-grade plutonium for fueling a new generation of large breeder reactors that it hoped to build, but the interaction with Washington and independent scientists in the 1990s has apparently persuaded it to take seriously the use of excess weapon-grade plutonium as MOX in existing light-water reactors. In this regard, in addition to the government-to-government route, there was created a U.S.–Russian Independent Scientific Commission on Disposition of Excess Weapons Plutonium, of which Garwin was a member.[8] Recall that each ton of plutonium is enough to make some 200 nuclear weapons, and a ton of highly enriched uranium about 50 nuclear weapons. So the total of excess material in Russia would provide something like 10,000 plutonium weapons and 60,000 uranium implosion weapons. Securing this material is truly a daunting task.

Immobilization

CISAC proposed that the United States not make an early choice between immobilization in glass with fission products and MOX fabrication and burning in reactors, but that both processes be developed and used. The amount of weapon-grade plutonium going to MOX or immobilization could then be adjusted according to the cost of the processes. This would also increase the likelihood that the United States (and Russia) would have at least one process operating for this important task in case of unexpected problems, either technical or political. The United States has adopted this approach; it is pursuing a "dual track" program, although the Bush Administration apparently wants to abandon the vitrification approach.

The actual disposition, after the methods are chosen and development is complete, will take at least 10 and probably 20 years, during which time the material will need to be stored securely and safely. To this end, the United States is cooperating with and partially funding a storage facility at Ozersk, Russia, to store the tens of thousands of containers that will hold the disassembled parts of nuclear weapons before these are finally disposed of. A faster pace is required, and can be achieved, if the problem is recognized by Congress and the leadership of the other industrial countries and resources are supplied. At their June 2000 Moscow summit, Presidents Clinton and Putin signed an agreement for each side to dispose of 34 tons of weapon plutonium; the Russian program will need funding from the G-7 nations. All of this excess Russian Pu will be burned as MOX, as will 25.5 tons (75%) of the U.S. stock; 8.5 tons of the U.S. plutonium will be disposed of by immobilization.

DESTRUCTION OF SURPLUS NUCLEAR WEAPONS

Just as we have discussed in the case of the fuel cycle for a nuclear reactor, or as automobiles are built on an assembly line, so there is a set of steps that must be followed for the destruction of surplus nuclear weapons. They can be demilitarized by removing the batteries for the arming and fuzing system, the containers for the tritium boost gas, the arming and firing systems, and the like, so that they could not readily be used. Then the bomb or missile can be disassembled, so that in the end one is left with what is euphemistically called the "physics package," the nuclear explosive itself. For a thermonuclear weapon, this would include the primary and the secondary. The primary contains the high explosive to implode the plutonium or uranium.

In the United States, disassembly takes place at the DOE Pantex plant in Amarillo, Texas. After mechanical disassembly, the primary has its detonators removed, and then the high explosive must be extracted. The result of the process is a pit (that is, the metal-encased shell of fissile material), bare of high explosive, with the flexible metal tritium fill tube attached.

Disassembled nuclear weapons emerge from Pantex in three streams—high explosive, fissile materials, and miscellaneous metal and plastic parts. The miscellaneous metal and plastic are for the most part crushed and sold for scrap, while the high explosive is burned or recycled for nonweapon uses. The fissile materials are shipped to storage locations—many plutonium pits are temporarily in storage containers at the Pantex plant. The highly enriched uranium parts and any other uranium from the weapon are shipped for storage to Oak Ridge, Tennessee, where the enrichment was initially performed. The process has gone full cycle at enormous expense. The excess weapon uranium is being stored for future use in submarine reactors.

Disposition of Excess Weapon Plutonium

Plutonium is a bigger problem. Both the MOX route of disposition by burning in light-water reactors and the immobilization route of vitrification in glass with fission products require plutonium oxide, and not plutonium metal. The plutonium in U.S. nuclear weapons contains about 1% of an alloying element, gallium, the purpose of which is to stabilize at the weapon storage temperature what is known as the delta phase of the metal. Plutonium has a complicated metallurgy, with at least five known crystal structures. The so-called alpha phase has a density 19 times that of water (twice the density of copper), while the delta phase has a density 15 times that of water (twice the density of iron). The delta-phase weapon before implosion is farther from criticality than would be an alpha-phase weapon, and most nuclear weapons are delta-phase plutonium for enhanced safety.

For converting the plutonium to oxide, the United States will probably employ a process under test at Los Alamos National Laboratory. In a glove box, the metal-enclosed plutonium pit is mounted in a lathe with a vacuum holding device, and the enclosing shell of stainless steel or other material is cut with a tool. It is also possible to cut the inert enclosing material with a laser. The two hemishells of the pit are then mounted in a chamber, plutonium side facing down, and hydrogen gas is admitted to the chamber. It reacts with the plutonium to form plutonium hydride powder, which falls to the bottom of the chamber. The hydride can be heated, and the hydrogen gas that is thus regenerated then travels to the plutonium hemishell and reacts with more plutonium. In this way the plutonium can be converted to powder or to hydride powder. The powder is transported to another portion of the enclosure, where

it can be made to react with nitrogen, thus liberating the hydrogen. The plutonium nitride can then react with oxygen to form the desired plutonium oxide.
Or the hydride can be oxidized in one step.

For use as MOX, the plutonium oxide needs to be heated to a high temperature to get rid of the gallium, which, if it remained, would react chemically
with the thin zirconium metal sheath of the fuel rod. For the immobilization
approach, the gallium can remain with the material.

In the immobilization route, plutonium oxide could be fed as powder into
the stream of coarse glass powder as the latter is being mixed with stored fission
products and poured into a melting chamber, where it is thoroughly melted
and occasionally drained or poured into a stainless-steel container. We have
already discussed this process in connection with the disposition of high-level
(fission-product) waste in the reprocessing-and-recycle approach to fueling
light-water reactors. The only difference here would be the plutonium oxide
added to the fission-product stream as it goes into the melter. This process must
be carried out by machinery in a "hot cell" shielded by several meters of
concrete, in order to protect plant personnel from radiation from the fission
products.

The United States plans to produce in this way some 6000 glass logs, each
weighing about 1.5 tons (2.5 tons including the stainless-steel canister), from
the fission-product waste left over from the manufacture of the weapon-grade
plutonium. If seventeen hundred of these each held about 2% plutonium, the
entire stock of 50 tons could be immobilized.

At present, the immobilization process, which is being studied by the
Lawrence Livermore National Laboratory, is planned to use a "can in canister"
approach, with the excess weapon plutonium oxide being fabricated into
ceramic "pucks" based on titanium and zirconium oxide—Synroc—that are
welded into cans, 28 of which would then be assembled onto a structure inside
the stainless-steel canister before the molten glass loaded with fission products
is poured into the canister. The Livermore team argues that the difficulty in
obtaining plutonium from the ceramic, combined with the difficulty of freeing
the plutonium cans from the highly radioactive glass, results in this approach
satisfying the so-called spent-fuel standard. In November 1998, the DOE asked
a panel of the National Academy of Sciences CISAC to study this point. One
specific concern is that one of the glass logs could be attacked with explosive
strips designed to cut the stainless-steel container and shatter the glass, allowing the cans containing the plutonium to be fished out within seconds and carried away unprotected by the intense gamma radiation from the fission
products, except during the very brief time of the attack itself. The CISAC
panel report of November 2000 judges that the can-in-canister approach would

meet the spent-fuel standard only if further analysis and actual trial bear out the Livermore claims that the 28 cans each containing 1 kg of Pu cannot readily and quickly be freed of the radioactive glass. The panel, incidentally, judged that the irradiation of MOX in CANDU reactors, in the normal CANDU fuel element with a weight of only 24 kg, would not meet the spent fuel standard.

SAFE, SECURE INTERIM STORAGE

The economics of the process (i.e., the discounted present value of the program cost is less if necessary expenditures are delayed without postponing the final outcome) would probably persuade the United States or Russia to perform the oxide conversion later rather than sooner. Since the dismantling of the nuclear weapons is also not likely to take place all within the next few years, it would be highly desirable to make the weapons unusable even if captured or stolen. This could be achieved by "stuffing" the pit—the fissile shell—with inert material that would prevent the implosion from reaching criticality, even if someone managed to provide high-voltage pulses to the detonators with appropriate timing. Inserting a substantial mass of wire into the pit is a well-known means of making U.S. nuclear weapons safe against a nuclear explosion. That is, some of the devices are built as hollow pits, with a tritium fill tube, and hundreds of grams or even kilograms of wire are spooled into the pit through the fill tube to form a heavy bundle of wire that would look like a ball of yarn. When the weapon is to be used, the wire is extracted from the pit by a motor, and the implosion can then proceed. For the disabling of nuclear weaponry, one could employ the same approach, but with brittle wire fed into the pit. After it emerged from the fill tube into the interior of the pit, rather than coiling into a ball from which it could later be extracted, it instead would break off into crumpled bits of wire. This or a similar scheme could presumably be contrived so that the wire could not be removed, even bit by bit. Of course, nothing prevents someone from dismantling a weapon, removing the plutonium pit, opening it, melting down the metal, and casting from it a more primitive nuclear weapon component. But pit stuffing certainly improves safety and eliminates much concern about the weapons in transit.

EVALUATING OPTIONS FOR PLUTONIUM DISPOSITION

Detailed analysis of the cost and the specific hardware required for the two plutonium disposition options can be found in the CISAC report and in a voluminous literature of the Department of Energy; to carry out either immobilization or the MOX fuel route will cost somewhere in the neighborhood of $1 billion to $4 billion. Since our nominal reactor producing one thousand megawatts of electrical power consumes about one ton per year of fissile material, about three reactors would be needed to burn the 50 tons of excess weapon plutonium in 17 years. But there is a small trick that reduces the cost. Since it costs more to fabricate MOX fuel than to buy low-enriched uranium fuel, the overall Pu disposition cost is less if one does not actually burn all of the plutonium but simply contaminates it heavily with fission products while it is in the reactor. So the cost can be reduced if instead of having approximately 4% Pu in the MOX, one has something like 7.5% Pu; only half as much MOX fuel needs to be fabricated to dispose of the stockpile of plutonium. Two reactors will do the job handily, each one being fed about 1.8 tons of weapon plutonium each year. The job will be accomplished in 30 reactor-years, or about 15 years of calendar time. Proper operation of the reactor with such high loadings of fissile material requires a "burnable absorber" such as gadolinium in the ceramic fuel, just as light-water reactors often operate with neutron absorbers in the water moderator when they are loaded with fresh fuel. It seems likely that MOX will constitute only one-third of the reactor core, thus reducing the disposal capacity of each reactor by a factor of 3. Four commercial reactors have signed up to do this.

We have already noted that in France some 20 reactors are already consuming MOX fuel to the extent of one-third of their load. That MOX was fabricated with plutonium separated from fuel that has been burned in French power reactors, but the use of weapon-grade plutonium in MOX has been analyzed and seems perfectly practical. In order to ensure that a U.S. reactor will operate safely with MOX, it needs to be licensed by the Nuclear Regulatory Commission. Several of the American reactors were specially built to a design that anticipated that they would use MOX from the reprocessing and recycle of spent uranium fuel. It is expected that these reactors can be licensed by the NRC without difficulty for the use of MOX fabricated from excess weapon plutonium.

Compared with the 50 tons of excess weapon plutonium, to consume the 500 tons of excess highly enriched uranium to be purchased from Russia over 20 years is a bigger job, but fortunately the fuel can be sold at the same price as virgin low-enriched uranium, and the reactor operator need make no adjustments in the operation of the plant. One might expect that 500 reactor-years would be required to consume 500 tons of highly enriched uranium that has been blended down, but another small detail must be taken into account. Russian highly enriched uranium has too much U-234 (an alpha-particle-emitting isotope) to satisfy international standards when it is blended to form low-enriched uranium for making reactor fuel, so the 500 tons is bought from Russia blended with 1.5% U-235 Russian LEU to form 4.4% LEU. Thus it requires about 760 reactor-years to consume the 500 tons because of the additional U-235 used in the blending. Forty of the 100 U.S. reactors, operating for 19 years, would consume the 500 tons.[9] No question here about loading more than the optimum 4.4% uranium-235 and discarding much of the fissile material. Other customers would pay for the enriched uranium, so it has a market price and one wants to use no more of it than necessary.

The CISAC report considered various degrees of elimination of the excess weapon-grade plutonium, including successive exposure in reactors, reprocessing and recycle, until the plutonium was very largely consumed. The report emphasized that such vigorous "elimination" made no sense, because the excess weapon plutonium is a small fraction of the plutonium present in the spent fuel discharged from U.S. reactors. Instead of elimination of plutonium, CISAC set the goal of achieving the spent fuel standard, so that after disposition, the excess weapon plutonium would be no more attractive or useful than a similar quantity of the much greater amount of plutonium in spent fuel.

It is clear that spent MOX fuel meets the spent fuel standard. It is protected against casual theft by the intense radioactivity of the fission products, for 10, 30, and one hundred years.

The plan at present is to form the ceramic for the can-in-canister approach so that the cans in each canister will hold a total of about 20 kg of plutonium and about 60 kg of depleted uranium; any uranium extracted after plutonium decay would be about 25% uranium-235 and thus not useful at all for making explosives. But in a few hundred years or a thousand or ten thousand years, the fission products would be almost nonradioactive and the plutonium could be extracted without much difficulty from the cans. Protection at that time would be provided by the three-hundred-meter burial depth of the canisters and by a ban on mining enforced by whatever government was in existence at that time.

NUCLEAR ENERGY WITHOUT NUCLEAR WEAPONS?

A general aversion to all nuclear activities, civil or military, as if they were inseparably linked, is all too common. It is as if the use of fire for domestic or industrial purposes were rejected under the pretext that it could be a terrible weapon. Without fire, humanity would be reduced to a quasi-animal existence. It is possible that without nuclear energy, billions of people soon to arrive on earth, whether we like it or not, will be denied a decent way of life. In hoping for a "decent" life we do not at all demand or even desire a servile copy of opulent societies whose wastefulness must and will one day come to an end—they cannot always behave as if they were masters of the universe. The ten or fifteen billion souls just over the horizon will require at least a minimum of energy resources, and it is possible that nuclear energy offers the least polluting and most available energy source in the long run.

Does that mean that the world has to resign itself to living with terrifying nuclear military forces? A brief look at the fits of rage that from time to time have beset mankind leads us to conclude that carelessness with respect to the proliferation of nuclear weapons could bring on the worst-imaginable catastrophes.

Clearly, the need to prevent clandestine access to explosive fissionable materials is today a major preoccupation of political leaders the world over. It was on the agenda of a meeting of experts held in Paris in October 1996. This problem has been a major agenda item for the so-called Gore-Chernomyrdin Commission (more formally, the United States–Russian Binational Commission) that met ten times through March 1998.

For proliferators with billions of dollars to spend, it is always tempting to try to procure material that has been optimized for military use, or else to divert civilian reactors from their intended purpose and to reduce the period of irradiation of natural uranium to limit the creation of undesirable isotopes of plutonium; but smaller fry may be content to buy what is most readily available—even if that is civilian plutonium more difficult to make into weapons.

Even with access to perfect military material, sophisticated installations are still necessary, so that terrorists and other blackmail artists may be more tempted by biological or chemical weapons, like the Japanese sect Aum Shinrikyo, which injured thousands of people and killed a dozen on the spot with a gas attack in the Tokyo subway. States that promote terrorist activities, often with substantial financial and logistical means, might not hesitate to sponsor large-scale adventures using these easily concealed weapons if they could be sure that the international community would not react against the sponsors. Aum Shinrikyo had also a program for acquiring fissile material and building nuclear weapons, but apparently was far from success.

MILITARY ACTION FOR PREVENTING PROLIFERATION

The 1981 eradication of the Osirak reactor by the Israelis was an extreme measure. In a world in which economies are more and more interdependent, intense economic pressure would suffice to prevent certain countries from toying with such plans, but not all. If the rescue of Kuwait deserved an expedition under the aegis of the United Nations, other adventures that should be resolutely cut off by the international community can be identified, even if they do not threaten the supply of oil to the Western world: Kosovo was perhaps the first, and the clandestine manufacture of weapons of mass destruction, even nonnuclear, should certainly be among them.

To establish a legal basis for action against a significant threat, that of chemical weapons, a large majority of the world's nations signed the Chemical Weapons Convention (CWC), which makes illegal not only the use of chemical weapons but also their development and possession. The convention provides for a strict system of inspection and verification. It would be desirable in this respect to strengthen the Biological Weapons Convention, although we should not delude ourselves that verification of such a convention can be effective to the same extent as the CWC. Only by making chemical and biological weapons illegal for those who already possess the technology and hope to prevent their use can the treaty become universal and apply to nations and individuals that may not willingly subject themselves to it.

VERTICAL PROLIFERATION AND DISARMAMENT

During the last thirty years, whenever the problem of nuclear weapon proliferation was raised, a number of nonnuclear states, led by a quasi-nuclear state, India, protested that there is a much more important problem, "vertical proliferation" by the so-called Nuclear Club. They claimed that a few hundred more nuclear weapons in the world would be nothing compared to the tens of thousands developed and produced by the United States and the Soviet Union or the hundreds in the possession of France, China, and Great Britain. The authors have a good deal of sympathy for this point of view, and we stress in this book, and have stressed long before, the need to decrease the American nuclear force from its high point of 33,000 warheads in 1967 and that of the Soviet Union of 45,000 in 1986. And there have already been major reductions. The American government has now limited itself to an "enduring stockpile" of only about eight types—two bombs, one type of cruise missile warhead, and five types of warhead for strategic ballistic missiles. Weapons designed for many specific missions have been eliminated, such as nuclear depth charges, air-to-air missiles, atomic demolition devices, and nuclear interceptors for ballistic missiles.

The objectives of tactical weapons can now be achieved with nonnuclear weapons that take advantage of the revolution in computer technology. The future of strategic weapons is a much more complicated problem.

NUCLEAR MEGATERRORISM?

In Chapter 3 we have described the various approaches to making military-quality nuclear weapons: gun-assembled U-235; implosion systems using U-235 or military-quality Pu (less than 10% Pu-240); and implosion systems using plutonium from civil reactors of perhaps 24% Pu-240. Although no nation has used so-called civil plutonium to fabricate its stock of nuclear weapons, we have noted that its use in an ordinary implosion device with necessary modifications would result in an assured yield of one or two kilotons, and much of the time in a much larger yield. We note also that an advanced design using civil plutonium could be assured of having a yield considerably greater.

It is not our purpose to aid those who might contemplate nuclear terrorism, but to indicate what would be involved. Terrorists have destroyed airplanes in flight, with the loss of life of all those aboard. The April 19, 1995, truck bomb in Oklahoma City killed 168 people. A single nuclear weapon in Hiroshima or Nagasaki killed some 100,000. Is such a weapon within reach of terrorists? Would they want it?

After the Second World War, much of the world was putting itself back together. Reconstruction, the advent and ripening of the Cold War between the Soviet Union and the West, and decolonialization marked this era. The use of force by nation against nation was limited in principle by the Charter of the United Nations, except in self-defense. But violence and threat of violence were not restricted to the interaction between nations; they were very much alive in the service of political causes or to exact revenge or retribution.

For many decades it was argued at those high levels of government concerned with terrorism in the United States that the facts showed that terrorists did not wish to kill many people, and in many cases wished to kill no one at all. They wanted simply to demonstrate their power and to obtain political concessions from those holding power.

Warnings that a bomb containing a few kilograms of explosives was emplaced and would go off at a certain time, allowing emergency evacuation of the neighborhood, proved this point. And even in Northern Ireland, "the troubles" killed a few people here and a few people there, while totally disrupting civil life. All told, about 3200 died in Northern Ireland. Many more would have been killed had the object been mass slaughter.

The vulnerability of large transport aircraft was clear to those contemplating the feasible scale of destruction, but it was argued that it was not in the interest of terrorist groups actually to kill a lot of people. Whether it was the persistence of long-standing resentments and internal conflicts or more recently the advent of how-to handbooks of terrorism in print or on the Internet, the ability to cause large-scale damage and some confidence in that ability spread to individuals or to tiny groups not controlled even by the political leadership of a substantial terrorist organization. At the same time, the ubiquity of modern consumer electronics led to the replacement of the ticking timer or modified alarm clock by the microprocessor or timing chip. The destruction agent of choice was high explosive, either stolen or diverted from commercial blasting sites or homemade of the same materials used in most commercial blasting—ammonium nitrate (widely available in bags for use as fertilizer) mixed with fuel oil or sugar.

Rather suddenly, hundreds of people could be killed at one blow by a pound of high explosive detonated in an aircraft in flight, or by a few tons of high explosive in a rental truck that collapsed much of the Federal Building in Oklahoma City.

The millions of Jews, gypsies, homosexuals, and others who died in Hitler's holocaust were murdered in a planned genocide. This was not terrorism, with the purpose of spreading terror; it was mass murder on an industrial scale. Similarly with the million Cambodians killed by the Khmer Rouge or the half million killed in recent years in Rwanda. But definitions sometimes get in the way of reasonable action. If 500,000 are killed tomorrow in a U.S. city by the detonation of a concealed nuclear weapon, and it comes out that the perpetrators were some fringe group with a stolen Russian device who simply believed that folks living in Chicago are evil, that might not be terrorism, but the results would be the same—a half million dead and terror everywhere.

On the other hand, it is not constructive to focus entirely on the threat of the moment, just because this is a book about nuclear weapons and nuclear energy. Naturally, one should be concerned with preventing the enormous damage that nuclear weapons can do if used by nations in war or by anyone else, whether politically motivated or not. And one should be concerned also as to whether the establishments of the nuclear power industry add significantly to the vulnerability of society.

CONTAMINATION AS A TERRORIST WEAPON

Before 1945, a curie (37 gigabecquerel) of radium was regarded as a very strong radioactive source, but in recent decades kilocurie sources of cobalt-60 or cesium-137 are commonly available for radiography or for sterilization of medical devices. Such intense gamma-ray sources could be dispersed with high explosive, and would contaminate a substantial area. No one would be killed immediately (except perhaps by the explosive itself), but costly and time-consuming cleanup would be required. The industrial use of such sources involves heavy shielding for transport and to protect personnel when the sources are not in use. Even when shielded, the sources may be detectable, and more so if terrorists were to store or transport them with improvised shielding.

The accidental distribution of radioactivity can have damaging and lethal effects, as noted in UNSCEAR 1993. In September 1987, a cesium-137 source of strength 51 trillion becquerels (51 TBq or 51×10^{12} disintegrations per second, or 1600 curies) was removed from a radiotherapy unit and dismantled by junk dealers, resulting in localized contamination of an inhabited part of the town of Goiania, Brazil. The cesium glowed, and townspeople applied it to paint luminous patterns on skin. Hands contaminated with cesium and used in eating resulted in heavy internal exposures. Fifty-four persons were hospitalized and four died. According to UNSCEAR, "In the course of the decontamination program, seven houses were demolished and large amounts of soil had to be removed. The total volume of waste removed was 3100 cubic meters"— about 4000 tons.

Large regions would need to be evacuated if they could not be decontaminated. For instance, even according to the standards in place in 1986 at the time of the Chernobyl catastrophe, areas contaminated to greater than 1.48 TBq per km^2 provoked general evacuation of the population. Thus this 1600-curie source, uniformly distributed, could impel evacuation and resettlement of population over an area of 30 square kilometers.

The question of terrorist use of plutonium, explosively dispersed, has often arisen. From the technical point of view, plutonium, in the form of powdered plutonium nitrate or plutonium oxide, in a sealed container presents no health problems to the person carrying it, and by the same token is far less apparent to remote radiation detectors than are the gamma-ray emitters used for radiography or therapy. As we have seen, the United States produced about one hundred tons of plutonium for its military program, and the Soviet Union considerably more. Furthermore, every nominal power reactor produces some 200 kg of plutonium per year, and in France and Britain, particularly, the spent fuel is reprocessed to yield plutonium oxide welded into individual containers containing 2 kg of plutonium. Contract reprocessing has resulted in tons of such plutonium being shipped to Japan. Because only a tiny amount of plutonium dissolves in water, the greater hazard by far is inhalation of fine dust.

The United States has had experience with explosively dispersed plutonium, in several accidents. On January 17, 1966, an Air Force KC-135 tanker and a B-52 bomber collided during an in-flight refueling operation 30,000 feet above the Mediterranean coast of Spain, near the town of Palomares. According to Randy Mayhew, who took part in the effort to retrieve two intact nuclear weapons:

> The collision killed all four tanker crew members and three of the seven
> bomber crew members. Four thermonuclear weapons fell from the aircraft;
> two of them struck the ground at high speed; their explosive detonated, and

hundreds of acres of farmland were contaminated with plutonium dust. A third bomb's parachute opened, and the nuclear weapon was recovered in good condition, where it fell. The fourth bomb, incidentally, was lost into the water, and was eventually located with the help of an undersea vehicle and recovered intact, at a cost of some $50 million.[10]

The radiologic consequences of the Palomares accident are detailed in UNSCEAR 1993.[11] The cloud contaminated 2.26 km^2 of uncultivated farmland and urban land. In the areas of highest contamination the vegetation and 10 centimeters of soil were collected and disposed of as radioactive waste. With lower contamination, arable land was irrigated, plowed to a depth of 30 centimeters, harrowed, and mixed. Of 714 people examined up to 1988, only 124 showed concentrations of plutonium in urine greater than the minimum detectable.

UNSCEAR estimates that the "collective effective dose due to acute inhalation" of plutonium immediately after the accident is about 1 person-Sv. As we have seen in Chapter 4, the International Committee for Radiation Protection assumes 0.04 lethal cancer per Sv. Taking into account resuspension of the plutonium in the soil, UNSCEAR estimates a total of 3 person-Sv for the ingestion of plutonium in the Palomares accident, leading to about 0.1 lethal cancer. With a population density one hundred times higher, the number of lethal cancers, everything else being equal, would thus be about ten from such an accident. In no way do the authors want to minimize the human and economic cost of such an event, but it would be far easier to kill a thousand people in a sports stadium by conventional means than these ten, years later, by explosive dissemination of kilograms of plutonium.

Because plutonium is highly insoluble, contaminating a water supply with plutonium oxide would produce a radiation dose far below normal background; one kilogram of plutonium dispersed in this way would kill a very few people.[12] The Sutcliffe paper discusses specifically the hazard due to the (hypothetical) explosive dissemination of a kilogram of plutonium in a city (Munich, Germany, in particular) and uses an estimate of 12 cancers per milligram of plutonium inhaled. It also points out: "The largest speck of plutonium that can be readily inhaled is about three micrometers in diameter and has a mass of about 0.14 millionths of a milligram." About 600 such particles would need to be inhaled to add a 0.1% chance of dying of cancer. With the average population density of Munich of about 4300 per km^2, and if very still air allowed the dust cloud to remain over the city for 12 hours, about 120 additional deaths from cancers would eventually be expected. Similar estimates have been published in 1990.[13] Sutcliffe referred to several field experiments.

In 1959 200 grams of plutonium was burned in open desert in Australia. At 200 meters from the burning plutonium (assuming 1 kg source) no person would have inhaled more than 0.1 microgram of plutonium, leading to about 0.1% increased risk of cancer. In experiments at the Nevada Test Site, explosive dispersal was tested, with the result that a person 300 meters directly downwind would have inhaled less than 0.1 microgram of plutonium, again increasing the risk of cancer by less than 0.1%.

NUCLEAR EXPLOSIVES AS A TERRORIST WEAPON

On March 20, 1995, the Aum Shinrikyu religious cult released Sarin nerve agent in the Tokyo subway system. Twelve people were killed and 5500 injured. Extensive investigation showed that the group had a program for obtaining biological and nuclear weapons. Although the Sarin episode was one of incredible ineptitude, the group's malevolent intent was clear. This should be taken as evidence that there are at least some groups that would use nuclear weapons if they could make them or otherwise acquire them.

However, the acquisition of a modern nuclear weapon does not bring with it the ability to detonate it. Thoroughly integrated with the weapon are various features that ensure that it does not detonate unless it has sensed the proper sequence of "environments" appropriate to its use. For instance, an artillery-fired projectile would need to feel the enormous impulsive acceleration in the barrel of the artillery piece, followed by a relatively quiet deceleration. The warhead of an ICBM would need a sustained acceleration and velocity gain, and a long period of coast before it would allow itself to be armed and ready to explode. In addition, U.S. nuclear weapons have so-called Permissive Action Links (PALs) integrated with the warhead; unless the correct code is inserted, the warhead will not explode. It is not known to what extent the warheads of other nations approach those of the United States in these features that augment safety and largely prevent unauthorized use.

While a PAL is not designed to the criterion that a competent nation should be unable to make use of a captured nuclear weapon for months, defeating a PAL could be a daunting and delicate task for a small group.

One could imagine the capture of a nuclear weapon pit from a modern disassembled weapon—the inert metallic case with its plutonium shell that needs to be fitted with an appropriate high explosive system to be made critical. It might in reality be easier for a small group with access to plutonium metal or to a pit to make its own more elementary nuclear weapon, with the solid sphere in a large amount of explosive, like the first weapons at Trinity, New Mexico, and at Nagasaki, Japan. Or a hollow shell approach might give better efficiency. The necessity of beginning with plutonium oxide would not add significant complexity, since the oxide can be reduced to the metal by simple means that were communicated in the 1950s and 1960s in conjunction with the Atoms for Peace Initiative that was part of a bargain in the creation of the nonproliferation regime.

A terrorist weapon would presumably not be designed to be necessarily "one-point safe"—that is, safe against giving a nuclear yield if detonated at a single point on the high explosive. It would not need to withstand the rigors of dropping from a bomb bay. Still, the task of actually fabricating a nuclear explosive, once the design is fixed, is not trivial. It could be done, but not on a tight schedule and not with high confidence.

PROTECTION AGAINST NUCLEAR TERRORISM

On the favorable side of the ledger, the usual tools against nuclear proliferation impede nuclear terrorism. These are to deny knowledge and to deny access to fissile materials. But the denial of knowledge, as we have made clear, is less and less effective with the passage of time. More information is made available, and the old information does not totally vanish. A chronology of the official release of information related to nuclear weapons has been provided by the Department of Energy Office of Declassification and is available on the Internet.[14] For instance, one can find in RDD-4 that "a nuclear test was conducted using reactor grade plutonium" that "it successfully produced a nuclear yield (77–4)," and that "DOE announced on June 27, 1994 that the event occurred in 1962 and gave a yield less than 20 kilotons."

Of course, the calculations that were performed by hand (with pencil and paper or with mechanical or electrically driven mechanical calculators in 1943 and 1944) are trivial with the simplest personal computer. Not only is the computing capacity higher by a factor of millions, but millions of people are accustomed to the convenience of computing and its accessibility. So restricting access to fissile material is more important than ever, since it is the major barrier to the acquisition of nuclear weapons by those who are driven to obtain them. Here it has not helped for plutonium, at least, to have gone in the last decade from a material that was highly prized and protected at all costs against theft or access by another superpower, to one that is a waste—in the sense that the cost of disposition is greater than the value to anyone who might buy it—except to those who want to make nuclear weapons.

A general remedy to most terrorism is to reduce the level of conflict in society and also the tolerance for violence. Finally, in the interests of the security of all, nations can and should implement a system by which individuals can confidentially provide information about activities that violate the law, including the law of treaties.

NUCLEAR MEGATERRORISM IN CONTEXT: BIOLOGICAL WEAPONS

Defining megaterrorism as 40,000 deaths or more caused by a small group, it appears that the only competitor to nuclear devices is a biological weapon. Biological weapons that were routinely developed for use in wartime, in contrast to chemical weapons, were never used in the modern era.

Writing on the Op-Ed page of the *New York Times* on April 8, 1998, Jessica Stern begins, "We've heard a lot recently about the horrors of biological weapons," and she goes on to challenge the assertion that "anyone with a biology degree, a crop duster and a grievance could kill you in your sleep along with millions of your neighbors." In early 1998, Secretary of Defense William Cohen appeared on television holding a five-pound bag of sugar and warning that that amount of anthrax, dispersed properly, could kill half the population of Washington, D.C. True, with a dose of some 8000 anthrax spores weighing a total of less than a microgram being fatal nearly 100% of the time within 5 to 7 days. To kill a million people would require them to inhale a total of less than one gram and not two kilograms, so Cohen is not imagining that the bacillus is fed individually to the population; but it is not easy to disperse such an agent "properly." A fact that appears on both sides of the ledger is that Iraq has formally admitted that it had produced before 1991 some 8.5 tons of concentrated anthrax and had loaded five Scud missile warheads each with hundreds of kilograms of anthrax.

Aum Shinrikyu, according to court testimony in Japan, made botulinum toxin and sprayed mists in central Tokyo, and also near the U.S. Navy installation at Yokohama and at Yokosuka, before moving on to Narita Airport.[15] No illness was reported from the clouds of visible mist. They then grew a starter culture of anthrax—*Bacillus anthracis*, a favorite germ warfare agent because its spores can survive in the ground or in storage for centuries. Inhaled, the spores in a few days convert to the "vegetative" microbial state, in which they multiply. Untreated pulmonary anthrax can have a lethality of 90%.

In June 1993 the leader of Aum Shinrikyu ordered an attack with anthrax. The members of the cult sprayed a mist of anthrax suspension from the top of the Aum building. Next month they then used a truck to spray anthrax around central Tokyo, near the Imperial Palace, but there were no reported illnesses. According to Japanese authorities, Aum Shinrikyu then attacked a subway station in Tokyo with botulinum toxin on March 15, 1995, again without harm. It is thought that one of the cult members may have sabotaged the assault.

But these episodes from "the gang that couldn't shoot straight" are no guide to the true potential of biological warfare or even biological terrorism.

Because anthrax is a spore-forming bacillus, it is highly stable against environmental influences and can be stored a long time and dispersed without being deactivated. It lies in the soil for decades, ready to infect the unwary who inhale the spores. But anthrax is endemic in large portions of the world—for instance, in Russia—and although infectious to humans is not contagious. That is, anthrax does not spread from person to person, unlike smallpox, which is contagious and does so readily. Earlier in the twentieth century, smallpox killed millions of people annually, and could destroy the majority of a society if introduced to a colony that did not have immunity. In one of the darkest blots in the history of North America, smallpox-laden blankets were given to native Americans in the early days of European settlement, since the disease was not present in North America and there was no immunity to it. As the result of his observation that milkmaids rarely caught smallpox, an English physician, Edward Jenner, some 200 years ago deliberately infected a young boy with cowpox—a relatively harmless disease—and then exposed him to smallpox. That vaccination (*vacca* means cow in Latin) was very effective and was routinely used to immunize children against smallpox. The United Nations World Health Organization (WHO) mounted a campaign of smallpox vaccination throughout the world, and eventually organized teams to track down cases wherever they occurred in the late stages of the campaign. By 1980, WHO declared smallpox "eradicated" from the world, as it had no animal hosts and did not survive in any durable form.

Since 1980, the only legitimate stocks of smallpox virus have been one in Moscow and one in Washington, held for scientific research. Although it was proposed to eliminate those stocks, it was hoped to retain them until the DNA of smallpox was completely sequenced, so that the virus could in principle be re-created if necessary; a decision on destruction of these stores was put off until 1999, when it was decided that the virus should not yet be destroyed. It is probable that additional stocks of smallpox were maintained by people or groups with the intent of reserving its use as a weapon.

Smallpox is not as contagious as, for instance, measles. In order to eradicate measles from a local population, 95% of the individuals need to be immunized in order to achieve "herd immunity." For smallpox, herd immunity is achieved with 80% prevalence of vaccination. These figures are far from precise.

To revive smallpox by its use in warfare or terrorism would be a crime against humanity of enormous magnitude. With the rapid dispersion of the virus by air travel of infected people, it is not at all clear that the infection could be contained without the loss of many millions of lives. The situation is entirely different with a contagious disease like smallpox than it is with an infectious (but not contagious) disease like anthrax.

Other types of biological agents are neither infectious nor contagious; in fact, they are not biological at all but are "toxins" produced by biological systems. Two of the commonly cited toxins for use in biological warfare are ricin from the common castor bean, and botulinum toxin, made by a bacterium that grows in the absence of oxygen.

The United States had an active program in biological weapons (BW) until it was terminated by President Nixon in 1969, when the United States unilaterally gave up development, research, manufacturing, possession, and of course use of biological weapons or toxins. A convention banning biological and toxin weapons and mandating their destruction was signed in Washington, London, and Moscow in April 1972 and entered into force in March 1975. The 1969 Nixon executive order and later Biological Weapons Convention banned biological and also toxin weapons; the toxins are effective in substantially smaller doses than the best chemical weapons. The Soviet Union, for one, continued an active military BW program and Russia may still have one to this day. In 1979 an outbreak of anthrax occurred in the Soviet city of Sverdlovsk, in the Ural region, which killed sixty-four people. Russia has now admitted that the epidemic was caused by the accidental release of anthrax from a military biological warfare facility. A 1999 book by Jeanne Guillemin, *Anthrax: The Investigation of a Deadly Outbreak*, purveys a wealth of detail.

A 1970 report of the World Health Organization, "Health Aspects of Chemical and Biological Weapons," estimates that the release of 50 kilograms of anthrax by an aircraft along a two-kilometer line upwind of a population center of 500,000 would kill 95,000 people and incapacitate 125,000; with the release of germs of Q fever, only 150 would die, while 125,000 would be incapacitated.

The United States in 1997 began an active program in which "first responders" have now been trained in 120 cities, with connection to a dedicated reserve force of physicians and scientists who are available to provide advice and analysis in support of field operations should there be an epidemic, a BW incident, or information on a BW threat. In early 1998, President Clinton signed a secret Presidential Decision Directive on the threat of terrorism, and of biological agents in particular. As a result, at least some $300 million is to be made available for stockpiling medicines to respond to epidemics caused by terrorism, and to put in place teams and equipment to identify the agent.

In fact, much can be done to render society less vulnerable to BW attack or to terrorism. It is relatively easy to provide either a commercial or residential building with "collective protection" in the form of slight pressurization with filtered air, which would reduce the hazard to occupants by a large factor—perhaps 100 or more. Furthermore, radio and television, properly used as a communication medium, could instruct the public in hygiene and in the fashioning and use of expedient facial masks, which could reduce the spread of a contagious agent so that a single focus might cause only a few additional cases and not thousands or hundreds of thousands.

A BW terrorism is real and serious, and probably more likely and dangerous than a terrorist nuclear weapon. But as regards national armories, even among small nations, nuclear weapons seem to be desired.

WHAT IS NEEDED TO MAKE A TERRORIST BOMB

To make a working nuclear explosive requires a design, materials, and fabrication. The many approaches attempted during the Second World War Manhattan Project for producing the first U.S. nuclear weapons are described exhaustively in a book that was written with access to the archives and people at the Los Alamos National Laboratory and was then edited to remove classified material.[16] Emphasizing the work performed at Los Alamos in actually building the bomb, rather than in obtaining the fissile materials, the book nevertheless introduces the various means that were proposed and indeed used for enrichment of natural uranium to the weapon-grade highly enriched uranium (some 95% U-235, compared with the 0.71% U-235 in natural uranium). The primary method was gaseous diffusion through a porous "barrier" material. But so-called electromagnetic separation was also used, in which uranium atoms are evaporated into a vacuum and ionized by an electron beam; the resulting positively charged uranium ion is accelerated and then bent by a steady magnetic field, just like the electrons in a television cathode-ray tube. Finally, some small contribution was made by thermal diffusion. The use of high-performance centrifuges operating with uranium hexafluoride did not contribute during the war. Modern uranium enrichment plants use either gaseous diffusion (United States and France) or centrifuge (Europe and Russia) or a vortex-tube technology (South Africa). There are many other possibilities.

One final approach of recent years to uranium enrichment is the laser isotope separation (LIS) approach, either the atomic vapor LIS (AVLIS) or the molecular vapor LIS (MLIS). The United States Enrichment Corporation provided some $75 million per year to the Lawrence Livermore National Laboratory to continue to perfect their AVLIS process, but it has now ceased its support. It is not yet clear whether this approach will ultimately be competitive with gas centrifuge or gaseous diffusion. South Africa used an indigenous "Helikon" process to produce hundreds of kilograms of uranium enriched to 80% U-235 or more, for fashioning its six gun-type nuclear weapons. This "vortex tube" or stationary-wall centrifuge approach uses uranium hexafluoride (UF_6) in hydrogen and requires far more electrical energy than does the centrifuge approach.

As we have discussed, the other option for nuclear weapons is the far easier chemical separation of a fissile material. But the plutonium created in a reactor fueled with natural uranium and with pure graphite or heavy-water moderator must be separated from its surrounding contaminants of uranium, intensely radioactive fission products, and the like.

It is absolutely essential to have fissionable material to produce a nuclear weapon. The original implosion weapon used some 6 kg of Pu that was at least 95% Pu-239. The weapon—"Fat Man"—weighed 10,800 pounds (including its tail fins)—4900 kilograms. It had a yield of some 22 kilotons—22,000 tons of high explosive such as TNT.

The second nuclear explosion at Hiroshima—"Little Boy"—was gun-assembled uranium-235. It weighed some 4080 kilograms and had a yield of about 13 kilotons.

Now the dominant option open to terrorists for the acquisition of fissionable material is to steal it, buy it, or have it given to them. We assume that it is acquired by one of these routes and that uranium is obtained in the form of highly enriched uranium metal, while plutonium is either in the form of weapon-plutonium metal or oxide of either weapon grade or "reactor grade." If the group has managed to receive U-235 metal (more precisely, uranium enriched in U-235 to perhaps 90% content), then it might wish to cast that metal into a spherical shell, or into a cylinder for gun assembly. This can be accomplished in a high-temperature furnace, preferably with an atmosphere of inert gas such as argon, capable of reaching the uranium melting point of 1130°C. It should not be assumed that terrorists or other groups wishing to make nuclear weapons cannot read, and they could have access to the vast literature in any university library, many public libraries, or now on the World Wide Web. Once cast to the rough shape, uranium would be machined, with an ordinary lathe, to the desired shape.

Should the material available be plutonium metal, it would need to be cast into a mold at a temperature above the melting point of Pu—639°C (to be compared with the melting point of aluminum at 660°C). If the material is plutonium oxide, it would need to be reduced with a reactive metal in a coated graphite crucible. This and the ensuing machining could be done in an improvised "glove box" for protection of the operators against inhalation of the plutonium dust. The high-efficiency particulate air (HEPA) filters that were developed by the Manhattan Project are now commonly available, retaining all particles 0.3 microns in size or over, with 99.97% efficiency.

The other design choice that needs to be made is that of an implosion weapon or a gun-assembly weapon. With plutonium, no choice is possible, as we have explained, since the large neutron background of plutonium would prevent any significant yield from a gun-assembly-type weapon.

The Implosion Route

The groups would need to acquire hundreds or thousands of pounds of high-performance explosive and decide whether to cast it into enormous hemispheres or to use precision molds and machining to make sector charges such as were employed in the first U.S. implosion weapons.[17] In either case, it would be necessary to arrange for launching a spherically converging detonation wave in the explosive—a task that was accomplished in 1945 by "lenses" containing fast and slow explosives. Later evolution of lenses reduced the size and mass and simplified construction and test—"ring lenses" and "air lenses."

Because ordinary detonators and electrical detonation systems for blasting explosives allow delays of microseconds or milliseconds after the switch is closed before the explosive is detonated, they are entirely unsuitable for nuclear weaponry. To obtain the spherically converging detonation wave that is needed to assemble and compress the fissionable material, the initially expanding detonation wave from each of the detonators must be generated at the same instant so that it enters the lens and then forms part of the converging spherical wave in the main explosive charge. Since the detonation speed of "Composition B" explosive (a mixture of RDX, TNT, and wax) is 7,900 meters per second or 0.79 cm/microsecond, a timing discrepancy of 10 microseconds between the start of the detonation wave on one side of the explosive sphere and that on the other would lead to an offset of 7.9 cm near the center, resulting in one side of the ball of fissionable material being struck while the other side is still untouched. This is not a recipe for maximum compression.

The problem was solved at Los Alamos in 1944 by the use of very-high-energy electrical pulses applied to the tiny "bridge wires" in the electrical detonators. Many thousands of detonators were fired and measured to achieve the required timing accuracy.

With a U-235 implosion weapon and with weapon-grade plutonium, the implosion assembly requires an "initiator" to provide neutrons at the requisite time. Many modern nuclear weapons use an "external initiator"—a tiny particle accelerator in which deuterium and tritium are caused to react by a high-energy electrical pulse. Such systems are in common use for down-hole diagnosis of oil wells. But the initiator needs to be timed properly in order to provide its pulse of neutrons at the time of maximum compression of the fissionable material.

The alternative is to use the original solution of the Nagasaki implosion bomb—an internal initiator in which a small sphere of beryllium is employed, together with the alpha-particle-emitting radioactive element polonium-210 in order to produce neutrons. But neutrons too early in this design would result in a yield reduced by as much as a factor of 10 from the design yield of 20 kilotons. Accordingly, the polonium and beryllium are separated by a thin layer of material, which is disrupted by the shock that penetrates through the plutonium core to the initiator at its center. In this type of initiator, many curies of polonium are required in order to be sure of having a neutron in a fraction of a microsecond, and the polonium (usually obtained by neutron capture on large amounts of bismuth in a nuclear reactor) has a half-life of four months and must be replaced every six months or so.

The bomb must have some kind of case or frame, and if it is to be delivered or moved by truck, it must be sufficiently sturdy to withstand the jolts of potholes. If it is to be dropped from an aircraft, then it needs extensive tests to ensure that it would withstand the various stresses. Detonation in the hold of a ship or in a cellar is a less stressful environment.

Gun-type Weapons

U-235 could be made into an effective nuclear weapon like that detonated at Hiroshima, with a yield of some 13 kilotons, by gun assembly. Here the projectile of an ordinary gun is replaced by a slug of highly enriched uranium, which constitutes less than a critical mass while it is in the gun barrel. The projectile can be fired into a set of U-235 rings that also constitute less than a critical mass. Together, assembled, one can thus have actually more than two critical masses. The material is not compressed, so the efficiency of the gun is of the order of 1%, compared with a number more like 20% for the implosion weapon.

The development of a gun-type weapon is far less conspicuous than that of an implosion weapon, since whatever testing is to be accomplished can be achieved in a near-laboratory environment (aside from accidents). The Hiroshima bomb was designed to retain the uranium projectile, but in order to ensure a nuclear yield even if the projectile did not stop within the matching rings, a polonium-beryllium neutron initiator was used in this case as well.

Because the fissionable material is not compressed, about 60 kilograms of enriched uranium was required in the gun, compared with less than half that much if U-235 were to be used in an implosion weapon, as was the case with the first Chinese devices.

Weapons Recently Tested by India and Pakistan

On May 11, 1998, India detonated at least one nuclear weapon underground at its test site at Pokhran, 560 kilometers southwest of New Delhi. The Indian government stated that it had tested three weapons, simultaneously detonated, the largest with a yield of 43 kilotons, a second with a yield of 12 kilotons, and the third of sub-kiloton yield. On May 13, New Delhi stated that it had detonated two additional nuclear weapons of 200 tons and 800 tons.

There is considerable doubt about these yields, since seismic measurements from a seismometer nearby in Pakistan indicated a total yield of about 10 kilotons from the May 11 explosions. India had previously tested a nuclear weapon underground in 1974 with a yield measured as somewhat less than 10 kilotons. No seismic signals were observed from the May 13 tests, with a sensitivity more in the range of 10 tons than of one hundred tons.

On May 28 Pakistan responded with "five nuclear tests," and another announced on May 30, at its Chagai test site in southwestern Pakistan, 30 kilometers from the Afghanistan border. Dr. A. Q. Kahn, a leader of the Pakistan nuclear weapons effort, is credited with the creation of the Pakistani uranium enrichment program and the development of the Ghauri missile (tested April 6, 1998, within Pakistan, reportedly with a range of 1300 km and a payload of 700 kg—widely regarded as similar or even identical to the North Korean No Dong missile). He stated that the largest explosion on May 28 was 30–35 kilotons and that the four tests were small, low-yield weapons. On May 30, an additional test was stated as having a yield of 15–18 kt, but the seismological data led Western analysts to estimate the yield as 2 kt. In any case, Pakistan has made its nuclear weapons thus far with enriched uranium, while India has used plutonium separated from spent fuel from its nuclear reactors.

Iraq's Nuclear Program

After the Gulf War in 1991, it was discovered that Iraq had had an active nuclear weapons program that included a multiple-technology effort to obtain the capacity to enrich uranium. In January 1991, at the beginning of the military operations against Iraq, the latter possessed somewhat more than 10 kilograms of unirradiated uranium enriched to more than 90%, and somewhat more uranium enriched to perhaps 80%, that had been irradiated in a reactor. This enriched uranium had been supplied from foreign sources and was under the accounting safeguards of the International Atomic Energy Agency, but Iraq apparently had a plan to seize the uranium, purify it, and perhaps enrich it slightly to make a single nuclear weapon.

Such plans were ended by the defeat of Iraq and the imposition of UN sanctions. In June 1998, IAEA announced that it had no evidence that Iraq had not (as regards its nuclear weapon program) complied with the requirements to make a full, final, and complete disclosure (FFCD) of its nuclear weapon activities, and accordingly the specific sanctions in regard to the nuclear program could be lifted. Nevertheless, IAEA inspections would continue to ensure that the nuclear weapon program was not restarted.

There are several lessons to be drawn from the experiences with India, Pakistan, and Iraq. The first is that making nuclear weapons is not trivial, even for a country with substantial wealth and scientific resources. The second is that the acquisition of fissionable material is the critical step in producing nuclear weapons.

CURRENT ISSUES: THE STOCKPILE STEWARDSHIP PROGRAM

Democratic societies go astray when propaganda replaces reasonable discussion. One can see this excess in the arguments both for and against any major program. Thus the U.S. stockpile stewardship program is both vilified and sanctified by partisans of either side. It should be supported not only to keep remaining nuclear weapons safe and reliable, but because the program is essential for the United States to maintain the Comprehensive Test Ban Treaty. In order to gain acceptance of the CTBT, President Clinton agreed with the nuclear weapons establishment to support a science-based stockpile stewardship program that would include major new facilities. Among them are advanced pulsed radiographic machines (flash X-rays) that would allow improved imaging of high-explosive implosions, which would be accomplished with natural uranium replacing fissile uranium-235, or some simulant replacing plutonium, or at reduced scale so that there could not possibly be any nuclear yield.

The largest single instrument is the National Ignition Facility (NIF) at Lawrence Livermore, to cost somewhat more than $2 billion. This machine is to employ 192 powerful lasers, whose light is directed through apertures in the two ends of a small hollow gold cylinder a few millimeters across. The light is converted into soft X-rays, simulating at considerably lower energy density the soft X-ray radiation from a nuclear weapon primary that can "drive" a secondary fusion charge. Nevertheless, one of the principal goals of NIF is to "burn" tritium with deuterium in tiny glass shells imploded within the gold cylinder, just as a thermonuclear weapon secondary is imploded within the weapon's radiation case. But the X-ray energy in NIF is on the order of a megajoule (10^6 J) in comparison with a primary yield of more than a kiloton (4×10^{12} J); this factor of 4 million between the NIF energy and that of a nuclear weapon primary means that the volume of material that can be compressed in the machine is about 4 million times smaller than in a weapon secondary—a sphere smaller in diameter by a factor of 150. The National Ignition Facility does not in any way faithfully simulate a real nuclear explosion. On the other hand, the enormous pressure of soft X-rays can perform weapon-related experiments at this small scale. Furthermore, the possibility of obtaining breakeven (more fusion energy out from the tiny secondary capsule than the amount of X-ray energy—or some similar definition of breakeven) is of interest in a non-military program to explore obtaining electrical energy from fusion.

Nevertheless, detailed comparison of experimental results from this facility with the theoretical prediction for complex tiny fusion capsules could help to maintain teams at the same scientific level that they held during the arms race. One of the explicit goals of the American nuclear stockpile maintenance program is to preserve the capacity to design new weapons, to guarantee their feasibility, and to be able to test them if necessary, but obviously not while the Comprehensive Test Ban Treaty is in force. This is a useful goal, and one compatible with the CTBT, but it should not be justified as directly necessary for maintaining a safe and reliable stockpile. It is, perhaps, indirectly necessary as a way to maintain competence of those who might respond to defects observed in aging weapons. In late 1999, it was discovered that the National Ignition Facility was far behind schedule and over budget—in contrast with the upbeat reports of the laboratory management both to Congress and to the Secretary of Energy. U.S. nuclear weapon development laboratory personnel have repeatedly stated that in an era in which other countries no longer detect seismic waves from U.S. nuclear test explosions, it is the achievement of ignition in NIF that conveys the credibility of our nuclear deterrent. It is wrongheaded to relate the credibility of our nuclear weapons to the totally distinct question of reaching breakeven in a tiny laser-driven capsule of deuterium and tritium gas.

It is possible to preserve the safety and reliability of weapons over a number of decades by a much simpler program of inspection and remanufacture when necessary. A country with a nuclear weapons arsenal can choose either to retain and expand scientific and technical competence by an American-type program based on research or to adopt a more conservative attitude based on maintenance and frequent replacement. It is likely that Russia, China, and perhaps Britain will choose this latter, more conservative approach.

The scientific program has the auxiliary objective of being able to resume the development of nuclear weapons in case of a failure of the CTBT. It is possible that the test ban regime will collapse a decade or two from now, and it is understandable that the nuclear powers, as well as the laboratories and the experts, would like to maintain their competence. But excessive activism on the part of the nuclear experts could actually lead to a decrease in the reliability and safety of the weapons stockpiled, if they are unable to resist the temptation to proceed with improvements that they will not be able to test. There is another potential problem: we heard from a weapons laboratory official recently that those whose daily work is to oversee nuclear weapons in the laboratories are worried that the investment and the excitement of new installations will effectively distract the engineers and physicists and divert them from the essential task of maintenance and inspection.

With lasers or with pulsed power installations, one can generate X-rays, in a small volume, similar to those emitted by the primary nuclear charge, and can study certain aspects of their effects on the assembly of the secondary charge. As a result, lasers don't help solve problems of the primary charge, but they can be of use for studying details of the secondary charge configuration. In this way, for example, they could provide answers to questions on corrosion, which would determine whether or not new secondary components would have to be remanufactured frequently. But the cost of precautionary replacement without verification might be less than the cost of the laser program. Countries that don't possess high-powered lasers would, therefore, not be at a disadvantage. They might even benefit: a recent study on the survival rate of cardiac patients shows that the best diagnostic technique, which uses a catheter inserted into the heart to provide the basis for minute-by-minute decisions on treatment, was associated with a higher rate of mortality than were the cases of patients who were spared this intervention.

THE BALLISTIC MISSILE THREAT

In 1997, Congress created the nine-person Commission to Assess the Ballistic Missile Threat to the United States, which legislation was signed into law by President Clinton. Garwin served on this commission, chaired by Donald H. Rumsfeld, former (and current) Secretary of Defense. Among the other members were R. James Woolsey, former Director of Central Intelligence in the Clinton administration; Paul Wolfowitz, now Deputy Secretary of Defense; General Lee Butler, first head of the U.S. Strategic Command and thus manager of all U.S. strategic nuclear forces; and General Larry D. Welch, former Chief of Staff of the U.S. Air Force and president of the Institute for Defense Analyses. As required by the law, the commission was granted the highest security clearances by Director of Central Intelligence George J. Tenet; it worked assiduously from its first meeting on January 14, 1998, to produce its report as mandated six months later on July 15, 1998. Its conclusions are presented in the 30-page unclassified Executive Summary.[18]

The highly classified full report of 200 pages is available to members of Congress and to those in the executive branch. Elements with still higher security classification were briefed to the leaders of the Congress on July 15, 1998.

The Rumsfeld Commission judged that of three countries—North Korea, Iran, and Iraq—that currently have an intense enmity toward the United States, North Korea and Iran have substantial programs for the development of long-range ballistic missiles. Based on the 300-km-range Soviet Scud missile, of which hundreds were used in the war between Iraq and Iran, and which are ubiquitous as a matter of export initially from the Soviet Union, North Korea has built a force of No Dong missiles of some 1300-km range. It has sold many of these to other countries, and on July 22, 1998, Iran tested such a missile with range allowing it to strike all of Israel, all of Saudi Arabia, most of Turkey, and some of Russia. Pakistan had tested a similar missile in April 1998. Despite various efforts to prevent the commerce in missiles of medium and intermediate range, North Korea in June 1998 issued a statement to the effect "Our missile export is aimed at obtaining foreign money we need at present." So it regards these missiles as a way of earning currency and supporting (at least) the military sector. Although the Missile Technology Control Regime (to which the United States and Russia belong) forbids the export of such missiles, North Korea is not a member of the MTCR, and there is no legal barrier nor agreement that restricts its actions in this regard.

The Rumsfeld Commission report concluded that with short-range ballistic missiles (e.g., Scuds) based on ships, several countries could threaten the continental United States with payloads of nuclear or biological weapons. It judged that North Korea has one or two nuclear devices that could be used in this way, and despite the agreement under which Pyongyang has shut down its 20-megawatt graphite reactor and ceased construction on a two-hundred-megawatt graphite reactor, North Korea has continued to pursue a nuclear weapons program in all other aspects. In addition, there is the concern that North Korea might obtain weapon uranium from Pakistan, in return for additional missiles and missile technology transfer.

Although the study of cruise missile attack on the United States was not within its primary charge, the Rumsfeld Commission nevertheless reported that such missiles (small, pilotless aircraft) launched from ships and carrying a biological weapon or nuclear weapon payload also posed a threat to much of the 50 United States, and that this threat was too little acknowledged.

In the development of ballistic missiles, the purchase of technology and consulting services from Russia and China has played an important role, as has the transfer of obsolete technology from the United States to several of these countries. Finally, as regards weapons with which North Korea, Iran, or Iraq could directly strike America, the commission judged that if Iran or North Korea established a well-supported high-priority program to acquire ICBMs, they could have a few such missiles within five years, and because of the measures these countries take to hide their activities from U.S. satellites and other observation, the United States might not know for several years of the existence of such a program. Iraq, of course, since 1991 has been under UN sanctions, which have been adequate to preclude a large program for the production of weapon uranium or plutonium and also impede work on long-range missiles. Nevertheless, under UN rules, Iraq has been permitted to work on missiles of range 170 km or less, which has allowed it to retain a core of scientists and engineers to have a running start when sanctions are lifted. The commission estimated that Iraq could possess an ICBM within ten years of its ability to begin work on such a program, and in August 1998 revised its estimate to five years, in view of the likely lifting or erosion of UN sanctions, as largely happened with the expulsion of UNSCOM, the United Nations Special Commission, in October 1998.

The creation of the Rumsfeld Commission followed years of frustration on the part of what had been for a long time the Republican minority in the House Armed Services Committee, which with the 1994 elections giving the party a majority in the House of Representatives was renamed the House National Security Committee. Already in 1991, Congress considered the Missile Defense Authorization Act, the purpose of which was to authorize the

deployment of a missile defense of the United States. Garwin and Hans Bethe wrote to the chairman of the House Armed Services Committee, Les Aspin (later to become Secretary of Defense in the Clinton administration), and the chairman of the Senate Armed Services Committee, Sam Nunn, arguing that no missile defense under consideration would protect the United States against the thousands of Soviet nuclear warheads, and further, that such a defense would not secure the United States against ship-launched cruise missiles or short-range ballistic missiles, and in fact would not even detect them.[19]

Further, the letter added, the defense would offer no protection against ICBMs of emerging missile powers armed with biological weapon payloads, if (as maximizes military effectiveness) these missiles were equipped not with a single massive warhead of hundreds of kilograms of biological warfare agent, but with one hundred or more "bomblets," each equipped with its own shield against the heat of reentry. The bomblets would separate from the missile as soon as it reached maximum velocity at the end of the boost phase of flight. Finally, as for defense against an ICBM armed with a nuclear warhead, the letter pointed out that any country that could develop an ICBM could far more readily equip its warheads with a large enclosing balloon 30 meters in diameter. The balloon would be obvious to any missile defense radar, but the defensive weapons — the "hit-to-kill interceptors" that the missile defense was supposed to use — rely on actually hitting the target at extremely high speed in order to destroy it. The interceptor would very likely be able to strike the balloon and punch a hole in it, but very unlikely to strike the warhead hidden within the large enclosing balloon.

At the turn of the century, the Clinton administration was completing a three-year program to develop a national missile defense, and in the summer of 2000 was to make the decision whether or not to deploy that defense by the year 2005. But the missile defense under development has precisely those deficiencies noted in the Bethe-Garwin letter of 1991. Nevertheless, many members of the House National Security Committee called for the urgent deployment of the missile defense system, and even among some of the former opponents of such systems there was little recognition that it would do no good at all. On March 17 and 18, 1999, the House and the Senate respectively passed somewhat similar bills, stating the policy of the United States to deploy a national missile defense. Compromise legislation was signed on July 23, 1999, by President Clinton, who highlighted portions of the law that subject the national missile defense program to the regular authorization and appropriation process and also support negotiated reductions in strategic arms. Clinton stated that he would decide in July 2000 on the basis of four criteria — the long-range ballistic missile threat; the technological readiness of a defense; the cost;

and the effect of a national missile defense system in furthering or inhibiting U.S. national security goals such as nonproliferation of nuclear weapons or the reduction of Russian nuclear weaponry. In September 2000, Clinton decided to defer the deployment decision—an action approved both by those who felt the proposed NMD was unneeded and those who felt it inadequate.

THE PROPOSED NATIONAL MISSILE DEFENSE SYSTEM

On August 26, 1999, Garwin spoke at the Second Annual U.S. Army Space and Missile Defense Conference in Huntsville, Alabama.[20] He argued that the proposed National Missile Defense (NMD) system with its 20 or so deployed interceptors (or even one hundred) would provide no significant protection against even 5 ICBMs from North Korea, in view of the likelihood that any biological warfare payload would be loaded into bomblets dispersed just after the rocket boost, and that a nuclear weapon payload would be hidden from the interceptor inside a large enclosing balloon as the warhead fell through space. He proposed instead that the United States deploy a boost-phase intercept system that would destroy North Korean ICBMs during the 250 seconds or so of their powered flight, just after their launch. This could be done with interceptors very similar to those under development for the NMD system, but with homing sensors that are much less sensitive, since they need only detect the rocket flame and not the subtle heat from a trash-can-sized reentry vehicle at 100 km distance. No radars would be necessary, since the ICBM launch would be detected on the so-called Defense Support Program (DSP) satellites that the United States has maintained since the 1970s in geosynchronous orbit, looking at the entire world.

The authors propose that such a system be built jointly with Russia and deployed first on the strip of land south of Vladivostok, which abuts North Korea. In fact, an interceptor that reaches ICBM speed in one hundred seconds could be based on a U.S. military cargo ship placed in the international waters of the Sea of Japan, 500 km off the North Korean shore, and still carry out the intercept. This seems to us a much better approach to protection than the difficult task of midcourse intercept. A similar approach could be taken to ICBMs from Iran or Iraq, if their missile programs reach that stage before relations improve with the United States; sites in eastern Turkey could negate ICBMs from Iraq.

In reality, nations that might wish to threaten the United States with missiles carrying biological warfare agents or nuclear warheads could much more readily accomplish that from ordinary merchant ships. In the harbor of a major United States city, such a ship could explode a nuclear weapon, dispense BW

agents into the breeze, or launch cruise or short-range ballistic missiles as indicated above. Furthermore, nuclear or especially biological weapons could be
covertly deployed within the United States and employed without the return
address that an ICBM inevitably provides.

In April 2000, a joint study of the Union of Concerned Scientists and the
MIT Security Studies Program issued a report on countermeasures to the
national missile defense; Garwin was a coauthor.[21] This report goes into substantial detail on the feasibility of the bomblet threat for the delivery of biological weapons—bomblets released on ascent would be unstoppable by the
proposed NMD system. Countermeasures suitable for protecting nuclear
weapons against intercept in their fall through space and analyzed in detail are
balloon decoys with a similar balloon around the warhead, shrouds cooled
with liquid nitrogen to reduce the infrared emissions to an undetectable level,
and a large enclosing balloon to foil the hit-to-kill interceptor.

Political leaders of any stripe are reluctant to identify threats to the security
of the people whom they lead, unless they have at the same time either the
existence or a promise of a defense. Insurgent politicians in an election campaign have routinely charged deficiencies in national security and threats to
the population, as did John F. Kennedy in his 1960 campaign. Kennedy
accused the President of having permitted a "missile gap," when in fact Eisenhower and a few in his administration knew from the first satellite photographs
of the Soviet Union that the U.S. missile development program was substantially in advance of Moscow's effort. In a democracy, the people should have
the information, unpleasant though it may be, in order that they and their
leaders can protect themselves.

THE BUSH ADMINISTRATION MISSILE DEFENSE PROGRAM

On March 1, 2001, and again on May 1, President Bush emphasized his commitment to missile defense. No longer would National Missile Defense be
treated separately from Theater Missile Defense. And the President specifically included comments on the virtues of destroying missiles in boost-phase as
well as after they have reentered the atmosphere—in addition to the Clinton
NMD emphasis on mid-course hit-to-kill intercept. Bush also emphasized that
the program would not be limited by the strictures of the ABM Treaty as it
stands, but did not indicate expressly that the treaty itself would have to be
abandoned, rather than being amended or reinterpreted. The authors are concerned that an inevitable bureaucratic momentum will lead to continued
expenditures on the mid-course system, with its vulnerability to countermeasures, while shortchanging the more technically promising and poten-

tially effective boost-phase intercept. And there is always the hazard of any program—making a decision narrowly focused on the program itself without regard to its proper ranking among threats to national security and their potential solutions.

Against missile threats from emerging powers, the main protection of the United States is still deterrence—the threat of retaliation against such an attack. And a commitment to deterrence (unpleasant though it may be and of challenged morality) can also serve to defer the investment of scarce resources in the building of an ICBM threat to the United States.

If such a threat nevertheless emerges, deterrence in all likelihood can prevent its use; in addition, the potential adversary would always need to fear a preemptive strike from the United States that would disarm it of its ICBM capability.

In order to deter such development or use, and especially to carry out a disarming strike, the United States needs to some extent the tacit approval of the vast majority of nations of the globe, and this can be obtained only if America minimizes the number of its nuclear weapons and emphasizes that their purpose is to ensure the security not only of the United States but of other likeminded nations. The United States has clearly emerged as the leader in the revolution in military affairs, typified by bombs and cruise missiles armed with high explosives and guided to meter accuracy by laser beams or the global positioning system (GPS), satellite imagery and digital data bases that enable such targeting, and other capabilities. But nuclear weapons of the types available to the United States more than 40 years ago would themselves pose a threat to civilization—the more so because of the global freedom of trade and travel that are so important to economic growth and well-being. The responsibility of those who have built and operated nuclear weapons (and those who have paid for them and prized them for the security they have brought) is not discharged until stocks of nuclear weapons have been reduced to the minimum possible, and whatever benefits they convey provided to the society at large. This is more than a duty; it is self-interest and interest in the security of our families and our societies.

CHAPTER 13

Can We Rid the World of Nuclear Weapons?

PROLIFERATION of nuclear weapons is a cause of grave concern. In the present-day world, the priority of priorities is to escape the current situation in which neither the size of the stockpile of nuclear weapons nor the uncertain and unstable conditions for their possible use are acceptable. We shall discuss the possibilities for worldwide control, possibly leading to the elimination of weapons whose possession offers today only a delusion of security — a delusion that can be life-threatening to their owners as well as to the rest of the world.

We live in a world that, in the course of the present century, has known military convulsions whose destructive level has grown in parallel with the level of scientific development. Over the span of human history, and to the present day, science has, in large measure, been supported because national leaders have been convinced that it would be a source of knowledge and tools for the national defense — in short, weapons.

After each of the two world wars, society created institutions, the League of Nations and the United Nations, that it hoped would prevent a repetition of the cataclysm from which it was emerging. The fear of a nuclear exchange on the battlefield was shared by political leaders who came to power after the Second World War. Despite belonging to opposing camps, they had no illusion that nuclear weapons would be confined to the battlefield, and that is why they resisted the temptation to use them in times of crisis.

This innate caution prevented a massive recourse to military force during the Cold War between the Soviet Union and the United States. It no doubt saved Europe, which was the most exposed region, from the total destruction that would have been created by the large number of nuclear weapons stationed there for use on its soil.

But the inhabitants of the planet have not changed. We often see them seized with violent destructive rages, turning to massacres with machetes or

361

submachine guns. We watch them with horror, but passively, on our television screens. They add a real-life vision of the unbearable to the romanticized scenes of violence with which we are saturated from childhood, in the name of entertainment.

We should try to find the reasons for the explosions of savagery that devastate our planet while some of its inhabitants intoxicate themselves with the fruits of scientific progress. They indulge, like drunkards with unlimited resources, in their own destruction, consuming these costly wines in ever-increasing quantities, instead of savoring them wisely in moderation.

It would take an entire library to contain the description of the horrors perpetrated by men in wartime. But there is another aspect of human nature. We should not forget those who love themselves enough to love others, those whose service to others is a monument to their genius for the centuries to come, those who can sometimes be martyrs without becoming executioners—in short, those to whom we owe the beauty and greatness of our world. But at

"It is this childlike, innocent quality among scientists that I always find so touching."

all latitudes, policed by regimes of all kinds, imbued with the most attractive ideologies, the most generous religions, human society can suffer uncontrollable convulsions, and in this soil nuclear weapons can germinate if they are sown into the wind by a lack of political will on the part of the international community.

By a caprice of history, nuclear weapons may be acquired because of the corruption of the leaders of a great power or the anarchy of a regime in ruin, in association with huge sums of dirty drug money or clean oil money to pay for them. That could happen if the prevention and countering of nuclear proliferation does not receive our undivided attention. The traumas of history have bequeathed an inheritance of inextinguishable hatred in the world as it is today. It is simple madness and irresponsibility to blow up airplanes and buses transporting innocent victims with the idea of serving God or a cause worthy of the sacrifice of one's own life. Just as it is urgent to disarm children playing with submachine guns and grenades, it is essential to prevent nuclear weapons from falling into irresponsible hands. But it is also necessary to respond to the security needs of those who seek nuclear weapons for their protective effect as a shield.

How to go about it? We shall suggest a realistic program to achieve a desirable goal: to rapidly and massively reduce the stockpile of nuclear weapons while respecting the legitimate concern for security of all nations, those that have nuclear weapons and those that don't. First we shall examine the efforts thus far in place to reduce the nuclear arsenal and to avoid the proliferation of

weapons. We shall see, from the difficulties encountered by the Comprehensive Test Ban Treaty, how divergent interests have appeared among the powers themselves, but also within countries, between opposing groups.

REASONS TO REDUCE TO VERY LOW LEVELS

The nuclear arsenal had reached such an insane level that even by reducing to zero the number of tactical weapons and to a third the number of strategic weapons planned for the year 2003, life on the planet would remain threatened in case of a global exchange of fire. There are, in the world, 2300 cities with more than 100,000 inhabitants. That means that the United States or Russia could destroy them all and each retain thousands of bombs in reserve. Over a long period of time, one cannot exclude the prospect of an unauthorized launch, of a madman coming to power with a group of totally devoted supporters, even if one took the trouble to provide the system with all the safety features imaginable.

A reduction to 2000 strategic warheads for Russia and an equal number for the United States has the attraction that it is much lower than present levels but not different in kind; hence it can be achieved rapidly. A maximum of a few hundred nuclear warheads altogether would be amply sufficient for deterrence. The world was not drawn into war when deterrence was based on numbers of that order. It is urgent for reasonable minds to work toward a reduction

in the stockpile of weapons to a level that no longer threatens the lives of hundreds of millions of totally innocent people, if the weapons were ever to be used. A more ambitious goal would be the establishment of a system of collective security that would reduce this stockpile, even to zero. To reach this goal, we shall have to involve others, and not only physicists and military personnel.

For skeptics, it is utopian to hope to go to zero at this time. To do so, they argue, would give an unchallenged tool of conquest to those who might keep a few nuclear weapons in hiding, or to newcomers to the club, because the potential victims would then have no means of retaliation—hence no deterrence. But this is a problem to be solved, not an impassable obstacle; we advocate going to zero nuclear weapons only if it *can* be solved.

A simple calculation illustrates the destructive power of even a small number of nuclear weapons.[1] Nearly 25 million people, or about a sixth of the Russian population, live in cities of more than a million inhabitants. Their total area is about 2500 km^2. The explosion of a single 475-kiloton nuclear warhead, launched from a submarine, would destroy 100 to 150 km^2. This figure corresponds to the destruction by the blast effect alone—millions of additional deaths might result from firestorms, radiation, disease, and starvation. Twenty nuclear warheads exploding on their targets would be enough to ensure the death of 25 million. In Russia, such an attack would lead to the loss of nearly a quarter of its industrial potential. In China, whose industry is more concentrated, seven bombs would be enough to assure the same slaughter.

In a strategy of deterrence, for which these threats are amply sufficient, bomb stockpiles a thousand times greater are simply unjustifiable. One might have thought that the great powers would have decided to bring the overall stockpile of nuclear weapons to a level where it would be impossible for a global nuclear exchange between the two largest proprietors to threaten all life on the planet. Before the fall of the Soviet system, only one button had to be pushed to trigger hell on earth. If radar had indicated that a thousand rockets were heading toward an adversary, probably toward its strategic launchers to neutralize them by a first "surgical" strike, then this button would have launched an immediate response: thousands of nuclear warheads targeted on the aggressor's vital centers.

With "detargeting," it would seem at first sight that more than one button is needed. Since rockets are no longer being directed toward predetermined targets, a first button would activate launching programs to load into the missile guidance systems the precise objectives, which wouldn't take very long but would give strategists a little time to think. The second button would then release the fire of hell. In reality, a single button could first authorize the loading of the targeted objectives in the programs of the missiles and then automatically launch them; one wouldn't have to push two buttons, one after the other. Still, agreements to detarget are useful, unless they are taken as a substitute for more serious limits.

CONCRETE PROPOSALS FOR DISARMAMENT

The authors propose a disarmament process in which most nuclear warheads would be rapidly dismantled and the parts put in storage so that initially they could be reassembled in a few months (but not a few minutes), while retaining a minimal stockpile of weapons ready to be launched. Perhaps more acceptable to military personnel are other possibilities, with most of the weapons no longer in a state of alert. For example, land-based silos could be covered over with a thickness of twenty meters of earth; even with power shovels it would take several hours to reopen them. As for missile-launching submarines, they could be sent to patrol in the southern hemisphere, so that their targets would be out of missile range, verifiably limited by the insertion of inert steel dummies in place of the warheads withdrawn in accordance with the START II agreement of January 1993. START II eliminates multiple warheads from U.S. and Russian land-based missiles, but allows multiple, independently targeted warheads on submarine-launched ballistic missiles. If the sub-launched missiles had some of their warheads removed, they could have long enough range so that they could strike targets from anywhere in the world—hence the desire to add dead weight to limit missile range. It would be necessary to verify adequately that the inert reentry vehicles remain in place. Such measures would help to avoid the danger of an apocalypse while continuing to deter any conceivable aggressor, who could not emerge from nowhere in a few weeks armed with thousands of nuclear weapons adequate for wiping out the nonalert retaliatory force.

The 450 French warheads would be more than enough for the whole world. We defy anyone to prepare a reasonable list today of 450 cities or sites as targets. Reductions to a few tens of one-megaton nuclear warheads would make less likely the order to wipe from the face of the earth millions of people, the vast majority of whom bear no responsibility for the behavior of their leaders.

During the crisis of October 24–25, 1973, in the last stages of the Yom Kippur War, the American military were put on an all-forces alert, as one of a set of crucial actions taken by the White House to respond to the Soviet threat of intervention in the Middle East. Secretary of State Henry Kissinger chaired the session. Kissinger was National Security Advisor as well. Spiro Agnew had recently resigned as Vice President, but Gerald Ford was not yet confirmed as his successor. Kissinger agreed with White House Chief of Staff Al Haig that President Nixon, burdened by his involvement in covering up the Watergate break-in, was so "distraught," "agitated and emotional" that he should not be informed of the meeting.[2] Some Russians claim that the first attack on Chechnya was decided by Russian generals toward the end of a drinking bout. Thanks to emotion, alcohol, and drugs, spiritual or moral inhibitions grow blurred so that orders are more easily given, and this is particularly dangerous when the order, once given, is carried out automatically.

PREVENTING ACCIDENTAL LAUNCH AND INSTABILITIES

During the Cold War both the United States and the Soviet Union regarded deterrence provided by their force of strategic nuclear arms as the solution to the otherwise insoluble problem posed by the confrontation of ideologies. Today, with Russia a struggling democracy, the nuclear forces themselves are far more of a threat than is the possibility of purposeful attack. It is time for statesmen to give highest priority to reducing the unintended nuclear danger. Some of the tools available were created to solve analogous problems under very different circumstances. Political leaders and their staffs, not entirely indifferent to the obvious danger of this accumulation of apocalyptic weapons, raised a serious question: Is it possible to avoid instabilities that would lead to the untimely use of nuclear weapons and precipitate actions impossible to control? Direct telephone lines called "hotlines" or "red telephones" (actually teletypes or fax or computer communications) were installed between heads of state of the United States and the Soviet Union. The U.S. and Russian Presidents, as well as their counterparts in the other nuclear powers, never go anywhere without a military officer always nearby, carrying a "football," or briefcase, containing a coded radio transmitter controlled by a key always in his or her possession that could enable the head of state to play the essential role in triggering a nuclear strike.

ARMS CONTROL AND DISARMAMENT AGENCY

Those in the U.S. Arms Control and Disarmament Agency (ACDA) who worked on the arms control and nonproliferation treaties and on the analysis of security systems like the hotlines between heads of nuclear states must sometimes have felt like mice lost among elephants. Their annual budget was $40 million, against $240 billion for the Defense Department. For every $6000 spent on weapons and military personnel, $1 was allocated to ACDA activities, which include negotiation of arms control agreements and studying protective systems to prevent weapons development from leading to catastrophes beyond recall.[3] On April 1, 1999, the Arms Control and Disarmament Agency ceased to exist. In a tactical concession to win Senate approval of the Chemical Weapons Convention, the Clinton administration agreed to the absorption of ACDA into the State Department; President Clinton named the director of ACDA, John D. Holum, to the post of Under Secretary for Arms Control and Nonproliferation, as well as Senior Advisor to the President and the Secretary of State on arms control and nonproliferation issues. It is too soon to judge how long the work of ACDA can be performed in its new structure as four of a score of bureaus of the State Department after Holum left the department at the end of the Clinton administration.

The United States and other nations should devote much larger sums to reshape the future by negotiation or in cooperation on disarmament. In reality, some of the Defense Department's funds are expended on improving security by means other than weapons. We have already noted the department's "Cooperative Threat Reduction" program paid for under the Nunn-Lugar legislation, conducted in association with the former Soviet Union.

DETERRENCE IN ONE NATION'S HANDS?

Obviously, it will not be possible to persuade the Russians or the Chinese or the French to leave to the United States alone the responsibility for managing a deterrent force, even of modest size. The French long maintained that an American politician would never exchange the survival of Detroit for that of Paris if there were a standoff with a nuclear-armed Soviet Union. They hold the same view about the United States facing even a small state with nuclear weapons. France will, therefore, not easily give up its own deterrent. Only when the world arrives at a situation in which nuclear weapons are entrusted to an effective international organization responsible for defending objectives defined by the international community might they be persuaded to change their minds.

Furthermore, it is clear that we still have a long way to go when we see the ineffectiveness of the United Nations during the several years when specialists in ethnic cleansing were able to carry on their genocide in the heart of Europe. But it is not at all certain that the United States would sacrifice Paris in order that Detroit survive. That would depend, evidently, on who is president. U.S. leadership in military action in Kosovo and Serbia responding to the atrocities in Kosovo showed a willingness to intervene where American interests are not immediately at stake, but influential voices were raised in Congress against this involvement; in a different administration, such views might be adopted by the White House.

The objective of the Comprehensive Test Ban Treaty is to establish a regime that will be sustained by the vast majority of countries opposing not only nuclear proliferation, but also the dominance of those who historically have possessed nuclear weapons. The support of the great majority of nations is a significant weapon against those who would break this agreement. We should try to arrive at a stage where guarantees would be provided by national nuclear forces, under the control of a collective international organization; after all, the nuclear protective shield that the United States made available to its allies during the confrontation with the Soviet Union was taken very seriously by the Kremlin leaders even though, to some Americans, like Henry Kissinger, National Security Advisor to President Nixon and later Secretary of State, it was not always apparent that this shield was more than psychological. In a speech in Brussels in 1979, Kissinger declared: "Our European allies should not keep asking us to multiply strategic assurances that we cannot possibly mean or, if we do mean, we should not execute, because if we execute we risk the destruction of civilization." But most people and most nations did not trust the United States to be as logical as Kissinger, and deterrence held.

A SINGLE, MUCH-REDUCED NUCLEAR FORCE

We should seriously consider the goal of establishing an international consensus for sharing a single nuclear force and of creating the needed organization.

It is true that the Soviets rejected the Baruch plan in 1947 by which the United States proposed to share nuclear secrets with them, in return for giving up national sovereignty over these weapons. But times have changed. Must we wait for the next catastrophe before setting up an effective security system? For the moment, it is important that the CTBT go into effect, that international agreements forbidding the use and possession of chemical and biological weapons be universally accepted and their violation punished, that substantial reductions in nuclear weapons arsenals be encouraged, and that nuclear deterrence be maintained by arms that are not capable of use within minutes or hours.

The entire world and each nation individually would enjoy greater security if we were able, during the next decade, to bring down the present American and Russian stockpiles of more than ten thousand warheads to a thousand each, while the British, the Chinese, and the French would limit themselves to three hundred each. These numbers would include the totality of weapons, whether mounted in launchers or held in reserve. The need to arrive rapidly at these reductions derives from the goal of gaining the support of the great majority of non-nuclear-weapon states toward terminating the threat of proliferation of nuclear weapons into the hands of additional countries or terrorist groups. Such reductions would also limit the global effects of nuclear weapons if ever they were to be used.

It is easy to imagine a further transition toward a world with two hundred nuclear warheads controlled by an international security organization, and with no such weapons in the hands of any single nation. Perhaps this arrangement should initially coexist with a much smaller number of nuclear weapons in national stockpiles. And the problem of command and control—decisions as to when, or if, nuclear weapons should actually be used—clearly requires an organization in which no participant would have a veto. The UN Security Council with its permanent membership expanded beyond the traditional nuclear five, and with a voting system that permits action against the will of one of the permanent members, has not yet been achieved, but it would seem a necessity for such a transition to be realized.

Proposals for massive reductions in nuclear weaponry and for measures to reduce the readiness to use nuclear arms have moved close to the mainstream in the past decade. The dissolution of the Soviet Union and the transfer of all of its nuclear weapons to a newly democratic Russia should, after all, be celebrated as a great achievement and should serve as the foundation for further progress in the control of nuclear arms.

For some, this major reduction would only be one step toward further reductions—leading to the elimination of nuclear weapons in national arsenals. What better objective, they ask, than to arrive at a world devoid of such an arsenal, in which large-scale conflicts, which would undoubtedly lead to their resurrection and eventual use, would be eliminated?

THE CANBERRA COMMISSION

Australia was very much involved in the realization of this objective and has played an active role in disarmament negotiations, in particular on banning nuclear testing. It set up and financed the seventeen-person Canberra Commission, to propose realistic steps toward a world without nuclear weapons, including consideration of the problem of maintaining stability and security during the transition period as well as after the goal had been attained. In August 1996, the commission, presided over by Richard Butler, at the time Australian ambassador to the United Nations, submitted its report.[4] Among its members were Joseph Rotblat, president of Pugwash; General George L. Butler, who was commander of the American strategic forces until 1992; Michel Rocard, former Prime Minister of France; Jacques-Yves Cousteau; and Robert McNamara, former American Secretary of Defense.

The report urgently requests the five nuclear powers to undertake initial steps toward the elimination of nuclear forces and to begin immediately the necessary negotiations. It also asks them to set up unilateral measures limiting the danger of accidental use of weapons, such as the de-alerting (i.e., reducing the readiness) of all nuclear devices.

WORLDWIDE CONTROL: AN IMPERATIVE

The control of nuclear weapons will complete the formal regulation of all weapons of mass destruction resulting from scientific development. The category "weapons of mass destruction" is commonly taken to include not only nuclear but also chemical and biological weapons, which are the subject of individual international agreements—the Chemical Weapons Convention and the Biological Weapons Convention. Unlike nuclear devices, chemical and biological weapons, both possession and use, are totally banned by these agreements; but BW agents are unfortunately easy to produce in facilities normally devoted to making beer or pharmaceuticals. The Chemical Weapons Convention is a detailed treaty that defines the materials that are banned and provides verification measures administered by an international organization specific to the convention. To ensure compliance with these agreements, each nation must ensure that all on its territory or under its control respect these agreements, under penalty of national law. It would be desirable to make use of this fact by supplementing the conventional verification measures—inspections, detection devices, etc., with "societal verification."[5] This would involve each nation's passing appropriate domestic legislation which would encourage (or even require) that nation's residents and citizens to report to the verification organization any violations of the treaty. Such reporting could be aided by the widespread availability of the Internet and e-mail, and could be effectively anonymous and encrypted, and explicitly encouraged by the nation as a signatory to the treaty.

There are seven crimes that are universally punishable—among them piracy, interference with diplomatic personnel, aircraft hijacking, and the like. It would help to bring individual violations of the BW and CW conventions under this same code.

This typifies the work that must be performed seriously to limit not only genocide committed by traditional means but also new crimes feasible only because of the progress of science and technology. There is the obligation to work out new missions for organizations like the United Nations. Realism demands, however, that one go beyond the creation of obligations on paper. How can the community of nations counter even those who intentionally and openly violate these agreements and these norms? Who will take on the role of sheriff? Can there be a body to judge the need for intervention? Can it act sufficiently rapidly? Is there anything that can replace the sheriff?

THE UNITED STATES AS INTERNATIONAL AUTHORITY

The power of the United States makes it a prime candidate for the world's sheriff, although few of its politicians openly advocate that role. It is clear that if America were to assume this responsibility, it would have to take into account the interests of other countries and show itself capable of making decisions independent of electoral pressures or the influence of certain lobbies. If not, it would soon be discovered that instead of the protection of a sheriff, the ungrateful citizens would prefer that of their local mafia. The decline of relations with Russia, in the wake of the spring 1999 NATO bombing of Kosovo and Serbia, typifies this reaction.

It would indeed be a major political advance if the world were to have a just, effective, and, in particular, elected sheriff. It would first be necessary that this basic constitutional change be accepted by most nations and that the limits of the mission be clearly defined. It is possible that this will only occur after a planetary catastrophe, which might lead instead to an attitude of everyone for himself, to a dispersion of centers of authority or centers of civilization. Before coming to a world sheriff, it would be wise to reduce the potential danger of the existence of the stockpiles of weapons of mass destruction made available by technological progress, and not just nuclear devices. In the case of chemical and biological weapons, we are well on our way to a legal structure ensuring the elimination of all stockpiles of these agents, but the problems with Iraq show some of the difficulties that will be encountered on the way.

OUR PROPOSALS FOR REDUCTIONS

For our recommendations, or those of the Canberra Commission, to be accepted, the problem is to find an international security system that would decrease the probability of conflicts and in any case not let them degenerate into major war between nations. Until this objective is attained, the total elimination of nuclear weapons will remain wishful thinking.

Nuclear weapon designers and builders of the five powers are proud of their accomplishments, and most of those who have built and tested the airplanes, missiles, and submarines intended to carry these terrible destructive devices to their targets have accomplished their tasks with competence. In almost every case, however, the secrecy inherent in these activities and the barriers raised by the Cold War long isolated and deprived them of a wider vision of the possible options in the domain of international security. In particular, they have been cut off from those who participate in the same activities in other countries. To a limited extent, however, the Pugwash meetings since 1957 have brought together a few of the creators of these weapons, and more recently the activities of the Committee on International Security and Arms Control of the National Academy of Sciences (and its predecessors) have contributed in interacting with the national security communities of the Soviet Union (now Russia) and China.

The 1990s have brought many changes, and when we now speak with these builders of nuclear arsenals, they recognize the great excess of destructive power in the world and would support responsible leaders when they decide to undertake a massive reduction in the nuclear arsenal. Our societies should consider seriously the analyses and constructive proposals of groups such as CISAC, the Pugwash Conferences, the Union of Concerned Scientists, the Federation of American Scientists, and even Greenpeace.

The authors now summarize their proposals to effect a massive and rapid reduction in nuclear weapons stockpiles:

1. In 2002 and 2003, Russia and the United States would be able to reduce their arsenals to two thousand warheads each, and set up a bilateral system for keeping track of all weapons and stocks of military-grade fissionable materials over and above what is contained in the weapons permitted by the treaties. There is no reason why this could not have been done in the late 1990s, had it been given the priority it deserved. As soon as possible, the bilateral system should be transferred to an international

organization that would be responsible for the surveillance of warheads and materials, while the warheads to be destroyed would be sealed and rendered nonoperational, and the nuclear material that they contain allocated irrevocably to civilian industrial use. Uranium of military grade, which, as we have already seen, presents a major danger of proliferation, would be mixed, inexpensively, with natural uranium before being stored, so as to yield an enrichment in uranium-235 of 19.9%, the upper limit for so-called low-enriched uranium. This could then be sold to the civilian nuclear industry under normal IAEA safeguards. Plutonium removed from weapons should be stored in a secure fashion until such time as it is used in reactors or, depending on the preferences of individual nations, safely deposited in deep geological sites.

2. The five nuclear powers should adopt positions, commitments, and finally a treaty in which they reject first use of nuclear weapons.

3. In response, Britain, China, and France would reduce their stockpiles as soon as the United States and Russia had arrived at a thousand weapons each. At a subsequent stage, the United States and Russia should consider a reduction to fewer than three hundred warheads, while the other powers would reduce theirs to two hundred.

4. In collaboration with the United Nations, the five powers would provide nonnuclear states and signatories of the Non-Proliferation Treaty with negative security guarantees—that is to say, a promise not to use such weapons against them—and positive security, by assuring them protection, nuclear if necessary, in case they were subject to a nuclear threat.

5. The five powers, as well as the other nations that would like to associate with the process, should begin negotiations and conclude a treaty containing specific measures to arrive at total elimination of nuclear weapons. India, Pakistan, and Israel would eventually be brought into this regime. One step might be that the nuclear weapons in national arsenals be used exclusively with the authorization of an international authority; this stage could be followed by the transfer of some or all of the national nuclear arsenals to this authority.

Whereas neither the total elimination of nuclear weapons nor the transfer of a small number of them to an international authority can possibly take place unless nations succeed in setting up a better system of international security, the first stages toward this objective depend only on transparency and on the commitment of the parties concerned. These first four steps would be taken in the new context of near-universal adherence to a system that also forbids the use of biological and chemical weapons, as well as their development and possession.

These proposals would have been considered utopian in the 1980s, but they are now close to the views of many specialists on these questions. Nevertheless, it is clear that nations, so long accustomed to weapons and even war, do not yet have mechanisms that would allow the necessary changes to take place at the required pace.

For example, the UN Disarmament Conference seems to be able to negotiate only one treaty at a time, and at a snail's pace. We have already pointed out that in the United States the Arms Control and Disarmament Agency was granted only $1 for $6000 spent by the Defense Department; it appears obvious that the possibility to reduce external threats deserves the allocation of substantially greater resources. A June 2001 report, *Toward True Security: A Nuclear Posture for the Next Decade,* offers specific proposals for U.S. action on nuclear-weapons numbers and policy generally — proposals consonant in large part with the views expressed here.[6]

The authors of this book don't claim to have all the right answers or to be able to offer a full solution to these tremendous problems. Nevertheless we must not give up hope in our political leaders, some of whom are intelligent, of good will, expert at navigating in troubled waters, and capable of thinking of the future beyond the next election.

CHAPTER 14

A Turning Point in the Nuclear Age?

You, the reader, have journeyed with us in this book from the discovery of radioactivity and its origins to the revolution that nuclear energy in the form of nuclear fission has already wrought in the world. That world has lived through 40 years of intense antagonism between the United States and the Soviet Union, buttressed with tens of thousands of city-destroying thermonuclear weapons on each side, and has miraculously emerged into an era in which all but a few agree that a number of nuclear weapons in the low thousands is the maximum tolerable, and that U.S. security actually requires a move to a level of hundreds and even perhaps to zero. A mere 20 nuclear warheads like those carried on the U.S. land-based or submarine-based missiles exploding on their targets in Russia would kill 25 million people. If those missiles were directed against China, more would die. But the United States is and will remain vulnerable to similar weapons; even a few delivered by missile, aircraft, truck, or ship could kill millions in a tragedy of unprecedented monstrosity. The task for all who care about the world, and about civilization and themselves, is to prevent such a disaster this year, this decade, and indeed forever.

The release of heat from nuclear fission in reactors has enabled submarines and surface vessels to travel for years without refueling, their endurance limited only by the needs of their crew. This is the second application of nuclear reactors, which were initially built in the United States during the Second World War to produce the fissile artificial element plutonium for the nuclear weapon program. Now, more than 400 large reactors worldwide produce commercial electrical power.

Nuclear power has a potential importance for the world's energy future far beyond its present contribution. Long before the exhaustion of the fossil fuels—oil, natural gas, and eventually coal—the great increase of carbon dioxide in the atmosphere caused by the combustion of these fuels and the accompanying effect on climate will have adverse consequences on ecosystems. Nuclear power offers one of the few well-developed and affordable approaches to satisfying the world's energy needs without contributing to this problem.

We have emphasized the necessity of operating nuclear reactors and the entire nuclear fuel cycle in a responsible fashion, ranging from appropriate treatment of the ground-up rock from uranium mines as discussed in Chapter 7, to the design and safe operation of reactors, and to the disposal of spent fuel or of the radioactive materials resulting from reprocessing. The International Atomic Energy Agency will continue to have an essential role in monitoring civil nuclear operations to ensure that materials are not diverted or stolen for weaponry. Whether or not the United States rebuilds or expands its population of power reactors in the next 50 years, nuclear power will continue to operate in much of the rest of the world; in either event, the United States has an interest in moderating the rate of rise of carbon dioxide in the atmosphere. America also has a vital interest in preventing the proliferation of nuclear weaponry. If it is going to influence the world toward these goals, it will need to maintain its competence within the government, the engagement of its citizens, and the investment to produce new knowledge and options to solve these problems.

We conclude this book by summarizing our proposals, the basis for and the details of which have been stated in earlier chapters.

Regarding nuclear weaponry, international agreements such as the Non-Proliferation Treaty and the Comprehensive Test Ban Treaty greatly support American national security interests as well as those of all parties to these treaties. As the only states possessing such enormous nuclear armories, the United States and Russia must work urgently to reduce the number of nuclear weapons (including those deployed and in reserve, as well as the quantities of weapon-usable plutonium and uranium in the military sector) to the equivalent of one thousand weapons on either side, from the present levels that exceed 10,000.

At the same time, the United States and Russia should begin conditional discussions with Britain, China, and France to persuade them to reduce their holdings in a similar fashion to 300 each when the United States and Russia achieve the one thousand level. In support of these eventual reductions, rapid demilitarization techniques should be developed so that a weapon would give no nuclear yield if detonated either intentionally or by accident.

For instance, one can use the "pit stuffing" techniques described in Chapter 12, which have already been employed for other purposes, by inserting hundreds of grams of wire into the hollow plutonium pit of a deployed nuclear weapon.

On the political and diplomatic front, the United States should take the lead in providing security guarantees—both negative and positive—to nations not possessing nuclear weapons, which means (negative security guarantees) that they will have no reason to fear being threatened by a nuclear weapon state, and (positive guarantees) that they will benefit from the protection of such weaponry, available through international support, if attacked by others.

As nuclear weapons are reduced in number, and it becomes apparent that the remaining nuclear weapons serve the purpose of maintaining security for every nation, more attention should be given to coalitions for international security in general, and a role for the United Nations in supporting and strengthening nuclear security guarantees.

BREAKING THE LINK BETWEEN NUCLEAR POWER AND NUCLEAR WEAPONS

The authors hope that it will be possible eventually to prohibit nuclear weapons and to see them never reemerge. Banishing nuclear power is neither necessary nor sufficient to eliminate nuclear weapons. Insufficient, clearly, because nuclear weapons emerged in the Manhattan Project before there was any nuclear power; and with the general advance of technology the acquisition of such weapons through dedicated enrichment plants or plutonium production reactors will be much easier in the future. Unnecessary, because nuclear power activities conducted under appropriate national and international safeguards would under normal circumstances be guarded against diversion to weaponry—monitored by the IAEA and we hope buttressed by societal verification.

RECOMMENDATIONS FOR AMERICAN NUCLEAR POWER IN THE SHORT TERM

The authors see nuclear energy as a desirable source of electrical power for the world in general, and for the United States in particular—in the operation of existing reactors and the next generation of similar reactors. The U.S. Department of Energy in 1999 assumed the responsibility for removing spent fuel from commercial reactors to an interim aboveground storage site until the mined geologic repository at Yucca Mountain is ready to accept such fuel in a routine fashion. This largely resolved a dispute with reactor operators that began when DOE did not begin accepting spent fuel in January 1998, as it was committed to do in return for the one-tenth cent per kWh charge it has been collecting from those utilities operating nuclear reactors in the United States.

For the longer term, the industry needs to continue to be held to a high standard of openness and safety. Successor reactors to ones that reach the end of their safe operating life ought to be acquired on the basis of experience and demonstration, and not await the introduction of new reactor types. As the existing pool of light-water reactors in France were initially built under contract with Westinghouse, so new reactors in the United States should draw on the most experienced and most competent suppliers in the world. Nuclear power plants are a global market, and to contract with the best domestic supplier is often to settle for second best.

To serve goals of economy and ecology, for the U.S. nuclear power industry a safe and economical route is to continue to use the once-through, direct disposal approach, so that the spent fuel after the storage at the reactor, and then in interim storage, will go to the mined underground repository without reprocessing.

ON THE WORLD'S MANAGEMENT OF NUCLEAR POWER

In general, nations are now responsible for the ultimate disposal of their own spent fuel from power reactors, or for the highly radioactive fission-product waste (the steel-encased glass logs) from the reprocessing of spent fuel. This policy ought to be reversed. With the opening of competitive, commercial mined geologic repositories around the world, small countries like Korea, Lithuania, Switzerland, and Armenia would not have to perform the research and development to create their own repositories. Reactor operators would have a choice between direct disposal of spent fuel or reprocessing and disposal of the separated fission products, with the commercial repositories accepting either form. In an era of market success, it is an anachronism and a restraint of trade to require each nation to dispose of its own spent fuel. Such a restraint serves neither the economy nor the environment.

On the other hand, safety, environment, and nonproliferation concerns will be served by the requirement that mined geologic repositories all be regulated and monitored by the International Atomic Energy Agency, and that they accept only waste forms (packaged spent fuel and vitrified fission products) that have been approved by this agency. Under these circumstances, competition will benefit not only reactor operators but the commercial firms that operate the repositories, as well as the countries in which they are located. IAEA and the United Nations, as well as interested coalitions of nations, will need to be alert to potential violations of repository requirements, in order that the stores of spent fuel not serve as the source of nuclear weapons material. The "Pangea" proposal to build in Western Australia a commercial mined geologic repository for spent fuel and for vitrified fission products (as well as, perhaps, excess weapon plutonium vitrified with fission products), together with the required transportation system, is a good start on a program that will benefit the environment, the consumer, and the nuclear industry. If properly structured, it can play an important role in reducing the likelihood that material from disassembled nuclear weapons will end up in nuclear arms in less responsible hands.

The future of the world's nuclear power depends upon nations, reactor operators, and investors having confidence in a continued supply of fuel for their reactors. While the suppliers of uranium and enrichment services in the short run have no interest in their customers knowing that there is a thousand times as much uranium available at affordable cost, it is certainly in the public interest to learn whether two of the four billion tons of uranium in seawater could be obtained by industrialization of existing processes at a per-kilogram cost of $100 to $300, in comparison with a current cost of some $25 from the ores that are being exploited at present. Those who sell reprocessing services and argue that uranium reserves are limited, and those who advocate breeder reactors in the near term for the same reason, are proposing extending the fuel supply at very much higher cost—saving uranium at a marginal cost of some $500 per kilogram—which is why it is so important for public policy and for the world's energy future to establish upper limits to the cost of separating uranium from seawater.

REPROCESSING VS. DIRECT DISPOSAL

The authors have discussed in detail the merits and problems of reprocessing fuel for recycle of plutonium and reenrichment of the uranium, as opposed to the direct disposal of spent fuel, as is the practice in the United States. For the United States, direct disposal is cheaper and safer. The spent fuel will remain potentially accessible for hundreds or thousands of years if it proves to be beneficial and economical to reprocess it to obtain plutonium and uranium for recycle; its entombment under hundreds of meters of rocks protects it from theft and banditry, while keeping it available for retrieving the canisters by a mining operation if that proves desirable.

On the other hand, we do not propose that France stop reprocessing the uranium fuel from its power reactors (it seems to plan not to reprocess the plutonium fuel), or that Britain cease reprocessing fuel (although it has not a single reactor licensed to burn plutonium from reprocessing). The people and the governments of those nations must ensure that the reprocessing they adopt is acceptable from the point of view of public health and nonproliferation. Reprocessing is presumably profitable for those who manage these activities (given that the reprocessing plants exist), although it costs the reactor operators, and hence the public, more than would the direct disposal of fuel. Expansion or replacement of these plants is a decision that should be taken openly after a serious analysis of cost and benefits.

We emphasize in this book that the plutonium separated from spent reactor fuel is usable to make nuclear weapons and, indeed, to produce powerful and reliable nuclear weapons. Those nations whose fuel is reprocessed and who receive the separated plutonium must guard that material and account for it, and the International Atomic Energy Agency must ensure that this so-called "civil plutonium" is indeed protected just as if it were weapon-grade plutonium.

THE FUTURE OF FISSION REACTORS

Because of the large size, high cost, and inherent hazards of nuclear power reactors, evolution is not nearly so rapid in this field as it is, say, in the field of personal computers. Evidently, reactors should evolve so that automatic computer control replaces operators and operating manuals. Such evolution is to be encouraged, but it does not require a change in the type of reactor.

At present no commercial power reactor uses fast neutrons and plutonium fuel, although the French Superphénix provided some experience with a fast-neutron uranium-plutonium breeder reactor. Higher-cost terrestrial uranium resources and especially the uranium from seawater render breeder reactors unnecessary for many centuries. On the other hand, if fast-neutron reactors were to be cheaper and safer than light-water reactors, we would encourage their adoption.

For fast-neutron breeder reactors, reprocessing is a necessity, but the operation can be simplified to the extent of removing only the fission products and recycling *in toto* the heavy materials — uranium or thorium, plutonium, americium, curium, etc. — all of which are eventually fissioned in repeated recycling through the reactor.

This approach to recycling actinides would not convey significant benefit from the point of view of waste disposal, but it is less hazardous for proliferation of nuclear weapons than is the exquisitely complete purification of plutonium for recycle in light-water reactors. The plutonium and other long-lived actinides do not migrate in solution in groundwater, but are fixed in place, as we observed in the discussion of the Oklo natural reactor, so the removal of all plutonium from the vitrified waste is not important. What is significant in limiting access to plutonium in this potential fuel cycle is the intense radioactivity of the gross heavy-metal fraction, with penetrating radiation — not at all in the same category as the recycled plutonium for light-water reactors, which can be handled (in its thin steel container) with bare hands.

ACCELERATOR-DRIVEN FAST REACTORS

The authors have discussed the project to use a high-power proton accelerator to supply 2% to 5% of the neutrons required in breeding the fissionable material, uranium-233, from naturally occurring thorium. What appeals to us is the largely passive nature of this fast reactor, with its core immersed in a deep lead-filled well, with the transfer of heat from the reactor achieved by natural circulation of the heated lead. Its proponents claim that the cost of electrical power from such a system would be only 50% that from a light-water reactor—a claim of which we are not yet persuaded. On the other hand, since the accelerator demands on the order of 20% of the cost of the heat-generating element, and 10% of the electricity produced, doing without the accelerator would be cheaper still. For the foreseeable future, the neutron economy of the thorium/uranium-233 cycle could be supplemented by feeding the reactor instead with a small amount of plutonium or highly enriched uranium obtained from the dismantlement of surplus nuclear weaponry; we urge the serious consideration of such systems, spurred by the analyses already done for the accelerator-driven cycle.

MODULAR HIGH-TEMPERATURE GAS-TURBINE REACTORS

Both the development in Russia led by the General Atomics Corporation and that by ESKON in South Africa envisage the use of graphite as a moderator, employing enriched uranium fuel in the form of tiny pellets or capsules of highly refractory and durable carbon-containing material. The graphite would be processed into large prismatic blocks in the General Atomics approach, and into spheres the size of tennis balls in the ESKOM utility approach; and in both, high-pressure helium would transport the heat from the reactor to the gas turbine. These concepts are the basis for a new generation of passively safe reactors. They promise smaller staff requirements, and open the possibility of deployment of nuclear power to more remote areas, to provide both electrical power and heat to utilities in a a cogeneration plant, and both lower costs and higher efficiency than do the current generation of nuclear plants. If such reactors can be brought to commercial status, then there may be further variants on this approach.

GLOBAL WARMING: NUCLEAR POWER VS. COAL

Some 70% of the heat from fission in a power reactor is sent locally to the atmosphere or to a river or ocean, while 30% of it in the form of electricity is converted to heat by the user of the electrical energy—from electric lighting, friction in motors, etc. A coal-fired power plant rejects slightly less heat locally to the environment at the time of generation, but each year for almost a century forces the earth to retain an equal amount of heat received from the sun; this is due to the carbon dioxide emitted to the atmosphere by the combustion of coal. This is no minor matter, fossil fuels having contributed to a 40% increase in atmospheric carbon dioxide since the industrial revolution. Major increase in capital investment and operating cost of coal-fired plants (on the order of 50%) could capture and sequester the carbon dioxide, but unless that is accomplished, or less costly carbon-sequestering options developed, nuclear power has a great relative benefit: the waste heat from a year of operation of a nuclear plant is only about 1% of the heat ultimately committed to the earth by a year's operation of a coal-fired plant producing the same electrical output.

THE ABILITY TO LEAD AND DECIDE

In the past decade or so, the embrace of the market economy in the United States, Britain, and much of the rest of the world has led to an expansion of trade and to great accumulations of wealth. Nevertheless, important aspects of life—education, health care, government—are not fully provided by the market economy, and analysis and explicit policy setting must serve instead for these domains. For questions with large technical components, such as the future of nuclear energy, governments must marshal and use scientific and technological counsel.

In his farewell address to the American people, President Dwight D. Eisenhower warned against the "military-industrial complex," but, in a less widely quoted portion of the speech, also against the hazard of capture of public policy by a "scientific-technological elite." We interpret this as referring to Eisenhower's difficulty in choosing the appropriate policy in matters with a high technical component, and his vulnerability (like that of managers at lower levels) to plausible but deceptive or inadequately analyzed proposals. We have noted Eisenhower's reliance upon his Science Advisory Committee; the United States would benefit both from a stronger science advisory apparatus in the White House and from regular exchanges with similar strengthened advisory apparatus in support of other governments abroad.

In this complex world, full of perils and promise, we have pointed to a path that for many centuries can allow the world to profit from the benefits of nuclear energy while minimizing the threat posed by nuclear weaponry. It is well within the ability of governments and industry to achieve these goals. But it will happen only if an informed and concerned public pushes them to recognize and to solve these problems. It is the public that is, after all, likely to pay the price of poorly chosen public policy. It is that public for whom we have written this book and to whom it is dedicated.

NOTES

INTRODUCTION
1. Steven A. Fetter, "Climate Change and the Transformation of World Energy Supply" (Stanford: Center for International Security and Cooperation, May 1999), 95 pages, ISBN 0-935371-54-0. Available from http://www.puaf.umd.edu/papers/fetter/publications-climate .htm.

CHAPTER 1: ALL ENERGY STEMS FROM THE SAME SOURCE
1. With her husband, Ida Noddack had discovered the element rhenium in 1925. The two of them, unfortunately, also "discovered" the non-existent element "masurium." Her paper, "On Element 93," was published in the *German Journal of Applied Chemistry*, September 1934, p. 653.
2. See William Lanouette with Bela Silard, *Genius in the Shadows* (Macmillan, 1992). Bela spelled the family name differently than his brother, Leo.
3. Abraham Pais, *Niels Bohr's Times: In Physics, Philosophy, and Polity* (Clarendon Press, 1991). The Hahn-Strassman paper of January 6, 1939, is quoted on p. 453.

CHAPTER 3: NUCLEAR WEAPONS
1. Carl Sagan and Richard Turco, *A Path Where No Man Thought: Nuclear Winter and the End of the Arms Race* (Random House, 1990).
2. Donald Porter Geddes, Gerald Wendt, et al., eds., *The Atomic Age Opens* (Pocket Books, 1945), p. 21; quoted in Sagan and Turco, p. 106.
3. T. Akizuki, *Nagasaki 1945* (Quartet, 1981); quoted in Sagan and Turco, pp. 106–7.
4. Committee for the Compilation of Materials on Damage Caused by the Atomic Bombs in Hiroshima and Nagasaki, *Hiroshima and Nagasaki: The Physical, Medical, and Social Effects of the Atomic Bombings*, trans. Eisei Ishikawa and David L. Swain (Basic Books, 1981), p. 92.
5. See R. P. Turco, O. B. Toon, J. B. Pollack, and C. Sagan, "Global Consequences of Nuclear Warfare," *Eos*, Vol. 63 (1982), p. 1018.
6. Lillian Hoddeson et al. *Critical Assembly: A Technical History of Los Alamos During the Oppenheimer Years, 1943–1945* (Cambridge University Press, 1993).

CHAPTER 4: NATURAL RADIATION AND LIVING THINGS

1. United Nations Committee on the Effects of Atomic Radiation, UNSCEAR 1993 Report to the General Assembly, with Scientific Annexes (United Nations Publication), p. 65, Table 7.

2. B. Vogelstein and K. W. Kinzler, "The Multi-step Nature of Cancer," *Trends in Genet.*, Vol. 9 (1993), pp. 138–41.

3. Maurice Tubiana and Robert Dautray, *La Radioactivité et ses applications*, Que sais-je? No. 33 (PUF [Presses universitaires de France], 1996), p. 94.

4. *Problèmes liés aux effets des faibles doses de radiations ionisantes*, Académie des sciences, Report No. 34, October 1995, pp. 28, 41–42.

5. W. C. Hahn, C. M. Counter, et al., "Creation of Human Tumour Cells with Defined Genetic Elements," *Nature*, Vol. 400, No. 6743 (29 July 1999), pp. 464–68. Available at www.nature.com.

6. K. S. Crump, D. G. Hoel, C. H. Langley, and R. Peto, "Fundamental Carcinogenic Processes and Their Implications for Low Dose Risk Assessment," *Cancer Research*, Vol. 36 (1976), pp. 2973–79.

7. Richard Wilson, "Low Dose Linearity: An Introduction," and Bernard L. Cohen, "Test of the Linear, No-Threshold Theory of Radiation Carcinogenesis," *Physics and Society*, Vol. 26, No. 1 (January 1997).

8. (http://www.uilondon.org/about UI/index.htm).

9. *Problèmes liés . . .* , op. cit.

10. Wei Luxin (High Background Radiation Research Group, China), "Health Survey in High Background Radiation Areas in China," *Science*, Vol. 209 (August 1980), pp. 877–80.

11. Tao Zufan and Wei Luxin, "An Epidemiological Investigation of Mutational Diseases in the High Background Radiation Area of Yangjiang, China," *Journal of Radiation Research* (*Japan*), Vol. 27 (1986), pp. 141–50; Wei Luxin et al., "Epidemiological Investigation of Radiological Effects in High Background Radiation Areas of Yangjian, China," *Journal of Radiation Research* (*Japan*), Vol. 31 (1990), pp. 119–36.

12. Yuan (1997), reported by Radiation, Science, and Health, Inc., edited by J. Muckerheide, http://cnts.wpi.edu/rsh/Data_Docs/1-2/6/2/126211We97.html.

13. John W. Gofman, *Radiation-Induced Cancer from Low-Dose Exposure: An Independent Analysis* (Committee for Nuclear Responsibility Book Division, 1990). Available at www.ratical.org/radiation/CNR/RIC/.

CHAPTER 5: THE CIVILIAN USE OF NUCLEAR ENERGY

1. See Nukem Market Report, p. 13 (December 2000).

2. U.S. Department of Energy, "Viability Assessment of a Repository at Yucca Mountain," DOE/RW-0508, December 1998. Available at http://domino.ymp.gov/va/va.nsf/.

3. Ibid.

4. Nuclear Regulatory Commission, "Strategic Assessment Issue: 6. High-level Waste and Spent Fuel," Viability Assessment, September 16, 1996.

5. C. McCombie, "Nuclear Waste Management Worldwide," *Physics Today*, June 1997, pp. 56–61.

6. For information on Pangea see Charles McCombie et al., "The Pangea International Repository: A Technical Overview," Waste Management '99 Conference, Tucson, March 3, 1999. Available at http://www.uic.com.au/Pangea-tech.htm.

7. See Georges Vendryes, *Superphénix Pourquoi?* (Nucléon, 1997). Vendryes received the Enrico Fermi Award from the Department of Energy and the President of the United States in 1984.

8. Edward Teller, "Fast Reactors: Maybe," *Nuclear News*, August 21, 1967.

9. R. Barjon, "Nécessité et sûreté de la filière des réacteurs à neutrons rapides en général et de Superphénix en particulier" (Need for and safety of fast neutron reactors in general and of Superphénix in particular), *Bulletin de la Société Française de physique*, July 1996, pp. 15–17:

10. R. L. Garwin, "The Role of the Breeder," in F. Barnaby et al., eds., *Nuclear Energy and Nuclear Weapon Proliferation* (Taylor and Francis, 1979).

11. Press communiqué, "L'aval du cycle" (The back end of the nuclear fuel cycle), February 2, 1998. Now available as "Politique Nucléaire et Diversification Energétique, Orientations Gouvernementales (Nuclear Policy and Energy Diversity, Government Direction)," at http://www.premierministre.gouv.fr/spihtm/sig_nn4/texte/recherché_txt_cat.cfm (in French: search for "Politique Nucleaire") and in English at http://www.plan.gouv.fr/organisation/seeat/nucleaire/rapportangl.pdf.

12. Ibid.

13. *Le Monde*, July 18, 1996.

14. Committee on Separations Technology and Transmutation Systems (STATS), *Nuclear Wastes: Technologies for Separations and Transmutation* (National Academy Press, 1996). OECD/NEA, *The Economics of the Nuclear Fuel Cycle* (1994), available only on the Web at http://www.nea.fr/html/ndd/reports/efc.

15. Jean-Michel Charpin, Benjamin Dessus, and René Pellat, *Etude Economique prospective de la filière électrique nucléaire* (Economic forecast for nuclear electric power), Documentation Française. Available in French at http://www.ladocfrancaise.gouv.fr/fic_pdf/charpinnucleaire.pdf, and in English at http://www.plan.gouv.fr/organisation/seeat/nucleaire/rapportangl.pdf.

16. J. S. Choi and T. H. Pigford, "Effects of Transmuting Long-lived Radionucleotides on Waste Disposal in a Geological Repository," in R. L. Garwin, M. Grubb, and E. Matanle, eds., *Managing the Plutonium Surplus: Applications and Technical Options*, NATO, ASI Series 1, Disarmament Technologies, Vol. 1 (November 1994), pp. 171–84.

17. Committee on (STATS). *Nuclear Wastes.*

CHAPTER 6: A GLIMPSE OF THE FUTURE OF NUCLEAR POWER

1. Committee on Separations Technology and Transmutation Systems (STATS), *Nuclear Wastes: Technologies for Separations and Transmutation* (National Academy Press, 1996).

2. M. Salvatores, I. Slessarev, M. Uematsu, and A. Tchistiakov, "The Neutronic Potential of Nuclear Power for Long-Term Radioactivity Risk Reduction," *International Conference on Evaluation of Emerging Nuclear Fuel Cycle Systems*, Vol. 1, pp. 686–82. (American Nuclear Society Topical Meeting, 1995).

3. C. Rubbia, "Harmless Energy from Nuclei," Einstein Lecture at the Israel Academy of Sciences, March 15, 1998 (and other reports).

4. John P. Holdren and R. K. Pachauri, "Energy," in *ICSU: An Agenda of Science for Environment and Development into the 21st Century* (Cambridge University Press, 1992), pp. 103–18.

CHAPTER 7: SAFETY, NUCLEAR ACCIDENTS, AND INDUSTRIAL HAZARDS

1. NEI, *The TMI 2 Accident: Its Impact, Its Lessons*, (April 1998). A less-detailed version of the NEI report has replaced the 1998 document, and is available at http://www.nei.org/doc.asp?Print=true&DocID=&CatNum=3&CatID=294.

2. "Report to the American Physical Society by the Study Group on Light-water Reactor Safety," published in *Reviews of Modern Physics*, Vol. 47, Supplement No. 1 (June 1975); S. M. Keeny, ed., *Nuclear Power Issues and Choices* (Ballinger, 1977).

3. "Reactor Safety Study: An Assessment of Accident Risks in U.S. Commercial Nuclear Power Plants," U.S. Atomic Energy Commission, Draft of WASH-1400 (1974). This was the first massive attempt at Probabilistic Risk Assessment, also termed Probabilistic Safety Assessment.

4. The full and moving account of the Chernobyl accident is from Grigori Medvedev, *The Truth About Chernobyl* (I. B. Taurus, 1991). Quoted by permission.

5. Ibid., p. 20.

6. Ibid., pp. 22–23.

7. Ibid., p. 52.

8. Ibid., p. 46.

9. Ibid., p. 118.

10. Ibid., p. 167.

11. Ibid., p. 174.

12. Ibid., p. 193.

13. Ibid., pp. 193–94.

14. Ibid., p. 224.

15. Ibid., pp. 203–4.

16. Ibid., p. 139.

17. International Program on the Health Effects of the Chernobyl Accident (IPHECA), "Health Consequences of the Chernobyl Accident: Results of the IPHECA Pilot Projects and Related National Programmes," summary report (Geneva: World Health Organization, 1995.

18. Table from Ibid.

19. U.S. NRC: Review of the Tokai-mura Criticality Accident, April 2000 (SECY-00-0085); Available at http://www.nrc.gov/NRC/COMMISSION/SECYS/secy2000-0085/2000-0085scy.html

20. Table from United Nations Committee on the Effects of Atomic Radiation, "Sources and Effects of Ionizing Radiation," UNSCEAR 1993 Report to the General Assembly, with Scientific Annexes, United Nations Publication.

21. UNSCEAR 1993, pp. 106–7.

22. Director, Safety, Health and Environment, "Annual Report on Discharges and Monitoring of the Environment," BNEL, Risley, Warrington, Cheshire, WA3 GAS, Britain. 1997; letter, R. Coates (BNFL) to R. L. Garwin, April 6, 1999.

23. UNSCEAR 1993, p. 107, paragraph 90.

24. "American Medical Association Science News Updates," November 12, 1997, *Journal of the American Medical Association*, Vol. 278 (1997), pp. 1500–4): "If current smoking patterns persist, tobacco will eventually cause more than two million deaths each year in China."

25. "The European Pressurized Water Reactor," *Technologies France*, December 1995–January 1996.

26. Stephen Breyer, *Breaking the Vicious Circle: Toward Effective Risk Regulation* (Harvard University Press, 1993).

27. Charles Hailey and David Helfand, "In the TWA 800 Crash, Don't Rule Out Meteors," *New York Times*, September 19, 1996.

CHAPTER 8: REDUCING GREENHOUSE GAS EMISSIONS

1. Data from the Energy Information Agency of the U.S. Department of Energy; see http://www.eia.doe.gov/emeu/iea/.

2. John P. Holdren and R. K. Pachauri, "Energy," in *ICSU: An Agenda of Science for Environment and Development into the 21st Century* (Cambridge University Press, 1992), pp. 103–18.

3. Tadao Seguchi, paper presented at Tokyo University–Harvard University Workshop, Tokyo, May 23, 1998.

4. J. Foos, Document of June 1996 on Extraction of Uranium from Seawater, provided to COGEMA. Foos is director of the Laboratoire des Sciences Nucléaires, of the Conservatoire National des Arts et Métiers, reporting on work of his laboratory and the Laboratoire de Catalyse et de Synthèse Organique de l'Université Claude Bernard de Lyon.

5. "A World Oil Price Chronology: 1970–1996" is available at http://www.eia.doe.gov/emeu/cabs/chron.html.

6. *Nukem Market Report*, April 2000.

7. Data from "Federal Energy R&D for Challenges of the Twenty-first Century," September 30, 1997; see http://www.whitehouse.gov/WH/EOP/OSTP/Energy/, updated from Congressional Appropriation Conference Reports, various years.

8. Matthew Wald, *New York Times*, June 20, 1999.

9. H. R. Lyndon, "Alternative Pathways to a Carbon-Emission-Free Energy System," *The Bridge*, Vol. 29, No. 3, National Academy of Engineering (Fall 1999).

CHAPTER 9: COMPARING HAZARDS OF NUCLEAR POWER AND OTHER ENERGY

1. R. H. Williams, "Hydrogen Production from Coal and Coal-Bed Methane, Using Byproduct CO_2 for Enhanced Methane Recovery and Sequestering the CO_2 in the Coal Bed," paper prepared for the Fourth International Conference on GHG Control Technologies, Interlaken, Switzerland, August 30—September 2, 1998.

2. UNSCEAR 1993.

3. Steven A. Fetter, "Climate Change and the Transformation of World Energy Supply," May 1999.

4. A. H. Rosenfeld, T. M. Kaarsberg, and J. Romm, "Technologies to Reduce Carbon Dioxide Emissions in the Next Decade," *Physics Today*, November 2000, pp. 29–34.

5. Intergovernmental Panel on Climate Change, "Climate Change 2001: The Scientific Basis," WG I contribution to the IPCC third assessment report, World Meteorological Organization, Geneva. Available at http://www.ipcc.ch/.

6. Svante Arrhenius, "On the Influence of Carbonic Acid in the Air upon the Temperature of the Ground," *Philosophical Magazine*, Vol. 41, p. 237, 1896.

7. Adam Serchuk and Robert Means, "Natural Gas: Bridge to a Renewable Energy Future," at http://www.repp.org/articles/issuebr8/issuebr8.html.

8. Steven A. Fetter, "Climate Change and the Transformation of World Energy Supply," (May 1999).

9. S. M. Keeny, Jr., et al., *Nuclear Power Issues and Choices*, Ballinger Publishing Co., March 1997.

10. Available at http://www.whitehouse.gov/energy/National-Energy-Policy.pdf.

CHAPTER 10: MAKING BEST USE OF SCIENTISTS

1. Jean-Henri Fabre, *Les Inventeurs et leurs inventions* (Éd. Champion Slatkine, 1986).

2. Henri Vincenot, *L'Âge du chemin de fer* (Denoël, 1980).

3. Thomas B. Cochran, Robert S. Norris, and Oleg A. Bukharin, *Making of the Russian Bomb* (Westview Press, 1995), p. 13.

4. G. A. Litvinov et al., *"KVS" Power* (in Russian), 1997, Snezhinsk (Chelyabinsk-70). The original Soviet nuclear weapons laboratory is located at what was long a secret city—Arzamas-16 (now named Sarov). The second such laboratory was the core of another secret city—Chelyabinsk-70 (now named Snezhinsk)—near Sverdlovsk (now Ekaterinburg) in the Urals.

5. *Chinese Science News*, December 27, 1995.

6. John W. Gardner, *Excellence—Can We Be Equal and Excellent, Too?* (W. W. Norton, 1961).

7. Stephen I. Schwartz, ed., *Atomic Audit* (Brookings Institution Press, 1998).

8. See http://www.pugwash.org.

9. See Bernd W. Kubbig, *Communicators in the Cold War: The Pugwash Conferences, the U.S.-Soviet Study Group and the ABM Treaty*, PRIF Reports No. 44, Peace Research Institute, Frankfurt, Germany. Available at http://www.rz.uni-frankfurt.de/hsfk.

10. U.S. Atomic Energy Commission, "Reactor Safety Study: An Assessment of Accident Risks in U.S. Commercial Nuclear Power Plants," Draft of WASH-1400, 1974.

CHAPTER 11: FROM ARMS RACE TO ARMS CONTROL

1. See Bernard Brodie, ed., *The Absolute Weapon* (Harcourt Brace, 1946).

2. R. S. Norris and W. M. Arkin, "Nuclear Notebook," *Bulletin of the Atomic Scientists*, July/August 2000.

3. H. A. Bethe and R. L. Garwin, "Antiballistic Missile Systems," *Scientific American*, March 1968.

4. Henry A. Kissinger, *White House Years* (Little, Brown, 1979), p. 217.

5. These excerpts can be found, for instance, at http://www.cnn.com/SPECIALS/cold.war/episode/22/document/starwars.speech/

6. "Analyses of Effects of Limited Nuclear Warfare," Testimony Prepared for the Subcommittee on Arms Control, International Organizations and Security Agreements of the United States Senate Foreign Relations Committee, Testimony by R. L. Garwin, September 18, 1975.

7. Lee Butler, "Zero Tolerance," *Bulletin of the Atomic Scientists*, January/February 2000.

8. E.g., Fig. B-6 in the 1997 CISAC Report at http://books.nap.edu/html/fun/appb.html.

9. See http://www.nrdc.org/nrdcpro/nuclear.

10. Stephen I. Schwartz, ed., *Atomic Audit: The Costs and Consequences of U.S. Nuclear Weapons Since 1940 (Brookings Institution Press, 1998). See also http://www.brook.edu/fp/projects/nucwcost/weapons.htm.*

11. S. D. Drell et al., "Nuclear Testing—Summary and Conclusions," JASON Report JSR-95-320, August 3, 1995. (Available at http://www.fas.org/rlg.)

12. See http://www.fas.org/rlg. Choose "1990–1999" and search for "frontline."

13. The KNET project: http://yin.ucsd.edu/ANZA/home.html.

14. R. L. Garwin, C. E. Paine, and R. E. Kidder, "Discussions in Paris Regarding the Necessity of Nuclear Tests for Maintaining a Reliable French Nuclear Force Under a Comprehensive Test Ban," unpublished paper, Federation of American Scientists–Natural Resources Defense Council, December 1994.

CHAPTER 12: CURRENT NUCLEAR THREATS TO SECURITY

1. The report is available at http://www.fas.org/rlg.

2. W. K. H. Panofsky, *Management and Disposition of Excess Weapons Plutonium*, pp. 32–33 (National Academy Press, 1994).

3. American Nuclear Society, "Protection and Management of Plutonium," Special Panel Report, August 1995.

4. John Deutch, "The Threat of Nuclear Highjacking," testimony to the Senate Permanent Investigation Committee on Government Operations, March 20, 1996.

5. Monterey Institute of International Studies, together with Carnegie Endowment for International Peace, "Nuclear Successor States of the Soviet Union: Status Report on Nuclear Weapons, Fissile Material and Export Controls," May 1998. Available at http://cns.miis.edu/pubs/reports/statrep.htm (see p. 108).

6. *Washington Post*, October 23, 1996.

7. Joseph R. Biden, Jr., "Maintaining the Proliferation Fight in the Former Soviet Union," *Arms Control Today*, March 1999. Available at http://www.armscontrol.org/ACT/march99/bidmr99.htm.

8. See J. P. Holdren et al., "Final Report of the U.S.-Russian Independent Scientific Commission on Disposition of Excess Weapons Plutonium," 1 June 1997. Available at http://www.fas.org/rlg.

9. Committee on International Security and Arms Control, National Academy of sciences, "The Spent-Fuel Standard for Disposition of Excess Weapon Plutonium: Applications to Current DOE Options," National Academy Press (2000).

10. See http://www.sandia.gov/LabNews/LN01-19-96/palo.html.

11. "Sources and Effects of Ionizing Radiation," UNSCEAR 1993 Report to the General Assembly, with scientific annexes, p. 117.

12. W. G. Sutcliffe et al., "A Perspective on the Dangers of Plutonium," UCRL-JC-118825, Lawrence Livermore National Laboratory, April 14, 1995. Available at http://www.llnl.gov/csts/publications/sutcliffe/.

13. Steve Fetter and Frank von Hippel, "The Hazard from Plutonium Dispersal by Nuclear Warhead Accidents," *Science and Global Security*, Vol. 1, (1990), pp. 21–41.

14. "Drawing Back the Curtain of Secrecy," Restricted Data Declassification Decisions, 1946 to the Present (RDD-4), January 1, 1998, U.S. Department of Energy Office of Declassification, "Approved for Public Release," http://www.doe.gov/html/osti/opennet/opennet1.html.

15. Sheryl Wu Dunn, Judith Miller, and William J. Broad, "How Japan Germ Terror Alerted World," *New York Times*, May 26, 1998.

16. L. H. Hoddeson, *Critical Assembly* (University of California Press, 1993).

17. Richard H. Rhodes, *The Making of the Atomic Bomb* (Simon & Schuster, 1986), pp. 655–58.

18. D. H. Rumsfeld et al., Executive Summary of the Report of the Commission to Assess the Ballistic Missile Threat to the United States (July 15, 1998). To be found at http://www.fas.org/rlg.

19. H. A. Bethe and R. L. Garwin, Letter of September 3, 1991, to Cong. Les Aspin and Sen. Sam Nunn, with a technical appendix by Garwin. Available at http://www.fas.org/rlg.

20. R. L. Garwin, "Missile Defense Policy and Arms Control Issues," presented at Second Annual U.S. Army Space & Missile Defense Conference, Huntsville, Alabama, 08/26/99. A similar paper is available as "Cooperative Ballistic Missile Defense," presented at the Department of State Secretary's Open Forum, November 1999. See http://www.fas.org/rlg, choose "1990–1999" and search for "Cooperative."

21. A. M. Sessler et al., "Countermeasures: A Technical Evaluation of the Operational Effectiveness of the Planned U.S. National Missile Defense System." Available at http://www.ucsusa.org.

CHAPTER 13: CAN WE RID THE WORLD OF NUCLEAR WEAPONS?

1. See National Academy of Sciences, *The Future of U.S. Nuclear Weapons Policy* (National Academy Press, 1997), p. 43.

2. McGeorge Bundy, *Danger and Survival* (Random House, 1988), pp. 519–20.

3. Garwin chaired the Director's Advisory Committee of ACDA from 1994 to 1999, and then the Arms Control and Nonproliferation Advisory Board of the Department of State.

4. Canberra Commission on the Elimination of Nuclear Weapons, Report, August 1996. Available at http://www.dfat.gov.au/cc/cchome.html.

5. R. L. Garwin and R. Z. Sagdeev, "Verification of Compliance with the ABM Treaty and with Limits on Space Weapons," in F. Calogero, M. L. Goldberger, and S. Kapitza, eds., *Verification — Monitoring Disarmament* (Westview, 1991), pp. 23–44.

6. B. G. Blair, T. B. Cochran, T. Z. Collina, J. Dean, S. Fetter, R. L. Garwin, K. Gottfried, L. Gronlund, H. Kelly, M. G. McKinzie, R. S. Norris, A. Segal, R. Sherman, F. N. von Hippel, D. Wright, S. Young, FAS/NRDC/UCS (available at http://www.ucsusa.org).

FOR FURTHER READING

NUCLEAR WEAPONS

The Future of the U.S.-Soviet Nuclear Relationship. Committee on International Security and Arms Control, National Academy of Sciences, Washington, DC: National Academy Press, 1991.

Management and Disposition of Excess Weapons Plutonium, Committee on International Security and Arms Control, National Academy of Sciences, Washington, DC: National Academy Press, 1994.

Genius in the Shadows: A Biography of Leo Szilard, the Man Behind the Bomb, by William Lanouette, Reprint edition (November 1994) © 1992 Chicago: University of Chicago Press.

The Making of the Atomic Bomb, by Richard Rhodes. Paperback Reprint edition (August 1995) © 1986 New York: Touchstone Books.

Dark Sun: The Making of the Hydrogen Bomb, by Richard Rhodes. (August 1996) New York: Touchstone Books.

The Future of U.S. Nuclear Weapons Policy, Committee on International Security and Arms Control, National Academy of Sciences, Washington, DC: National Academy Press, 1997.

Atomic Audit: The Costs and Consequences of U.S. Nuclear Weapons Since 1940, by Stephen I. Schwartz (Editor) (June 30, 1998), Washington, DC: Brookings Institution;

The Nuclear Turning Point: A Blueprint for Deep Cuts and De-Alerting of Nuclear Weapons, by Harold A. Feiveson (Editor) (June 1999) Washington DC: Brookings Institution.

NUCLEAR POWER

Nuclear Power Issues and Choices, S. M. Keeny, Jr., et al., Cambridge, MA: Ballinger Publishing Co, 1977.

Nuclear Energy: Principles, Practices, and Prospects, by David Bodansky (Hardcover, May 1996) Woodbury, NY: American Institute of Physics.

CLIMATE CHANGE, INCLUDING CLIMATE-CHANGE MITIGATION

Laboratory Earth: The Planetary Gamble We Can't Afford to Lose, by Stephen H. Schneider (New York: Basic Books, 1997).

For Further Reading

Global Warming: The Complete Briefing, by John Houghton, 2nd ed. (Cambridge: Cambridge University Press, 1997).

Climate Change Policy: Facts, Issues and Analyses, by Catrinus J. Jepma and Mohan Munasinghe (Cambridge: Cambridge University Press, 1997).

Policy Implications of Greenhouse Warming, National Research Council (Washington, D.C.: National Academy Press, 1991) (available at http://www.nap.edu).

ENERGY OPTIONS & ENERGY FUTURES

Energy, The Next Twenty Years, H. H. Landsberg, et al. (Cambridge, MA: Ballinger, 1979).

Renewable Energy: Sources for Fuels and Electricity, Thomas B. Johansson, Henry Kelly, Amulya K. N. Reddy, and Robert H. Williams, eds. (Washington, DC: Island Press, 1993).

Renewable Energy: Power for a Sustainable Future, Godfrey Boyle, ed. (Oxford: Oxford University Press, 1996).

Global Energy Perspectives, Nebojsa Nakicenovic, Arnulf Grubler, and Alan McDonald, eds. (Cambridge: Cambridge University Press, 1998).

Introduction to Energy: Resources, Technology, and Society, by Edward S. Cassedy and Peter Z. Grossman, 2nd ed. (Cambridge: Cambridge University Press, 1999).

Prospects for Sustainable Energy: A Critical Assessment, by Edward S. Cassedy (Cambridge: Cambridge University Press, 2000).

INDEX

Page numbers in *italics* indicate figures.

Index

Glenn Amendment, 302

global warming, 205, 225, 232, 238–9, 240, 241, 246; nuclear power vs. coal, 382–3

Gofman, John W., 103–4

Gold, Thomas, 231

Goldberger, Marvin L., 266

Goldschmidt, Bertrand, 318

Gorbachev, Mikhail, 266–7, 284, 320

Gore-Chernomyrdin Commission, 336

governments, 257–9, 309, 383; use of experts, 250–1

Graham, John, 100–1

graphite, 43, 51, 127, 172; as moderator, 138, 382

graphite reactor, 113–14; graphite-water, 315

Gray, L. H., 81

gray (1 Gy), 81, 94

greenhouse effect, 5, 6, 108, 165, 201, 206, 215, 234–42

greenhouse gases, 5, 225, 234–5, 238, 246; reducing emissions of, 206–30, 237–8

Greenpeace, 193, 200, 373

Group of Seven, 178

Guillemin, Jeanne, 346

gun assembly, 59, 75, 313, 348, 350

Habiger, Eugene, 324

Hadès rockets, 67

Hahn, Otto, 23, 24

Haig, Alexander, 318, 367

Halban, Hans von, 26

half-life, 81; defined, 32

Hanford, Wash., 32, 33, 55, 59, 109, 117, 126, 143, 270, 319; Fast-Flux Test Facility, 163; pollution, 142

Harvard University, 264, 325; Belfer Center for Science and International Affairs, 269

Hatfield-Exon-Mitchell Act, 301

hazards, 169; comparing nuclear power and other energy, 231–47; industrial, 170–205

"Health Aspects of Chemical and Biological Weapons" (report), 346

health issues related to radiation, 4, 89–104

heat: nuclear reactors as source of, 110–11, 112, 113, 114–15, 116–17; waste, 382–3

heat exchangers, 115, 131, 222

heavy water, 43, 46, 51

heavy-water reactor, 111, 115–17, 127, 212, 315

Heisenberg, Werner, 34

helium, 18–19, 153, 167

Helms, Jesse, 282

Helsinki agreement, 297, 298

High Background Radiation Research Group, 101, 102–3, 104

highly enriched uranium (HEU), 117, 312, 313–14, 326, 347, 348; contraband, 323; excess, 327, 328, 329, 331; Russian, 321–2, 334–5

Hiroshima, 3, 33, 68, 69, 71, 72, 73, 108, 338, 348; fatalities at, 89–90; radiation, 192

Hiroshima bomb, 34–5, 48, 49, 59, 62, 64, 65, 186, 255, 273, 313, 350

Hiroshima victims, 93, 94, 104

Holbrooke, Richard, 280

Holdren, John P., 210, 217, 266

Holum, John D., 368

hotlines, 368

Hubble Space Telescope, 229

human activities: radiation from, 85–9; relative dangers of, 80

Hungary, 297, 321

hybrid vehicles, 225

hydrogen, 8, 9, 14, 15, 225, 232

hydrogen bomb, 62–5, 226, 228, 273

hydrogen bubble, 173–5

hydrogen tests, 301, 302

hydropower, 107, 207, 233

ICBMs, see intercontinental ballistic missiles (ICBMs)

ice ages, 234, 240

ICRP, see International Committee for Radiation Protection (ICRP)

Idaho National Engineering and Environmental Laboratory, 157

immobilization, 329, 330, 331, 332–3

implosion bomb, 59–61, 62, 75, 77, 117, 316

implosion system, 271, 313, 314, 315, 348, 349–50

incandescent lamp, 222–3

India, 215, 223, 299, 352; and CTBT, 304–5; and disarmament, 374; nuclear weapons, 268; tests by, 78, 252, 297, 302–3, 306, 350–1

Indonesia, 232

industrial energy, definitions for, xiv–xvi

industrial hazards, 170–205

inertial confinement fusion (ICF), 226–7, 228

Initiative for Proliferation Prevention, 327

inspections, 309, 353; by IAEA, 315–17, 321, 352; on-site, 300, 306

Index

monazite, 200
Monterey Institute of International Studies, 323
MOX (Mixed OXide) fuel, 51, 122, 137, 138, 139, 142, 148, 158, 164; cost of, 312, 317, 328–9, 330, 331, 332, 333–4; derived from excess weapon plutonium, 319; extracting Pu from, 318; spent fuel standard, 335
multiplication factor K, 54–5, 56
multiplicative model, 99, 104
muons, 80–2, 188

Nagasaki, Japan, 3, 20, 33, 68, 72, 108, 338, 342; fatalities in, 89–90; radiation, 192
Nagasaki bomb, 59, 60, 186, 255, 313, 349
Nagasaki victims, 93, 94, 104
National Academy of Sciences: Committee on International Security and Arms Control (CISAC), 266–8, 303, 318, 319, 328, 329, 330, 332–3, 335, 372; National Research Council, 97, 151, 156; National Research Council Committee on Health Effects of Exposure to Radon, 85
National Missile Defense system (NMD), 297, 358–9
NATO, see North Atlantic Treaty Organization (NATO)
natural disasters, deaths from, 203
natural gas, 224, 225, 241, 377; hydrates, 209; see also gas; methane
natural radiation, 79–106; vs. Chernobyl radiation, 188–9; coal in, 201; from the human body, 105; risk from, 199
natural radiation dose, origin of, 89
Natural Resources Defense Council, 289
natural uranium, 48, 49, 111, 122, 210; chain reaction, 26, 41, 45; fission, 27, 110; price of, 146–7; reactors with, 36, 43, 50–1, 52; toxicity, 159
naval propulsion, 108, 109, 117, 260
Naval Research Advisory Board, 260
"Nécessité et sûreté de la filière des réacteurs . . ." (Barjon), 133–4
neptunium, 22, 23, 123, 149
neptunium-239 (Np-239), 32
neutrino(s), 21–2, 82, 110, 111
neutron bombs, 65–7, 69
neutron capture, 22, 33, 41, 45–6, 129, 154, 159; in plutonium, 163; probability of, 42–3; by U-238, 32, 33

neutrons, 9, 13, 14, 15, 27, 48, 117, 196; absorption of, 36–43, 37, 38, 39; absorption probability, 43–8, 46, 47, 51, 129–30, 130; in atmosphere, 82, 83; causing death, 65, 68–9; discovery of, 20; in fission, 28, 31; heavy nucleus, 153–4; implosion bomb, 59–61; interactions, 33–5; as projectile, 20–7, 35, 42; properties of, 33–43
Nevada Test Site, 125, 253, 307, 342
Newton, Isaac, xiii
Nier, Alfred O., 24
nitrogen oxide, 232, 233
nitrous oxide, 235
Nitze, Paul H., 267, 269
Nixon, Richard M., 261, 277, 280, 295, 346, 367, 369
Nixon Administration, 262, 291
Nobel Prize, 22–3, 41, 79, 153, 263–4
nonnuclear weapons states, 107–8, 299, 300, 303; and CTBT, 306; and ending proliferation threat, 370; and NPT, 308, 321; and proliferation, 337
nonproliferation regime, 302, 303, 316, 343
Non-Proliferation Treaty (NPT), 263, 268, 291, 299, 302, 304, 312, 313, 315, 374, 377; CTBT and future of, 307–9; extension of, 303; former Soviet Union and, 321; Review Conference, 307
North Atlantic Treaty Organization (NATO), 66, 67, 258, 272, 273, 275, 281; bombing of Kosovo and Serbia, 372; expansion, 279, 297, 321; policy regarding nuclear weapons, 287
North Korea, 4, 279, 299, 324; ballistic missile threat, 355–6, 358; and CTBT, 304, 305; No Dong missile, 351
NPT, see Non-Proliferation Treaty (NPT)
nuclear age, turning point in, 376–83
nuclear binding energy, 27, 28, 29, 30
nuclear centers: pollution in, 142–3
Nuclear Cities Initiative, 326–7
nuclear energy, 3–4, 225, 240; Chernobyl and perception of, 193–5; civilian use of, 107–52; cost of, 165; in countries lacking technology, 315; discovery of, 20; fuel supply for, 248; and global warming, 246; generating, 27–30, 213; linkage between civilian and military applications, 311–12; without nuclear weapons, 335–6; separating myth from reality, 248–50
Nuclear Energy Agency, 136, 145

A NOTE ABOUT THE AUTHORS

Richard L. Garwin is Philip D. Reed Senior Fellow for Science and Technology at the Council on Foreign Relations in New York. A physicist, he has been Professor of Public Policy at Harvard University. He is a member of the National Academy of Sciences, the National Academy of Engineering, and the Institute of Medicine. He has been awarded 43 U.S. patents and published hundreds of papers. The U.S. government granted him the Enrico Fermi Award in 1996.

Georges Charpak, physicist, is a member of the Académie des Sciences and of the U.S. National Academy of Sciences. He started his career at the Collège de France (Paris) and has long worked at the European Center for Particle Physics, CERN (Geneva). He received the 1992 Nobel Prize in Physics for his invention of electronic detectors of ionizing particles, used widely in physics, industry, and biology.